The Ethics of Killing

Christian Erk

The Ethics of Killing

Life, Death and Human Nature

palgrave
macmillan

Christian Erk
University of St.Gallen
St. Gallen, Switzerland

ISBN 978-3-031-07182-9 ISBN 978-3-031-07183-6 (eBook)
https://doi.org/10.1007/978-3-031-07183-6

This Palgrave Macmillan imprint is published by the registered company Springer Nature Switzerland AG.
The registered company address is: Gewerbestrasse 11, 6330 Cham, Switzerland

Prologue and Acknowledgements

John Henry Newman (1982: 375) once said: "Reflect, Gentlemen, how many disputes you must have listened to, which were interminable, because neither party either understood either his opponent or himself". The ethical (im)permissibility of killing human beings is one of those interminable disputes. All of us accept that killing a human being is ethically impermissible in general. There is, however, widespread and continuing disagreement once we zoom in on particular cases of killing such as suicide, abortion and euthanasia. This book is not so bold or overconfident that it attempts to settle this debate for good.

When it comes to the problem of the ethical (im)permissibility of the above-mentioned and other kinds of killings one must realistically admit that the lines have been drawn for a long time. The essential arguments in favour or against the specific kinds of killing have also been on the table for a long time. And, unfortunately, the discussion has been going in circles for decades. While both sides can and do refine already existing arguments, neither side comes up with substantially new arguments.

What separates the camps of the proponents and opponents of these killing practices is not so much their conclusions, that is, the fact that they deem some practices ethically permissible or impermissible. What actually separates them are the premises from which the respective conclusions are derived. In ethical debate, it is not one's position as such that matters but why it is held and how it was reached; without a sound argumentative foundation, a position is merely an opinion. So, ethics—or as it is also called, moral philosophy—is not only about understanding the argument of one's counterpart but first and foremost an exercise in self-questioning and self-knowledge.

This book has made it its goal to help both sides of the debate understand either themselves or their counterpart better. Its main goal is to examine the ethical (im)permissibility of killing human beings. It does so by taking a deep dive and analysing the concepts that are needed to answer this ethical problem: What does it mean to act? What elements does an action comprise? What is the difference between a good or evil action and a permissible or impermissible action? How can we determine whether an action is good or evil? Is there a moral duty not to kill? Is this duty held by and against all human beings or only persons? What and who is a person? What is human dignity and who has it? What is it that is actually taken when somebody is killed, that is, what is life? And closely related to that: What and when is death? Without having seriously dealt with these questions first, it is virtually impossible to define the ethical limits of dealing with our own lives and the lives of other human beings. By answering each of these questions consecutively and integrating the answers into an argumentative architecture, the book offers a comprehensive answer to one of the most fundamental questions of mankind: Under which conditions, if any, is killing human beings ethically permissible?

Those familiar with ethical theories will see that the book answers these questions by developing an ethical theory that can be described as a "natural ethics" and that is congruent with what is usually called natural law theory, or, to be more precise, classical (also: traditional) natural law theory. Even if it might lack the glamour of (short-lived) novelty and the (shallow) thrill of unconventionality, natural law theory—to my mind—is still the most plausible and solid ethical theory that ethics has to offer. This is why this book prominently features thinkers such as Thomas Aquinas, David Oderberg and Edward Feser but also a range of other authors representative of this ethical theory. The book can, therefore, be seen as an endeavour to systematically ground a natural law ethics and consistently apply it to the topic of killing human beings.

Without realising it and albeit intermittently, I have been working on this book since 2006. This was the year I started my studies in Political Theory at the London School of Economics and Political Science (LSE). The first formative essay I had to write was about the ethical permissibility of one of the killing actions treated in this book. Sitting in front of a blank sheet of paper I realised back then that I was enjoying the comfort of opinion without having endured the discomfort—and sometimes pain—of thought. When it came to one of the most fundamental problems of mankind, I was living an only insufficiently and superficially examined life—and I have tried to remedy this situation in the meantime to the best of my abilities. Whether consciously or not, most of the academic work I have done since then has helped me to develop the argumentative puzzle pieces that can now be joined to an ethics of killing human beings. Certain parts of this book, therefore, draw on, and in part reproduce verbatim, material that has previously been published by me in Erk (2011, 2014a, 2014b, 2015, 2016, 2019). In the end, the formative essay I had to write (and hand in within four weeks) back in 2006 took me some 16 years to finish—and eventually turned out to be a full-fledged book.

Although it can be lonely sometimes, nobody writes a book alone. I am intellectually indebted to and would like to thank the many persons who have devoted some of their time as well as thoughts to me and were willing to serve as intellectual sparring partners. This exchange was infinitely valuable to sharpen my thinking and, where necessary, to correct it. Although not all of them might hold the same views I set out and defend in this book, I hope that they nevertheless find the result they have helped to shape of some worth. I am also grateful to the two anonymous reviewers of this book's manuscript for their careful reading and their insightful comments and suggestions.

Finally, some words on certain quotations and references might be in order: Whenever the original language of a quote is not English, I have taken the liberty of presenting the original quote as well as the English translation side by side. Although this might be distracting to some readers, it gives at least the opportunity to verify the proposed translation. For ancient or medieval authors such as Aristotle, Thomas Aquinas, Plato and Augustine I have used a condensed system of reference. References to their works begin with an abbreviation of the work's title, followed by numbers that correspond to the respective divisions of the work. The abbreviations and the corresponding full titles of the works are listed at the beginning of the references section at the end of this book. There the reader can also find the translations I have used as a basis, taking the liberty of adapting them, usually for the sake of precision and/or terminological consistency.

St. Gallen, Switzerland Christian Erk
July 2022

References

Erk, Christian. *Health, Rights and Dignity. Philosophical Reflections on an Alleged Human Right*. Frankfurt, Paris, Lancaster & New Brunswick: ontos Verlag, 2011.

Erk, Christian. "Das Eigentliche des Todes. Ein Beitrag zur Be-Lebung der Debatte über Hirntod und Transplantation." *Ethik in der Medizin* 26.2 (2014a): 121–135.

Erk, Christian. "The Diagnosis of the Absence of Life." *Bioethica Forum* 7.3 (2014b): 105–108.

Erk, Christian. *Rationierung im Gesundheitswesen. Eine wirtschafts- und sozialethische Analyse der Rationierung nach Selbstverschulden*. Berlin & Boston, MA: de Gruyter, 2015.

Erk, Christian. "(Brain) Death, (Brain) Life and the Value of Life." *International Journal of Bioethics and Health Policy* 1.1 (2016): 27–34.

Erk, Christian. *Moral Philosophy. Fundamental Concepts*. Second, revised and expanded Edition. Heusenstamm: editiones scholasticae, 2019.

Newman, John Henry. "Discipline of Mind." In: Newman, John Henry. *The Idea of a University*. Edited, with an Introduction and Notes by Martin J. Svaglic. Notre Dame, IN: University of Notre Dame Press, 1982. pp. 361–380.

Contents

Abbreviations

BHG	Basic human good
CR	Claim-right
D	Duty
DBB	Donation after brain death
DCD	Donation after cardiac (or: circulatory) death
DDR	Dead donor rule
EEG	Electroencephalogram
IVF	In vitro fertilisation
MTX	Methotrexate
PAS	Physician-assisted suicide
PDE	Principle of double effect
VRFF	Voluntary refusal of food and fluids
VSED	Voluntary stopping of eating and drinking

List of Figures

VII: The Special Ethics of Killing Human Beings

VIII: Postlude: Specific Questions at the Margins of Human Life

I: Prelude: Setting the Stage

Virtually all of us would agree that killing human beings is ethically impermissible in general, but virtually all of us would also agree that there are exceptions. The problem with the latter agreement is that there is wide disagreement on what exactly constitutes an exception to the rule and how. The greatest agreement is likely to exist with regard to killing in self-defence. But beyond that, unanimity quickly ceases. Some take abortion, that is, the killing of an unborn human being, to be another exceptional case of an ethically permissible killing; others deem euthanasia, that is, the killing of a human being to relieve his pain and suffering, to be an exception to the rule too. If we look at both the public and scholarly debate about the ethical permissibly of different kinds of killings, we are left with a confusing plurality of positions.

The problem with this situation is not the plurality of opinion as such. It is the fact that the differing positions on the ethical permissibility of the different kinds of killing actions are not only contrary but contradictory in nature. While contrary propositions ("contraries") such as "The sky is blue" and "The sky is black" cannot both be true, but can both be false, of two contradictory propositions ("contradictories") such as "The sky is blue" and "The sky is not blue" one must be true and the other false (cf. Erk 2019: 31ff). Something cannot be the case and not be the case at the same time and in the same respect. For example, as the latter excludes what the former affirms, the statements "Abortion is ethically permissible"

C. Erk, *The Ethics of Killing*,
https://doi.org/10.1007/978-3-031-07183-6_1

and "Abortion is ethically impermissible" are opposed in a contradictory fashion.[1] Therefore, only one of these statements can and has to be true.

It is the genuine task of the science of ethics—or as it is also called, moral philosophy—to establish which of these two contradictory ethical statements is true and which is false. For the moral philosopher, contradictory pluralism is a situation that is unsatisfactory from an intellectual perspective. It poses a challenge that wants to be taken on and should not be avoided by refuge in ethical relativism. The wide-ranging differences in the ethical evaluation of different kinds of killings, therefore, call for a systematic treatment of the ethical permissibility of killing.

1. Kinds of Killings

This book is about the ethics of life and death. As such it deals with the problem whether and, if so, why and under which circumstances it is ethically permissible to kill a human being.[2] Accordingly, the guiding question of this book can be stated as follows:

> Under which conditions, if any, is it ethically permissible for a human being to kill a human being?

[1] Correctly formulated, contradictories differ from one another only by the word "not". So, the statement that is opposed in a contradictory fashion to a less absolute position on abortion such as "Abortion is ethically permissible early in pregnancy, but not in the third trimester" is not the statement "Abortion is not ethically permissible" but the statement "Abortion is not ethically permissible early in pregnancy, but not in the third trimester". In the same fashion, the contradictory opposite to the statement "Abortion is ethically impermissible, save in cases of rape, incest and threat to the mother's life" is not the statement "Abortion is ethically permissible" but "Abortion is ethically permissible, save in cases of rape, incest and threat to the mother's life".

[2] In general, the impermissible killing of a human being is called homicide. The question underlying this book could therefore also be restated as asking whether every killing of a human being is an act of homicide.

Having introduced the term "homicide", it is also helpful to explain what is meant by two closely related terms, namely "murder" and "manslaughter". Murder is a kind of homicide, namely homicide in an unscrupulous manner and with malicious aforethought, in which the intention of the agent or the method of killing is particularly depraved. Manslaughter is a kind of homicide, too; it is homicide that lacks malice and is brought about in a state of extreme emotion that is excusable in the circumstances (e.g. provocation) or in a state of profound psychological stress.

Admittedly, answering this question is a somewhat ambitious enterprise. To be able to take up this book's mission it is helpful to concretise what is and what is not covered. For the book simply cannot and does not cover all kinds of killings. A first step in the concretion of this book's mission is to emphasise that it is an exercise in the *individual* ethics of *human* life and death.

- This book is about the *individual* ethics of life and death. Its focus is on the killing of a human being by a human being. It is, however, not an exercise in the social ethics of life and death. Albeit performed by individual human beings, too, the killings that are the focus of social ethics are socially authorised cases of killings. The prime examples of such kinds of killings are capital punishment and killing in the context of war. The person who kills in these cases does so representing and on behalf of a community of persons.
  So, this book does not address the question whether it is ethically permissible for a community of persons to impose the death penalty on one of its members who has committed a serious wrong.[3] Its focus is on killings which are committed by private authority. Nor does it touch upon the subsequent question whether and under which circumstances a human being is ethically permitted to carry out a death penalty—either on another human being or himself. The book is also not concerned about cases of public duels, that is, duels that are fought between two individual human beings on behalf of the common good (e.g. to end a war).
  The book also does not grapple with the question whether and under which circumstances it is ethically permissible for a state to resort to war (*ius ad bellum*). And it does not address the question whether and under which circumstances a human being is morally permitted to kill in the context of an armed conflict with another state or in a civil war (*ius in bello*).[4]
- Furthermore, this book is about the ethics of *human* life and death. As human beings are not the only beings possessing life, some might

[3] For comprehensive treatises on the problem of the moral permissibility of capital punishment, the interested reader is referred to Feser & Bessette (2017) as well as Pojman & Reiman (1998).

[4] As starting points for an in-depth examination of this topic, the interested reader is referred to Walzer (2015) as well as Regan (2013).

conclude that an ethics of life and death should not only cover the taking of human life but also animal life and plant life. This book only considers the life of human beings. Therefore, it does not deal with the ethical (im)permissibility of the killing of animals or plants—or the underlying question of the ethical (im)permissibility of the consumption of animals, animal products or plants.[5] The book only deals with cases of killings in which both the being who is killed and the being who does the killing are human.

Having narrowed down the ground to be covered, this book's exact scope and limits become even clearer if we bear in mind that the killing of a human being can occur in several forms. An act of killing requires a human being who commits the killing (the "killer") and a human being who is killed (the "killed"). The human being who is killed can either be the person who is doing the killing or another human being. So, there are killings in which the killer is identical to the killed and there are cases in which the killer and the killed are different human beings. While the former case is always tantamount to suicide, the latter case can come in various forms depending on (a) whether the killing takes place before (or right after) the birth or at a later stage in the life of the killed, (b) whether it is undertaken against the will, without a volition or at the request of the killed, and (c) whether the killing occurs in defence against an act of unjust aggression. Applying these distinctions yields the following kinds of killings: suicide, lethal self-defence, abortion (including infanticide) as well as voluntary, non-voluntary and involuntary euthanasia. It is the ethical permissibility of these kinds of killings that this book sets out to investigate.

Another topic that deserves ethical investigation is organ transplantation, or to be more precise, the explantation of vital organic material from a human being ("explantee") for the purpose of implantation into another human being ("implantee"). Organ transplantations are governed by the so-called dead donor rule, which states that organ donors must be dead before any of their vital organs may be procured for transplantation. The permissibility of the explantation of vital organic material therefore hinges on the definition of death used to determine whether a donor is already

[5] Such investigation would not only have to cover the killing of animals such as chickens, cows or pigs but also the killing of plants such as trees, lettuce or potatoes, as well as the killing of bacteria, fungi, algae and one-celled living organisms (protozoa) in general.

dead or still alive. Depending on the definition used, the transplant of vital organs can therefore constitute a killing action. Against this background it makes sense to also shed light on this topic from a killing perspective.

Although it does not constitute a case of killing as such, it is also useful and instructive to add another object of investigation to our list, namely assisted suicide. In assisted suicide, a human being—be it a physician or not—somehow assists in another human being's suicide. This raises the question whether and to what degree such cooperation is ethically permissible.

Given these concretions, the book will therefore focus on the following kinds of killings (cf. Fig. 1):

Given these kinds of killings, the general guiding question of this book can be broken down into a range of sub-questions. Four of them pertain to the ethical permissibility of killing an innocent human being, one deals with the ethical permissibility of killing an unjust aggressor, and one focuses on the ethical permissibility of cooperation in another human being's suicide:

- Under which conditions, if any, is suicide ethically permissible?
- Under which conditions, if any, is lethal self-defence ethically permissible?

Kinds of Killing	A kills A (A = killer = killed)	A kills another human being (B) (A = killer ≠ killed = B)				A cooperates in another human being's (B) suicide (A ≠ killer = = killed = B)
		at B's request and with B's express consent	without a volition on part of B	against B's will		
				without an attack by B	in defence against an attack by B	
before birth of the killee	not applicable	not applicable	Abortion (incl. infanticide)	not applicable	not applicable	not applicable
after birth of the killee	Suicide	Voluntary Euthanasia	Non-voluntary Euthanasia	Involuntary Euthanasia	Lethal Self-defence	Cooperation in Suicide ("Assisted Suicide")
		Organ Transplantation/ Explantation of Vital Organic Material				
	↓	↓	↓	↓	↓	↓
	Killing an Innocent				Killing an Unjust Aggressor	Cooperation in Suicide

Fig. 1 The kinds of killings covered in this book

- Under which conditions, if any, is abortion or infanticide ethically permissible?
- Under which conditions, if any, is euthanasia ethically permissible?
- Under which conditions, if any, is organ transplantation ethically permissible?
- Under which conditions, if any, is cooperation in suicide ethically permissible?

What all these actions have in common is that they are so-called killings or killing actions. But what makes an action a killing action? Before outlining the structure of the book, let us briefly explain what is meant when a particular action is characterised as a killing action.

2. Killing, Letting Die, Dying, Living

What is characteristic of a killing action is that at least one living human being dies as a result of this action, that is, changes his status from "alive" to "dead". This is what we mean by "killing". In its most general sense, killing can be defined as the taking or ending of a life. So, killing means being causally responsible for a living being's death.

> Properly speaking, killing is a human action in which a human being initiates a causal sequence, directing it to another person's death. (Jensen 2011: 179)

If A kills B, A puts an end to B's life; it is because of A's action—be it a commission or an omission—that B crosses the threshold from life to death. A kills B if A is "'the agent' of harm that befalls someone else" (Foot 1994: 281), that is, if A initiates the causal sequence that leads to B's death. In other words: If A was the cause of B's death, A has killed B.

In contrast, letting die is a failure to prevent death: "Someone allows another to die when (1) there already exists a causal sequence leading to the person's death, (2) he could intervene to stop or delay the sequence, but (3) he chooses not to" (Jensen 2011: 174). Letting die occurs in situations "where one has the opportunity to prevent or delay death—because some causal sequence is already leading to someone's death, or is at least believed to be doing so—but one chooses not to" (Jensen 2011: 179).[6]

So far, we have been using the terms "death", "life" and "killing" without further clarification. Although some of these terms shall be further elaborated upon in subsequent chapters, it makes sense to also shed a clarifying light on some important concepts:

- Living = possessing life; the term "alive" can usually be used synonymously but has an element of exhibiting activity to it.
- Life = the possession of which causes a human being to be alive; the term can also refer to the period of time during which a human being lives.
- Conception (also: fertilisation) = the beginning of the life of a human being, resulting from the fusion of two opposite-sex human haploid gametes (germ cells), namely an egg (also: oocyte or ovum; female) and a sperm (also: spermatozoon; male).
- Birth = the end of a pregnancy and the beginning of a living human being's life outside the womb.
- Dying = the process of losing one's life.
- Death = the end of the process of dying, that is, the complete loss of life; the end of the life of a human being; however, the term can also refer to the state of being dead.
- Dead = the state of having died, that is, having lost one's life.
- Killing = causing a human being's death, that is, causing a human being to lose his life and change his status from "alive" to "dead".[7]
- Living being = a being that possesses life; instead of the term living being, biology uses the term "organism".

[6] As letting die is not tantamount to killing, rescue case-style examples are not suited to determine the ethical permissibility of killing. Indeed, there is an ethical argument to be made (e.g. by reference to the so-called order of charity (cf. Erk 2019: 237ff)) for why it is at least ethically permissible to save one's son rather than a strange child, a five-year-old girl rather than a frozen embryo or a country's president rather than a regular citizen—provided, of course, that one cannot save both. But being ethically permitted to let somebody die by not rescuing him does not logically imply that one is ethically permitted to kill this somebody. The former is simply irrelevant to the latter. Or as Kaczor (2020: 85) puts it: "The principles governing triage in rescue cases are not the same as the principles governing which human beings should be intentionally killed".

[7] Strictly speaking, killing a human being means performing an exterior action that results in the death of a living human being. The term "exterior action" will be introduced and explained later on (cf. Chap. III, Sect. 2.2).

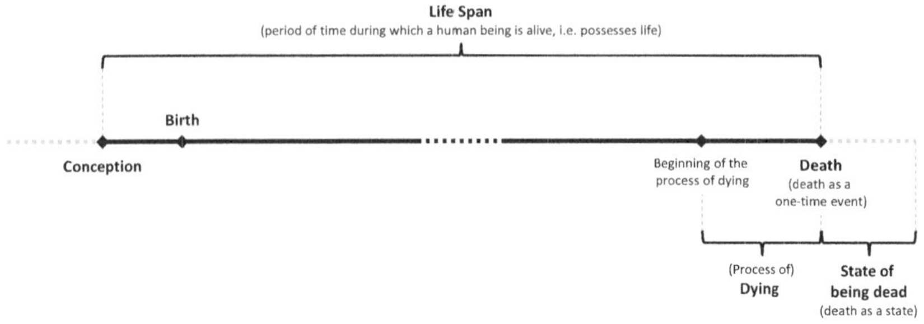

Fig. 2 Life, death, living and dying

To get a better overview, most of these terms can be ordered and plotted on a timeline (cf. Fig. 2):

As stated above, killing a human being means causing his death. This definition allows for the conclusion that every killing necessarily requires that somebody has died. If there is no one killed but "only" someone—more or less severely—injured than there has been no killing. A life-threatening assault is a serious wrongdoing, but it is not a killing.

3. The Structure of this Book

This book is devoted to the examination of the ethical (im)permissibility of killing human beings in general as well as selected kinds of killings in particular, namely suicide, lethal self-defence, abortion, euthanasia, organ transplantation and cooperation in suicide. So, this book is literally about matters of life and death. In order to not just scratch the surface, but to be able to answer the question with the seriousness due to such ethical investigation, a lot of preparatory work must be done. There are many concepts that need to be dealt with and understood before one can begin to define the ethical limits of dealing with our own lives and the lives of other human beings. The way to answer the guiding question of this book is to answer the following sub-questions:

- What does it mean to act?
- Which actions can be subject to ethical evaluation?
- What elements does an action comprise? Which of these elements are relevant for ethical evaluation?
- What makes an action evil? What makes it good?

- What is the difference between an action's goodness or evilness and its permissibility or impermissibility?
- What is a duty? What is a right? What is a moral right or duty?
- Is there a moral duty not to kill? Is this duty held by and against all human beings or only persons?
- What and who is a person? What is human dignity and who has it?
- What is it that is actually taken when somebody is killed, that is, what is life? What and when is death?

Without having seriously dealt with and answered these questions first, it is not only virtually impossible but also intellectually negligent to even attempt to answer the question of the (im)permissibility of killing human beings. This book addresses the above questions one by one and integrates the answers to them into an argumentative architecture that allows it to answer its guiding question. Accordingly, the book is divided into the following main parts (also cf. Fig. 3):

- **Human actions and their ethical quality**
  The first part of this book comprises Chaps. II and III. These chapters (a) familiarise the reader with the concepts needed to describe a killing action as a human action and (b) clarify how to determine the ethical quality of such action. Chapter II outlines the elements of which every action is composed and also describes the various degrees of voluntariness of an action. It then applies these findings to the action we call killing and by doing so defines the forms in which such action can take place.
  Chapter III explains how we can evaluate the morality and permissibility of a human action. It discusses the so-called sources of morality, that is, those elements of an action from which its morality derives. And it defines the terms "good" and "evil" and by doing so introduces the standard of morality that is needed to assess the ethical quality of an action, namely the essence or nature of human beings. Far from being simply neutral, good and evil have normative force. The chapter therefore continues to show how the nature of human beings grounds their most fundamental duties and rights, namely their moral duties and moral rights—and along with them the specific moral duty and right that are of interest to this book, namely the moral duty to live and the moral right to life.

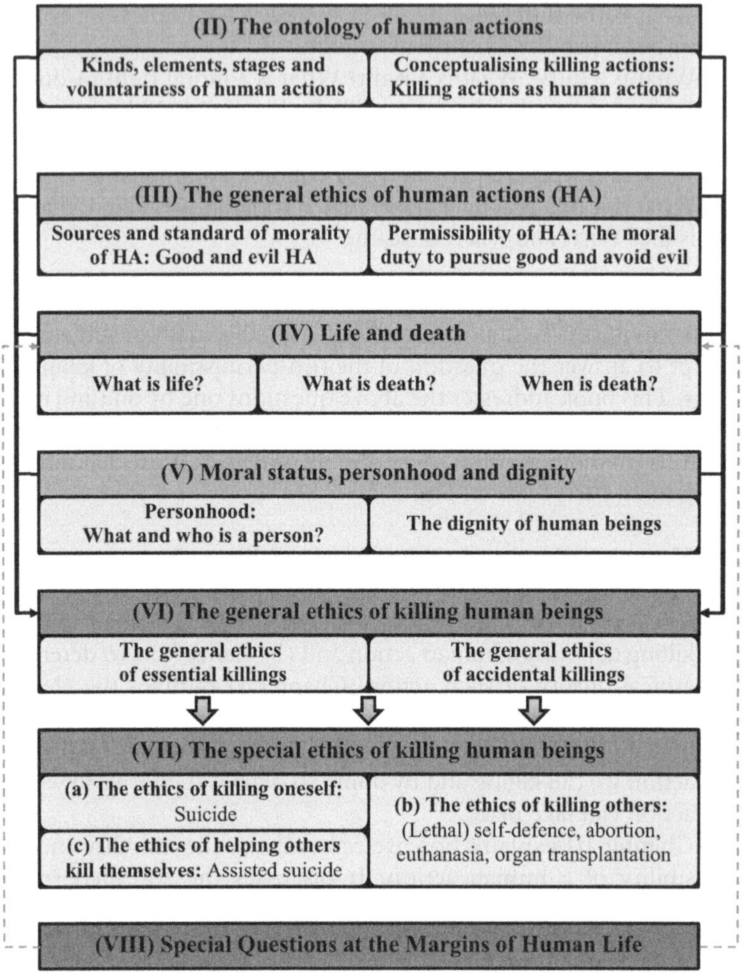

Fig. 3 The structure of the book

- **Life and death**
  In order to establish that an action is actually a killing and not something else, we need to be able to diagnose death. However, the tests for the diagnosis of death do not just fall from the sky but must be coherent operationalisations of an answer to the question of what we

mean by the term "death". Therefore, when thinking about the ethical (im)permissibility of killing, there is no way around dealing with the concepts of "death" and its contradictory opposite "life". Accordingly, Chap. IV—the second part of this book—is devoted to the definition of the concepts of "life" and "death". This chapter will answer the following questions: What is life? What is death? How do we know that a human being has died, that is, when is death?

- **The personhood, moral status and dignity of human beings**
  The third part of this book—Chap. V—contains a discussion of two of the most widely used but also most unclear concepts in moral philosophy, namely personhood and dignity. As moral rights and duties can only be possessed by beings who have moral status, that is, persons, it is vitally important to understand the relationship between the terms "person" and "human being". Therefore, the main tasks of this chapter are (a) to establish whether all human beings are persons and whether all persons are human beings and (b) to reflect on the sources of the dignity of human beings.
- **The ethical (im)permissibility of killing human beings**
  The fourth and penultimate part of this book comprises Chaps. VI and VII. These chapters apply the findings of the preceding chapters to the guiding question of this book, namely the (im)permissibility of killing human beings. This will be done by first outlining the general ethical limits of dealing with our own lives and the lives of other human beings. This general ethics of killing human beings is then used to evaluate the specific ethical (im)permissibility of suicide, lethal self-defence, abortion, euthanasia, organ transplantation and cooperation in suicide.
- **Specific Problems at the Margins of Human Life**
  The last part of this book comprises Chap. VIII and addresses several specific and somewhat technical questions concerning the beginning and end of human life and existence. It serves as an extension and deepening of the concepts introduced in Chap. IV. The first four parts of the book gradually develop a comprehensive ethical theory of killing and are therefore directed to a general audience interested in the (classical) natural law ethics of killing. The book's last part is directed at a readership interested in some important intricacies of the academic discussion within the natural law tradition, especially those pertaining to hominisation, ensoulment in general and ensoulment with a rational soul in particular.

Having said what needed to be said to set the stage, we can now go *in medias res* and start explicating the ethics of life and death.

Bibliography

Erk, Christian. *Moral Philosophy. Fundamental Concepts.* Second, revised and expanded Edition. Heusenstamm: editiones scholasticae, 2019.

Feser, Edward & Joseph Bessette. *By Man Shall His Blood Be Shed: A Catholic Defense of Capital Punishment.* San Francisco: Ignatius Press, 2017.

Foot, Philippa. "Killing and Letting Die." In: Steinbock, Bonnie & Alastair Norcross (Eds.). *Killing and Letting Die.* New York: Fordham University Press, 1994. pp. 280–289.

Jensen, Steven J.. "Killing and Letting Die." In: Jensen, Steven J. (Ed.). *The Ethics of Organ Transplantation.* Washington, D.C.: The Catholic University of America Press, 2011. pp. 170–191.

Kaczor, Christopher. *Disputes in Bioethics. Abortion, Euthanasia, and other Controversies.* Notre Dame, IN: University of Notre Dame Press, 2020.

Pojman, Louis P. & Jeffrey Reiman. *The Death Penalty: For and Against.* Lanham et al.: Rowman & Littlefield Publishers, 1998.

Regan, Richard J.. *Just War: Principles and Cases. Second Edition.* Washington, D.C.: Catholic University of America Press, 2013.

Walzer, Michael. *Just and Unjust Wars. A Moral Argument with Historical Illustrations. Fifth Edition.* New York: Basic Books, 2015.

II: The Ontology of Human Actions

As explained in the previous chapter, this book has set out to examine the ethical permissibility of certain acts of killing, namely suicide, lethal self-defence, abortion, euthanasia, organ transplantation and cooperation in suicide. To be able to enquire about the ethical (im)permissibility of these instances of killing actions, one must first understand their general nature as killing actions. So, the first step to take is to look at what we call a killing action and to describe as precisely as possible what kind of action we are actually dealing with.

To be able to do this, we must first acquire the necessary vocabulary. This is what the next four chapters are for. They provide us with the conceptual foundations that allow us to describe and categorise actions. A fifth chapter then applies these distinctions and categories to the action we call killing.

1. Kinds of Actions

In its broadest sense, an action is something someone does. Someone acts if he is carrying out an activity, and someone has acted if he has carried out an activity. As this initial characterisation reveals, an action requires an efficient cause, that is, something that brings it about. The efficient cause that produces an action is called an "agent". As this book is only concerned with killings that are committed by a human being, all agents relevant to this book's ethical problems are human beings.

C. Erk, *The Ethics of Killing*,
https://doi.org/10.1007/978-3-031-07183-6_2

When it comes to actions performed by human beings, we must, however, be careful with our choice of words. From the perspective of moral philosophy, the term "action" contains too many conceptual nuances to be used without further clarification. To be able to apply the term correctly, we need to therefore make several important distinctions.

The first distinction to be made is that between "human actions" and "actions of a human being". Although they are composed of almost the same words, they denote different things. While all human actions are performed by human beings, not all actions performed by human beings are human actions:

- A *human action*—or as it is also called, directly voluntary action—is an action that proceeds from a human being's deliberate and free will, that is, is freely willed on the basis and in the light of rational deliberation.
- In contrast, an *action of a human being* is an action which a human being happens to perform without having consciously controlled or deliberately willed it.

We can only speak of a human action if its agent is acting both knowingly, that is, knows what he is doing, and willingly, that is, freely wills what he is doing. Actions produced by a principle extrinsic to the agent (e.g. forced acts) or actions produced by an intrinsic principle but without influence of reason and will (e.g. digestion) cannot be said to be human actions. We can therefore give the following definition of human actions:

> A human action is a directly voluntary action of a human being, that is, an action caused and performed by its agent both knowingly and willingly.

The distinction between human actions and actions of a human being is relevant because the object of ethics is human actions only. Ethics—or as it is also called, moral philosophy—is "the science of moral good and evil in human acts" (Cronin 1930: 1). Strictly speaking, ethics can and does only evaluate the ethical quality of actions if and insofar as they will be, are or have been carried out knowingly and willingly. As will be detailed later, an agent can only be held morally responsible for having performed an action if and insofar as this action has been performed voluntarily, that is, is a human action. Only voluntary and therefore human actions are within the scope of ethics.

This general characterisation must, however, be rendered more precise right away. Given its etymological proximity to the word "active", some might be led to interpret the verb "to act" as involving a single observable bodily movement or physical activity of an agent. However, this is only partially true. Agents can act not only by actively doing something but also by refraining from acting, that is, by not doing anything. It is easy to understand why this is so if we bear in mind that one of the characteristics of an action is that it is freely willed. But this willing can be twofold: A human action can be willed in a positive sense (*voluntarium positivum*) or a negative sense (*voluntarium negativum*). The human action that results from the former kind of volition is called "positive human action"; the one that results from the latter is called "negative human action":

- A *negative human action* is characterised by the fact that its agent deliberately wills to do nothing. Such action can come in either of two forms:
  - In a *deliberately willed passivity (also: sufferance)* an agent deliberately wills not to react to something happening to himself. For instance, an agent might deliberately choose not to fight back when getting beaten up.
  - In a *deliberately willed inactivity (also: omission)* an agent deliberately wills to do nothing despite the possibility to the contrary. For instance, an agent might deliberately choose not to help a drowning child although he could do so.
- A *positive human action* is characterised by the fact that its agent deliberately wills to do something. Such action is also called a *deliberately willed activity* or *commission.* For instance, an agent might deliberately choose to help a drowning person.

So, doing nothing can be as much a human action as doing something. Whether a human being performs a human action does not depend on the presence of some bodily movement but on the movement of his will.

Within the class of human actions two further distinctions can be made. The first distinction concerns the observability of an action. Many of our actions involve a bodily movement on our part that can be observed by another person. However, the concept of human actions does not only extend to such *externally observable human actions.* There are also *internal human actions* (such as willing, loving and thinking) that are not necessarily accompanied by a bodily movement and therefore cannot be observed. Insofar as they are performed voluntarily, they are human actions, too.

The second distinction to be made at this point is that between basic human actions and non-basic human actions. This differentiation concerns the question whether a human action is a composite of several sub-actions or not:

- *Basic human actions* are actions we do not do by doing something else. Such action is simple in the sense that it is not made up of other basic or non-basic actions. A basic action is its own and only building block.
- *Non-basic human actions* comprise other (basic or non-basic) actions. Non-basic actions are an organised series of several distinct and discrete basic actions.

Given their nature, basic actions tend to be short-term occurrences; non-basic actions usually extend over a longer period of time—hour(s), days(s) and so on. Having breakfast is a typical example of a non-basic action: This action is performed by performing a variety of other (both basic and non-basic) actions ranging from drinking and eating in various forms and fashions to wiping one's mouth with a napkin at the end.

Having introduced these basic distinctions, we can now take the next step in our attempt to conceptualise the action of killing. This step consists in outlining the elements of which all human actions are composed. Knowing their elements allows us to describe killing actions more concisely than we have done so far.

2. The Elements of Human Actions

An action is a human action because and insofar as it is deliberately willed. But what exactly does the will actually will when causing and performing a human action? Posing this question means asking about the elements of a human action. As will be shown over the course of the following pages, every human action necessarily consists of four elements: (a) the end of the agent, (b) the exterior action and its natural end, (c) the circumstances of the exterior action and (d) the consequences of the exterior action.

2.1. The End of the Agent

The first and most fundamental thing which the will wills when causing and performing a human action is the *end of the agent* (also: purpose, motive; Latin: *finis operantis*). Human acts do not just happen; they are taken for a reason and with a purpose in mind. The end of the agent answers the question why the agent performs the act, that is, what he wants to see realised with his action.

> Every human action is for the sake of some end. That is, every human action is purposive in some way; it is done for the sake of attaining some goal or realizing some desired state of affairs. (Williams 2012: 199)

In voluntary acts, this purpose does not, however, just fall from the sky but must be deliberately willed. It is by the so-called *interior action (of the will)* (also: inward voluntary action) that an agent chooses the end he desires to see realised by his action.[1] At the end of the interior action there stands a consciously willed end of the agent that explains why the agent is taking some action.

When it comes to the end of an agent, we can draw several distinctions: On the one hand we can distinguish between primary and secondary ends. The *primary end* is the main motivating reason for the respective human action while the *secondary end* is a co-determining but only marginal and non-decisive motive for the respective human action. The primary end is the one that is above all and primarily intended, and which impels the

[1] As the name indicates, this action is something that happens within the agent. To be more precise, it is a mental process that involves the interplay of the intellect and will of the agent. Something can only be willed as an end if it has been apprehended as good by the intellect of the agent. The will depends on a judgement of the intellect; without such operation of the intellect, there can be no willing.

Of course, the judgement of the intellect can be wrong. Just because my reason judges something to be good does not automatically mean that this something is truly good; it just means that I am convinced that it is good.

But even in light of some particular good presented to our will, we remain free to will or not will that good. While the will cannot but strive for the good in general, no particular good can force the will to will it; with respect to all concrete goods the will possesses freedom of choice. This is due to the fact that all goods but the absolute and perfect good have at least one aspect to them that can be considered not good—if not in the good itself, then in the circumstances of and the opportunity costs related to its attainment.

For a detailed description of the interior action of will the interested reader is referred to Erk (2019: 82ff)

agent to act more than anything else. The secondary end is neither intended solely for itself nor moves the agent to act by itself alone. So, an action can be done with a single (and therefore necessarily primary) end or multiple (i.e. a primary and one or more secondary) ends in mind.

On the other hand, we can distinguish between an end that is pursued for the sake of itself and an end that is pursued for the sake of something else. An end that is pursued only for its own sake is called a *last end* (also: ultimate end, supreme end). An end that is realised not only for its own sake but also for the sake of something else is called an *instrumental end* (also: sub-end, mediate end). With respect to the last end we can distinguish between an absolute last end and a relative last end:

- A *relative last end* is an end that is a terminus for some action(s) but is still ordered towards a higher end; it is the last end of some action(s) but still related to an ulterior end. Strictly speaking, while being a last end in a certain order of ends, such a relative last end is an instrumental end in the grand order of ends.
- The *absolute last end* is the end which has no other end above it. The absolute last end is the end which is not subordinated to any other end; as such it is not striven for to reach another, higher end but sought for its own sake.

The class of instrumental ends can be further distinguished into superior (also: farther, remote, superordinate) and inferior (also: subordinate) instrumental ends:

- A *superior* (also: superordinate) *instrumental end* is an end that, although not a supreme end, is realised by means of other ends; a relative last end is a superior instrumental end.
- An *inferior* (also: subordinate) *instrumental end* is an end whose realisation helps to realise a superior instrumental or a supreme end. The supreme end can also be a superior end in relation to its subordinated instrumental ends. The lowest inferior end is also called the nearest or proximate end.

Certain instrumental ends can be both inferior and superior depending on whether they are looked at from the perspective of the supreme end or some subordinated lower-level instrumental end.

2.2. *The Exterior Action and its Natural End*

The second thing that the will wills when willing an action is causally related to the end of the agent. He who really wills the realisation of some purpose must necessarily also will the means suitable for the realisation of that purpose. Let us illustrate this with an example: If Peter is hungry, his intellect will cognise the satisfaction of his hunger as good and present this good to his will; his will will then will the satisfaction of his hunger. The satisfaction of his hunger is willed by Peter as his end of the agent or purpose. As Peter really wills the satisfaction of his hunger as his purpose, he also wills to take the appropriate means to realise this purpose by, for example, cooking a satiable meal and eating it. In consequence, Peter is cooking such a meal and eating it.

The means actually taken for the realisation of the end of the agent constitute the so-called exterior action (also: external action, outward action). Whereas the end of the agent is that for the sake of which something happens, the something that happens is the *exterior action* of the human action. This exterior action is the second element of a human action. Just like the end of the agent, the exterior action of a human action is consciously willed as the means to realise the end of the agent.

When it comes to the evaluation of a human action, careful attention must be placed on the concise definition of the exterior action. This definition can best be given by reference to the nature of the exterior action performed by the agent. What the body is moved to do in order to realise the end of the agent has a natural end that makes the exterior action the kind of action it is. This natural end immanent to the exterior action is the so-called (natural) end of the exterior action (Latin: *finis operis*)[2]. The natural end of the exterior action is that which the exterior action naturally—that is, regardless of what the agent wants to achieve with this action (*finis operantis*) and regardless of whether this or that agent performs it—results in and with the attainment of which it is usually completed. As such it is also called the objective or natural purpose of the exterior action. The end of the exterior action can coincide with the end of the agent; it can, however, also entirely disaccord with it.

For example, by its very nature the exterior act of eating is directed to the satisfaction of hunger and the procurement of the strength necessary

[2] From the perspective of the exterior action, the end of the exterior action is sometimes also called the exterior action's proximate end (*finis proximus*) and the end of the agent the exterior action's remote end (*finis remotus*).

for survival; this is its *finis operis*. By its very nature the *finis operis* of constructing a house is the house. But, obvious as the end of the action might be in these cases, how do we generally know what the natural end of some exterior action actually is?

Generally speaking, the end of an exterior action is the action's essential characteristic(s) that makes it the kind of action it is, that is, without which the exterior action would not be the kind of action it is but some other action. Its natural end is so intrinsic to an exterior action that the agent cannot will to perform the exterior action without also willing the realisation of this end; and neither can the agent alter or eliminate this end. As a side note: Usually, the overall human action is named after its exterior action.

The natural end of an exterior action can be determined by taking recourse to common experience (of life) and the usual or natural course of things. Its natural end is the outcome that results from a certain kind of exterior action in the majority of cases, that is, with some specified (very high) degree of probability. It is that which cannot be willed not to obtain if someone performs an exterior action. Or put differently: The *finis operis* is the end that must be intended by an agent if we abstract from the why of the exterior action, that is, if the agent would perform the action for the action's sake without a distinct *finis operantis*.

Before moving on to the remaining two elements of a human action, it might be helpful to sum up what has been established so far:

- In performing an action, every agent necessarily wills the realisation of two ends, namely the end of the agent (*finis operantis*) as well as the end of the external action (*finis operis*).
- The end of the agent (*finis operantis*) is the end that the agent seeks to realise by performing the external action, that is, by realising the end of the external action.
- The end of the external action (*finis operis*) is the natural end of the external action whose realisation the agent wills for the sake of the realisation of his *finis operantis*. Its natural end is what defines an exterior action, that is, makes it the kind of action it is rather than some other kind.

Having identified the two core elements of a human action we can now direct our attention to its third element, namely its circumstances.

2.3. The Circumstances

Human actions are not performed in a vacuum outside of time and space. Rather, every human action is performed under and accompanied by a variety of unique conditions that surround, that is, stand around (Latin: *circum stare*) it. These conditions are commonly called “circumstances”. To be more precise, the circumstances are not the conditions of a human action as such but of the exterior action of a human action. They do not detail the context in which a human action as such is performed; they detail the context in which the exterior action of a human action is performed. The circumstances are the accidental properties of an exterior action. Although they are not directly connected to its essence, that is, do not define the nature of the exterior action, they contain additional information that is relevant to understanding the exterior action.

Typically, the list of the circumstances of an exterior action comprises seven circumstances usually abridged—as perhaps familiar from the famous Latin hexameter—by the seven interrogative pronouns who (*quis*), what (*quid*), where (*ubi*), with what aids/instruments (*quibus auxiliis*), why (*cur*), how (*quomodo*) and when (*quando*). As the circumstance “what” has two components, the list of the circumstances actually comprises the following eight elements (cf. I^a^-IIae q. 7 a. 3 co.):

- **Who?—Circumstance of person**
  The circumstance “who” details the agent, that is, the efficient cause of the action. It specifies the special qualities or characteristics of the agent performing the action. Quite often it makes a difference who is actually performing an action: If a poor person donates the same amount of money as a rich person but in doing so gives away half of his wealth while the rich person only donates 1/10,000 of his fortune, the same action might be judged differently. An example of such information specifying the agent and his properties is his socio-demographic standing, such as his education, age, occupation, (societal) position and family status.
- **What?**
- The circumstance “what” has two aspects to it: It can either refer to (a) the quality and quantity or (b) the consequences of an exterior action. To distinguish these aspects this circumstance can be subdivided into the two circumstances “with respect to what/about what” (*circa quid*) and “what” (*quid*).

- **About what?—Circumstance of quantity and quality**
  The circumstance "with respect to what/about what" (*circa quid*) refers to and details the distinctive qualitative and/or quantitative characteristics of the object which undergoes the exterior action. To be more precise: While the *finis operis* describes the essence of the exterior action, the circumstance "with respect to what/about what" refers to the "merely" accidental properties of the exterior action, namely the quality and quantity of the object or matter affected by it.
  For example, it does not suffice to describe an action as "donating", that is, as the giving of money to a charity. It is also important how much money is donated, whether its one's own money that is donated and who the recipient of the money is. In the same fashion, an act of stealing needs to be concretised by indicating how much is stolen from whom. Similarly, an act of beating needs to be concretised by stating who is beaten—beating one's father is different from beating a stranger.
- **What?—Circumstance of effects of the act**
  The circumstance "what" (*quid*) refers to and details the unintended effects and consequences of the external action.
  While being an element of a human action in its own right (cf. chapter II.2.4), the consequences of an exterior action are also one of its circumstances. This is so because from the perspective of the exterior action its consequences are accidental to the exterior action; the same exterior action done by different persons at different times and so on can result in a variety of consequences. However, as this circumstance is already covered by a separate element and does not have to enter the ethical evaluation of a human action twice, we can exclude it from the list of circumstances.

- **Where?—Circumstance of place**
  The circumstance "where" requires us to specify the location where the exterior action is performed. This not only includes a geographical specification but also whether the exterior action has been performed alone and secretly or in the presence of others.
- **By whose aid? With what aids/instruments?—Circumstance of helps or influences**
  This circumstance refers to the instrumental causes of the act and requires a specification of the persons or tools with the help of whom or which the action is performed.

- **Why?—Circumstance of intention**
  The circumstance "why" refers to the personal purpose of the agent and explains what induces the agent to act. It is therefore tantamount to the end of the agent (*finis operantis*).
  We know that the end of the agent is an element of a human action in its own right. However, the *finis operantis* is also a circumstance of a human action's exterior action, because from the perspective of the exterior action the end of the agent for performing the exterior action is accidental to the exterior action; the same exterior action can be done for a variety of purposes. However, as this circumstance is already covered by a separate element and does not have to enter the ethical evaluation of a human action twice, we can exclude it from the list of circumstances.
- **When?—Circumstance of time**
  By answering the question of the "when" of an action we specify the point in time at which the exterior action was performed and/or the duration of the action.
- **How?—Circumstance of mode or manner**
  The circumstance "how" (also: in which way; in what way) denotes the mode or manner in which the exterior action was done. This circumstance especially asks about the inner state the agent was in when performing an action, that is, whether the action was done, for example, out of fear, out of ignorance or under the overwhelming influence of an emotion. This circumstance also asks how often the action is done (repeatedly or once) and with which intensity (lightly or forcefully).

The seven circumstances can be distinguished into three main classes, depending on whether they affect or modify the exterior action itself, its causes or its effect:

- The circumstances about what (*circa quid*), when (*quando*), where (*ubi*) and how (*quomodo*) affect and modify the exterior action itself.
- The circumstances who (*quis*), with what aids/instruments (*quibus auxiliis*) and why (*cur*; *propter quid*) affect and modify the causes of the exterior action.
- The circumstance what (*quid*) affects and modifies the consequences of the exterior action.

Having explicated this third element of a human action, we can now turn our attention to its last element, namely the action's consequences.

2.4. *The Consequences*

Every human action—or to be more precise: every exterior action—results in something. Every exterior action has, in other words, consequences. These consequences constitute the fourth element of a human action. According to its etymological roots, a consequence (derived from the Latin verb *consequi* which translates as "to follow after") is that which follows something. If something is said to be a consequence of an exterior action, this something is the result or outcome of that action; it is the effect consequent upon the action. Therefore, the consequences of an exterior action comprise all states of things, situations, results, subsequent events and/or effects that follow from the respective action.

When it comes to the consequences of an exterior action, we can distinguish the following main kinds of consequences:

- ***Potential consequences vs. actual consequences***
  A potential consequence is a consequence that can but has not (yet) come to pass. An actual consequence is a consequence that can and has come to pass. Before the performance of an external action there can only be potential consequences. While all actual consequences must have previously been potential consequences, not all potential consequences must necessarily become actual consequences. Only the actual consequences of an exterior action are relevant for the evaluation of the ethical quality of a human action.
- ***Direct consequences vs. indirect consequences***
  A direct consequence follows immediately from the external action. First and foremost, this immediacy is a logical one; it does not need to be a temporal one. The fact that a consequence has followed an external action with a certain time lag does not necessarily mean that it cannot be its direct consequence; it can still be the logically immediate result of the external action.
  An indirect consequence follows from an external action only mediately, that is, pursuant to one or more other consequences. It is not the result of the external action itself but of subsequent cascading effects. An indirect consequence is the consequence of an external

action's consequence. Usually, indirect consequences follow the external action with a time delay.

- ***Essential consequences vs. accidental consequences***
  An essential consequence is intrinsic to the exterior action, that is, results from it by necessity (or at least naturally and in most cases). The essential consequences of an exterior action are tantamount to its natural end (*finis operis*). Normally the natural end of the exterior action is not only an essential but also a necessary consequence of the exterior action. For if an exterior action fails to actualise a certain kind of end, then the agent might have tried but simply did not perform this but some other kind of exterior action.
  Although all non-essential consequences tend to be called accidental consequences, this book takes a more fine-grained approach and considers accidental consequences a special kind of non-essential consequence. In general, a non-essential consequence is a consequence the occurrence of which is not intrinsic to an exterior action. A non-essential consequence can but does not have to result from the performance of an external action and usually only occurs in exceptional cases or seldom. Strictly speaking, all consequences other than the natural end of the exterior action are non-essential to the exterior action. This not only includes the non-essential consequences of the exterior action as such but also the end of the agent (insofar as it has been successfully realised by the exterior action). The non-essential consequences are also called side effects, collateral effects or secondary effects. Following Thomas Aquinas' position according to which something is accidental if it is beside the agent's intention (cf. II^{a}-IIae q. 64 a. 7 co.), what sets accidental consequences apart from other non-essential consequences is that their occurrence is not intended; an accidental consequence is an unintended non-essential consequence.

When thinking about the consequences of an action, we must furthermore bear in mind that consequences are always consequences for somebody. The consequences of an action performed by some agent can affect the agent, one or more other persons, or both. A full account of the consequences of an exterior action must therefore also include the specification of the persons who are affected by the respective consequence(s).

2.5. Summary

Combining the insights gained so far, it can be stated that an individual action, that is, the action of an individual agent, comprises four elements:

- the end of the agent (*finis operantis*)
- the exterior action and its natural end (*finis operis*)
- the circumstances of the exterior action
- the consequence(s) of the exterior action

Put together, a complete human action can be depicted as follows (cf. Fig. 1):

Having outlined the four elements of a human action, we can now turn our attention to the next descriptive dimension of human actions. Some pages ago, human actions have been defined as voluntary, that is, deliberately willed actions. We now know that a human action consists of four elements. This allows us to conclude that an action is a voluntary and therefore human action to the extent that its elements are voluntary, that is, deliberately willed. But how do we know that an element is deliberately willed? This is the question addressed in the next chapter.

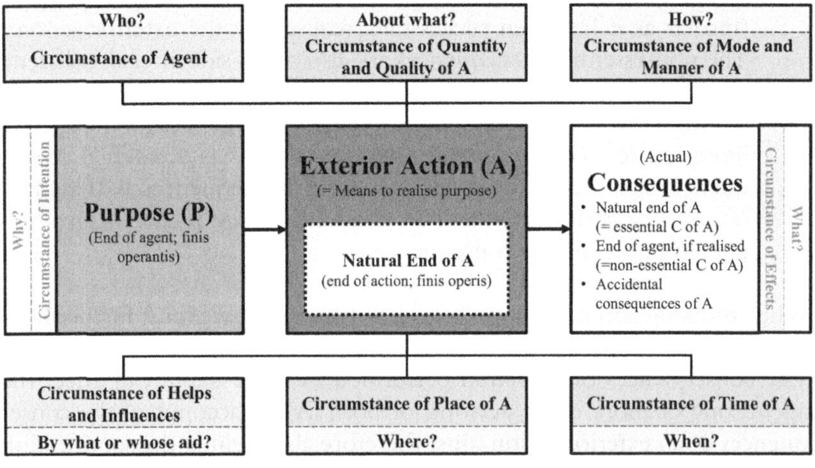

Fig. 1 The elements of a complete human action

3. Degrees of Voluntariness

In chapter II.1 we had a look at the factors that make an action of a human being a human action. As we have learned, human actions are directly voluntary actions of a human being, that is, actions that are caused and performed by their agents both knowingly and willingly. In addition, chapter II.2 has familiarised us with the four elements of a human action, namely the end of the agent, the exterior action and its end, the circumstances of the exterior action, and the consequences of the exterior action.

If we combine these insights, we can conclude that an action is a directly voluntary and therefore human action to the extent that its elements are directly voluntary (also: deliberately willed). But what is it that makes an element of an action a directly voluntary element?

As has been established, an action's degree of direct voluntariness is constituted by both its willingness and knowingness. Accordingly, it is the degrees of willingness and knowingness of the elements of an action that determine an action's overall degree of voluntariness. The degree of willingness and knowingness of the elements of an action can be determined by means of the following three distinctions:

- ***Foreseeability: Foreseeable vs. unforeseeable***
  The occurrence of something is foreseeable for an agent if it can be foreseen by him. In contrast, the occurrence of something that is unforeseeable cannot be anticipated, either in principle or given a certain set of circumstances.
  The measure of foreseeability is not what a perfectly informed being could have known but rather what a prudent and reasonable agent who has applied due diligence could have known under given circumstances.
- ***Foresight: Foreseen vs. unforeseen***
  Something has been foreseen by an agent if he knew that it would or at least could come to pass. In contrast, the occurrence of something has been unforeseen by an agent if he did not know that it would or at least could come to pass.
  As logic dictates, something can only be foreseen if it is foreseeable to begin with; unforeseeable occurrences are impossible to foresee. However, not everything that is foreseeable for the agent must also be foreseen by the agent.

- ***Intention: Intended vs. unintended***
  An agent intends something if he wills its occurrence, that is, if he chooses to bring it about.[3] In contrast, something is said to have occurred unintendedly if its occurrence was not willed by the agent when acting.
  Something can only be intended if it is foreseen (and in consequence foreseeable). It is a logical impossibility for something to be unforeseen but intended. But not everything that is foreseeable and foreseen by an agent must also be intended by the agent.

While intention constitutes willingness, foreseeability and foresight constitute knowingness. Something is willed if it is intended, and something is known if it is foreseeable and foreseen. If something has these three characteristics, it is directly voluntary. But as something can, for example, be foreseen but unintended, there must also be lesser degrees of voluntariness. Combining the possible expressions of the three dimensions of voluntariness yields the following degrees of voluntariness of the elements of an action and an action as a whole:

- **Directly voluntary**
  The occurrence of something is directly voluntary (also: voluntary in itself; premeditated) if it is foreseeable, foreseen and intended (cf. Prümmer 1960: 106).
- **Indirectly voluntary**
  The occurrence of something is indirectly voluntary (also: knowingly accepted; voluntary in cause) if it is foreseeable, foreseen but not intended (cf. Prümmer 1960: 106).[4]

[3] As Fagothey (1963: 32f) points out, we can distinguish different types of intention: "An actual intention is one that a person is conscious of at the moment he performs the intended action. The person pays attention not merely to what he is doing but also to the fact that he is here and now willing it. A virtual intention is one that was once made and continues to influence the act now being done, but is not present to the person's consciousness at the moment of performing the act. [...] A habitual intention is one that was once made and not retracted, but does not influence the performance of the intended act. [...] An interpretative intention is one that has not been made but presumably would have been made if the person were aware of the circumstances" (Fagothey 1963: 32f). For an action to be a voluntary action, virtual as opposed to actual intention is sufficient.

[4] Sometimes indirect voluntariness is enriched by the criterion of avoidability. On that account, something has been brought about in an indirect voluntary fashion if it has been foreseeable, foreseen, unintended and avoidable. For the purposes of this book, the criterion of avoidability will not be taken into further consideration.

- **Unknowingly accepted**
  The occurrence of something is unknowingly accepted if it is foreseeable but neither foreseen nor intended.

Figure 2 sums up the insights and gives an overview of the degrees of voluntariness that the respective elements of a human action can take on:

When it comes to human actions, the *end of the agent* must by logical necessity always be directly voluntary. To be more precise: Whenever a rational being moves itself to act, the action must have a directly voluntary final cause, that is, a directly voluntary reason why it is performed. There is a logical reason for this:

<table>
<tr><th colspan="3" rowspan="3">Element of Human Action</th><th colspan="3">Determinants of Voluntariness and their Expression in the Elements</th><th rowspan="3">Possible Degrees of Voluntariness of Element of Human Action</th></tr>
<tr><th colspan="2">Knowingness</th><th>Willingness</th></tr>
<tr><th>Foreseeable for agent?</th><th>Foreseen by agent?</th><th>Intended by agent?</th></tr>
<tr><td colspan="3">End of the Agent
(= purpose of A_{ext})</td><td>✓</td><td>✓</td><td>✓</td><td>→ Directly voluntary (by logical necessity)</td></tr>
<tr><td colspan="3">Natural End of the Exterior Action (A_{ext})</td><td>✓</td><td>✓</td><td>✓</td><td>→ Directly voluntary (by necessity)</td></tr>
<tr><td rowspan="5">Consequences of A_{ext}</td><td colspan="2">End of exterior action of A_{ext} (= essential consequence)</td><td colspan="4">If actualised, the natural end of A_{ext} is an essential actual consequence of A_{ext}. Its degree of voluntariness is already specified qua being the natural end of A_{ext}.</td></tr>
<tr><td rowspan="4">Non-essential consequences</td><td>End of agent</td><td colspan="4">If actualised, the end of the agent is a non-essential actual consequence of A_{ext}. Its degree of voluntariness is already specified qua being the purpose of A_{ext}.</td></tr>
<tr><td>Premeditated non-essential consequences</td><td>✓</td><td>✓</td><td>✓</td><td>→ Directly voluntary</td></tr>
<tr><td rowspan="2">Accidental consequences of A_{ext} (= Unintended non-essential consequences)</td><td>✓</td><td>✓</td><td>X</td><td>→ Indirectly voluntary (knowingly accepted)</td></tr>
<tr><td>✓</td><td>X</td><td></td><td>→ Unknowingly accepted</td></tr>
<tr><td colspan="3" rowspan="3">Circumstances of A_{ext}
(i.e. all circumstances other than the circumstances "why" (= end of agent) and "what" (= consequences of A_{ext}))</td><td>✓</td><td>✓</td><td>✓</td><td>→ Directly voluntary</td></tr>
<tr><td>✓</td><td>✓</td><td>X</td><td>→ Indirectly voluntary (knowingly accepted)</td></tr>
<tr><td>✓</td><td>X</td><td></td><td>→ Unknowingly accepted</td></tr>
</table>

Fig. 2 Degrees of voluntariness of a human action

> If the agent were not determined to a definite effect, it could not act and produce this effect rather than another. Therefore, it would not act at all. The determination of the agent to a definite effect is precisely what we mean by the end. (Renard 1948: 146; cf. I[a]-IIae q. 1 a. 2; Contra Gentiles, lib. 3 cap. 2)

Without a directly voluntary end of the agent, there can be no exterior action and therefore no action at all. As an exterior action is the means by which an end of the agent is realised, an absence of the latter makes the former obsolete. Without knowing the reason to perform an exterior action, there is no point in performing this rather than another exterior action. If his action did not have a directly voluntary *finis operantis*, the agent could not have performed this exterior action. So, the fact that an agent has performed an exterior action allows for the conclusion that his action must have had a directly voluntary end of the agent.

In reality, however, there is a problem with the end of the agent: In contrast to the exterior action as well as the circumstances and consequences of the exterior action, the end of the agent cannot easily be observed. The end of the agent is formed in the mind of the agent. Unless an agent reveals his end of the agent or unless we are dealing with an obvious case that allows for one conclusion only, we can only take a guess at an agent's *finis operantis*. We know that an agent must have had a directly voluntary *finis operantis* for performing an exterior action, but it is hard to have certain knowledge about its nature unless he tells us in all honesty.

While it is up to the agent to choose which *finis operantis* to pursue, *the end of the exterior action* is an objective given. The agent is not free to choose what his end of the action is. An exterior action just is what it is regardless of what the agent wants it to be. Every exterior action has a foreseeable natural end that its agent cannot not will to be realised when performing this exterior action. The end of an exterior action must therefore always be deliberately willed, too. For example, by its very nature the natural end of the exterior act of eating is the satisfaction of hunger and the procurement of the strength necessary for survival; unless one is suffering from a disease that prevents it, one cannot eat without intending to sate oneself. An agent who has performed an exterior action is always expected and held to have done so directly voluntarily.

As so often in life, there are exceptions: An agent can perform an exterior action without intending its natural end if he is coerced to do so; in that case it is, however, not so much the agent who performs the action but he who coerces the agent. It is also possible for an agent to perform an exterior action without foreseeing its natural end, that is, without knowing that this action will by its nature have a certain result. Such a case of acting out of ignorance is present if the agent in question is, for example, a child or a mentally ill person who does not yet understand what she is actually doing. These exceptions do not, however, make the rule but prove it.

4. The Stages of a Human Action

A human action does not simply happen. The performance of a human action is the result of a complicated interplay between reason and will. A condensed account of these processes can be construed as consisting of six consecutive stages (cf. Erk 2019: 309ff). Each of these stages comprises an act of the intellect (cognition) as well as an act of the will (volition). Every act of the will is preceded by a directive act of the intellect. So, every human action can be said to consist of 12 successive sub-acts that in their entirety form the process of a human action:

(I) Order of Intention (*Ordo Intentionis*) **(A) Acts of the intellect and acts of the will with regard to the end (intention)**		
(1a) Cognitio/(Simplex) Apprehensio ((simple) apprehension): By this act, the intellect apprehends some object or state of affairs as good in the agent's circumstances and, therefore, as a good.	**Act of the Intellect**	
	Act of the Will	**(1b) Voluntas/(Simplex) Volitio** ((simple) volition; inefficacious intention): As a consequence of this intellectual apprehension, by this act, the will desires the realisation of this good as an end; it wills, that is, wishes for, the end as such (as something not possessed).

(continued)

(continued)

(2a) Iudicium (judgement of (achievability of) end): By this act, the intellect determines that the end is attainable, that is, that its achievement is within the power of the agent. If the object of volition is not attainable it cannot become an object of intention and counsel. While we can wish for almost everything, we can only intend what can realistically be achieved. In the same fashion, we cannot deliberate about the means to something which is impossible to obtain.	**Act of the Intellect**	
	Act of the Will	**(2b) Intentio** ((efficacious) intention (of taking the means to attain the good)): By this act, the will desires the end not merely as a good as such, but as a good that can be attained by means of action; the will desires not only the end, but also the means leading to it, that is, the end through some means.
(II) Order of Choice (*Ordo Electionis*) **(B) Acts of the intellect and acts of the will with regard to the means** **(B1) Deliberation**		
(3a) Consilium/Consultatio (counsel; deliberation (and decision about means)): By this act, the intellect investigates and determines (or to be more precise: is moved by the will to investigate and determine) various means by which the intended end can be realised. As the end can only be realised by an external action, counsel is about investigating and determining the various options of external actions.	**Act of the Intellect**	

(*continued*)

(continued)

	Act of the Will	**(3b) Consensus** (consent (to means)): By this act, the will endorses those proposed means as desirable, or at least tolerable, means to the intended end.
(B2) Decision		
(4a) Iudicium practicum (seu discretivum mediorum) (practical judgement; discernment of fittest means): By this act, the intellect judges that and which one of the means that have received the will's consent is the best (all things considered and for the given circumstances) to realise the end.	**Act of the Intellect**	
	Act of the Will	**(4b) Electio** (choice): By this act, the will chooses the means the intellect proposes as best (and the will does so because the will already intends the end and must, therefore, also intend the means). Put differently, the will chooses the exterior action that is suited best to realise the intended end.
(III) Order of Execution (*Ordo Executionis*) **(B3) Execution**		
(5a) Imperium (command): By this act, the intellect orders that the elected means be carried out in action ("Do the best means!").	**Act of the Intellect**	
	Act of the Will	**(5b) Usus** (use): By this act, the will carries out the intellect's command by making use of the appropriate faculties of the human soul in the way that the intellect has prescribed so as to attain the intended end. In short: The will causes the act (= the means to achieve the intended end) to be performed.

(*continued*)

(continued)

(C) Fruition		
(6a) Contemplatio (contemplation): By this act, the intellect apprehends that the end has been attained and contemplates the goodness of the attained end.	**Act of the Intellect**	
	Act of the Will	**(6b) Fruitio/Delectatio** (enjoyment, fruition, delight, pleasure): By this act, the will rests in the enjoyment of the end, that is, enjoys/takes pleasure in the end insofar as the end is presented in thought as being possessed.

These six stages and 12 acts lay out the most elaborate form of a human action. However, in practice human actions may not involve all 12 elements that a complete human action comprises:

> If the means to the desired end is perfectly obvious, there will be no need for consilium to begin with, and the agent can proceed directly from intentio to iudicium. And the intellect's failure to stick to its command—a manifestation of what Aquinas calls inconstancy—can interrupt usus before it comes to completion in the exterior act by which the agent means to accomplish her end, thereby forestalling fruitio. (Williams 2012: 201)

Furthermore, neither do all of the above-mentioned acts need to be apparent to the agent's consciousness, nor do the acts of the intellect and reason need to be separated by any perceptible time.

5. Killing Actions as Human Actions

Now that we have acquired the vocabulary to describe actions, we can apply it to the kind of action that is of interest to this book, namely killing actions. But first, let us recapitulate exactly what we have learned so far with respect to the nature, elements and voluntariness of human actions:

- **Human actions:** A human action is a directly voluntary action of a human being, that is, an action caused and performed by its agent both knowingly and willingly. There are three basic forms of human actions: activity (commission), inactivity (omission) and passivity (sufferance).
- **Elements of a human action:** Human actions comprise four elements, namely (a) the end of the agent (*finis operantis*), (b) the exterior action and its natural end (*finis operis*), (c) the circumstances of the exterior action and (d) the consequence(s) of the exterior action. The end of the agent is the reason why the agent performs the exterior action. The exterior action is the means by which the agent seeks to fulfil his purpose for acting. It is by its natural end that an exterior action can be classified as a certain and not some other kind of exterior action. Usually, a human action is named after its exterior action. The circumstances are the specific conditions under which the exterior action takes place. The consequences are the effects that the exterior action actually results in. If realised, both the end of the agent and the end of the exterior action are counted among an exterior action's consequences, the first as a non-essential, the latter as its essential consequence. All other consequences are non-essential to the exterior action. An accidental consequence is an unintended non-essential consequence.
- **Degrees of voluntariness of human actions:** The elements of a human action can be directly voluntary, indirectly voluntary (also: knowingly accepted) or unknowingly accepted. Directly voluntary means foreseeable, foreseen and intended. Indirectly voluntary means foreseeable, foreseen but not intended. Unknowingly accepted means foreseeable but neither foreseen nor intended.

Let us now apply these distinctions to killing actions. This will help us get a better understanding of what we are talking about when we are talking about such actions. In Chap. I, Sect. 2 we have approached the action "killing" by broadly defining it as causing a human being's death: A human being (the killer) kills a human being (the killed) if he does something that causes a human being to lose his life. This definition has three aspects to it:

- A killing action requires a killed: For a killing to have occurred, a human being must have died. The death of a previously living human being is a necessary (albeit not sufficient) condition for a killing.

- A killing action requires a killer: The death of the killed must have resulted from an action performed by a human being (and not from, e.g., natural causes). This killer must be the efficient cause for the death of the killed.
- The killer and the killed can be identical.

As the above definition shows, a killing action is essentially defined by a certain consequence, namely the death of a human being. Without a dead human being there can be no killing action. In other words: In a killing action, the death of the killed must be a consequence of an exterior action performed by the killer.

The definition of a killing action makes no reference to the purpose of the exterior action that results in the death of a human being. The killer must have had some *finis operantis*; that much is certain. But our definition of killing actions does not require the presence of a specific end of the agent on the part of the killer. An exterior action that causes the death of a human being is a killing action regardless of the purpose which the agent pursues with the exterior action. This is not to say that the end of the agent has no influence on the overall ethical quality of a killing action; it just says that the why of an action does not determine whether this action is an actual killing action. Killings can occur for all kinds of purposes, but what makes them killings is not their purpose but the fact that a human being has died as a result of their exterior action.

We now know that a killing action has occurred if an agent has performed an exterior action that has caused a human being's death. But as we have seen earlier in this chapter, there are different kinds of consequences. We need to therefore specify what kind of consequence the death of the human being is that has resulted from the agent's exterior action.

- As should be obvious, the consequence must be an actual consequence. As long as death is only a potential consequence, no human being has actually died. And if nobody has died, no killing action can have occurred.
- The death of the killed can either be an essential or a non-essential consequence of the exterior action. If it is an essential consequence, death has resulted from the exterior action because it is its natural end. If it is a non-essential consequence, death has resulted from the exterior action, although it is not its natural end.

Natural end of exterior action (A_{Ext}) performed by agent A		**Actual consequence of A_{Ext}**		**Kind of killing action**
Death of B (exterior action = killing action)	→	**B is dead**	→	**Essential Killing** (i.e. killing action with death of B as essential consequence of A_{Ext})
Not death (¬death) of B (exterior action = not a killing action)	→		→	**Accidental Killing** (i.e. killing action with death as non-essential consequence of A_{Ext})

Fig. 3 Kinds of killing actions

So, in a killing action, the death of the killed can either come about as the natural end of the exterior action or it can come about as an accidental result of an exterior action whose natural end is not the death of the killed. We can therefore distinguish two basic kinds of killing actions (also cf. Fig. 3):[5]

- **Essential killing** (*also: directly voluntary killing*)
  In this type of killing action the death of a human being is the essential consequence of the killing action's exterior action. In essential killing actions the death of the killed has come about as it is the natural end of the exterior action.
  As the end of the exterior action is necessarily deliberately willed when performing the exterior action, this means that in essential killings the death of the killed is always brought about directly voluntarily.
- **Accidental killing**
  In this type of killing action the death of a human being is a non-essential consequence of the killing action's exterior action. In non-essential killing actions the death of the killed has come about as a result of the exterior action, although it is not its natural end.

[5] The distinction between essential killings and accidental killings is congruent with the distinction between what, for example, Bender (1951: 225) or Noldin (1935: 309) call direct killings and indirect killings. The expressions "essential killing" and "direct killing" are synonymous, as are the expressions "accidental killing" and "indirect killing".

> As will be shown in a later chapter (cf. chapter VI.2.2), a non-essential killing can only be ethically permissible if the death of the killed has resulted as an unintended non-essential, that is, an accidental, consequence. Anticipating this insight, the term that will be used for the purposes of this book will not be "non-essential killing" but "accidental killing". This terminology should also provide more linguistic clarity, as the expressions "essential" and "accidental" stand out more clearly from each other in the text than "essential" and "non-essential".

For the full description of a killing action, either of these types of killing actions must be further specified with regard to the circumstances in which its exterior action takes place. So, a full description of a killing action does not only require the specification of its type, that is, whether it is an essential or accidental killing action, but also the circumstances of its exterior action, that is, the circumstances of quantity and quality, time, place, mode or manner, agent as well as helps or influences.

Bibliography

Bender, Ludwig. "Occisio directa et indirecta." *Angelicum* 28.3 (1951): 224–253.

Cronin, Michael. *The Science of Ethics. Volume I: General Ethics.* Third Impression. Dublin: M.H. Gill and Son, 1930.

Erk, Christian. *Moral Philosophy. Fundamental Concepts.* Second, revised and expanded Edition. Heusenstamm: editiones scholasticae, 2019.

Fagothey, Austin. *Right and Reason. Ethics in Theory and Practice.* Third edition. Saint Louis: The C.V. Mosby Company, 1963.

Noldin, Hieronymus. *Summa Theologiae Moralis: II. De Praeceptis.* Editio XXIII. Oeniponte: F. Rauch, 1935.

Prümmer, Dominicus M.. *Manuale Theologiae Moralis Secundum Principia S. Thomae Aquinatis in Usum Scholarum.* Tomus II. Editio Quinta Decima. Barcinone, Friburgi Brisgoviae & Romae: Herder, 1960.

Renard, Henri. *The Philosophy of Being.* Second edition, revised and enlarged. Milwaukee: The Bruce Publishing Co., 1948.

Williams, Thomas. "Human Freedom and Agency." In: Davies, Brian & Eleonore Stump (Eds.). *The Oxford Handbook of Aquinas.* Oxford: Oxford University Press, 2012. pp. 199–208.

III: The General Ethics of Human Actions

This book is a treatise on the moral permissibility of selected kinds of killings, namely suicide, lethal self-defence, abortion (including infanticide), euthanasia, organ transplantation and cooperation in suicide. How can we go about finding an answer to this question? In order to know whether any of the above-mentioned killings is morally impermissible we need a yardstick that allows us to determine moral (im)permissibility of actions in a general, systematic and non-arbitrary fashion—and which can then be applied to the selected kinds of killings mentioned above. In ethics, this yardstick is called a "moral theory".

A moral theory is a reasoned account of what makes actions good or bad in general (cf. Erk 2019: 5); as such it provides us with the knowledge that is necessary to determine whether a particular action is good or bad. Moral theories can be categorised along two dimensions, namely according to (a) their stance on the "sources of morality" and (b) their "standard of morality".

- The sources of morality are those elements of an action, from which its morality (also: moral quality) derives (cf. Erk 2019: 112). By explicating its stance on the sources of morality, a moral theory sets out whether, to what extent and why which elements of an action influence its overall ethical quality.

C. Erk, *The Ethics of Killing*,
https://doi.org/10.1007/978-3-031-07183-6_3

- By its standard of morality, a moral theory tells us how to determine whether the elements of an action that have been defined as sources of morality and thus the action as a whole is good or evil. This requires a reasoned definition of the concepts of "good" and "evil".

The first of the following two sections details a moral theory that is based on a four sources theory of morality as well as an essentialist standard of morality.[1] This moral theory adheres to an essentialist standard of morality as it holds that goodness and evilness have to do with the essence (also: nature) of things. It is a four sources theory of morality as it takes all four elements of a human action—albeit to different degrees—into consideration when determining its ethical quality.

However, far from being simply neutral, good and evil have normative force. As the yardstick of goodness, the essence or nature of human beings grounds their most fundamental duties and rights, namely their moral duties and moral rights.[2] These duties and rights are possessed by all human beings of all times at all times and have validity independently of, and even in contrast to, all legal and conventional rights and duties. Therefore, as a second step, the second section details the general moral duties and rights that human beings possess and also grounds the special moral duty and right that are of interest to this book, namely the moral duty to live and the moral right to life. This knowledge of our moral duties and rights is important for the purposes of this book as an action that is morally impermissible is an action that an agent has

[1] When reading this chapter, the observant reader equipped with previous knowledge of ethics will notice that it presents an ethical theory that can be described as a "natural ethics" and that stands in the tradition of the moral theory commonly known as classical (also: traditional) natural law theory. This book's position is therefore based on a version of classical natural law theory.

As the name indicates, there is not only a classical but also a new natural law theory. The latter theory was developed by Germain Grisez in the 1960s in an interpretative article on Thomas Aquinas' first principle of practical reason (cf. Grisez 1965). For a useful brief overview of this theory and its proponents cf. Tollefsen (2008) as well as Lee (2019). New natural law theory differs from classical natural law theory in several crucial aspects. It has been severely criticised by traditional natural law theorists and others, such as Hittinger (1987) and Long (2013); a helpful compilation of additional works criticising new natural law theory can be found in Feser & Bessette (2017: 87 (FN 120)). I agree with Feser & Bessette (2017: 87) in that these criticisms "are reason enough to reject the theory".

[2] This is why the label "natural law theory" contains the term "law".

a moral duty to not perform; if there is no such prohibition-duty, the action is morally permissible. So, in order to know whether any of the killings mentioned at the outset of this chapter is morally impermissible we need to know whether there is a moral duty to not perform this particular action.

1. The Morality of Human Actions

As has been stated above, moral theories can be categorised according to their stance on the sources of morality and their standard of morality. The sources of morality refer to those elements of an action from which its morality (also: moral quality) derives. So, let's start the enquiry about whether, to what extent and why which elements of an action influence its overall ethical quality.

1.1. The Sources of Morality[3]

In Chap. II, Sect. 2 we have learned that a human action consists of four elements, namely (a) the end of the agent, (b) the exterior action and its natural end, (c) the circumstances of the exterior action and (d) the consequences of the exterior action. Interesting as this may be, knowing that human actions comprise these four elements does not, however, yet solve the problem of how to actually determine their ethical quality. To be able to do so, we need to establish (a) whether and to what extent the four elements influence the ethical quality of human actions and (b) by which yardstick to decide whether the respective elements of a human action and thus the action as a whole is good or evil. By devoting itself to an explication of the so-called sources of morality, this chapter addresses the first part of this task.

The ***sources of morality*** (also: principles of morality; Latin: *fontes moralitatis*) are those elements of a human action from which its morality (also: moral quality; ethical quality) and therefore its goodness (Latin: *bonitas*) or badness (also: evilness; Latin: *malitia*) derives. Explicating the sources of morality means explicating which of the four elements of a human action have to be considered to which extent when determining whether a human action is good or evil and how.

[3] This chapter is a revised and condensed version of and as such draws on, and in part reproduces verbatim, material published in Erk (2019: 112–129).

So, what are the sources of the morality of a human action? A look at the literature reveals that this a highly contested question. The disagreement centres around the sub-questions whether there is one or more source of morality and which element is (or which elements are) decisive for the evaluation of a human action's ethical quality. A first attempt to group the different opinions yields the following ideal-typical categories of moral theories:

- A first group of moral theories holds that it is only the end of the agent (*finis operantis*) that is decisive for the ethical quality of a human action. Proponents of this position adhere to the moral theory that can be named "motivism".[4] So, regardless of the moral quality of the external action, the consequences and the circumstances, motivism suggests that a human action is good if and because the end of the agent is good, and that it is evil if and because the end of the agent is evil.
- The moral theory named "deontology" advances the position that only the external action that is carried out by the agent is relevant for the evaluation of the ethical quality of a human action. The end of the agent, the consequences and the circumstances of a human action have no influence on its moral quality. A human action is good if and because its exterior action is good, and it is evil if and because its exterior action is evil.
  As the name "deontology" indicates,[5] deontologists usually hold that the goodness of the exterior action depends on the existence of a moral duty on the part of the agent. A human action is good if its external action is in accordance with, that is, does not violate, a moral duty possessed by the agent; it is evil if it violates a moral duty possessed by the agent.

[4] In earlier publications, I have used the term "intentionalism" to characterise this type of moral theory. This choice of term was, however, somewhat unfortunate. For when acting, the agent not only intends the end of the agent but also the end of the exterior action. Strictly speaking, the term "intentionalism" is too broad to only apply to the agent's motive or purpose for acting. In order to remedy this inaccuracy and align the term with the content, the term "intentionalism" has been substituted with the term "motivism".

[5] The English word "deontology" derives from the classical Greek word δέον (read: déon) which can be translated as "that which is binding" or simply as obligation or duty.

The ethical quality of a human action (HA) is derived from the ethical quality of one of the elements of a human action		
↓	↓	↓
End of the Agent (P)	**External Action (A_{ext})**	**Consequences (C)**
↓	↓	↓
Motivism	**Deontology**	**Consequentialism**
Only the end of the agent ("finis operantis"; P) is relevant for the moral evaluation of HA.	Only the external action (A_{ext}) is relevant for the moral evaluation of HA.	Only the consequences of the external action (C) are relevant for the moral evaluation of HA.
The ethical quality of HA is measured only against the ethical quality of P.	The ethical quality of HA is measured only against the ethical quality of A_{ext}.	The ethical quality of HA is measured only against the ethical quality of C.
Regardless of the ethical quality of A_{ext} and C (as well as the circumstances), HA is good if and because P is good and evil if and because P is evil.	Regardless of the ethical quality of P and C (as well as the circumstances), HA is good if and because A_{ext} is good and evil if and because A_{ext} is evil. A_{ext} is good if it is in accordance with the moral duties held by the agent. A_{ext} is evil if it violates a moral duty held by the agent.	Regardless of the ethical quality of P and A_{ext} (as well as the circumstances), HA is good if and because P is good and evil if and because P is evil.

Fig. 1 Single source moral theories: An overview

- According to the moral theory named "consequentialism", the ethical quality of a human action is measured only by the ethical quality of the consequences of the exterior action. Consequentialists deny that the goodness or badness of a human action directly depends on anything other than the goodness or badness of the actual consequences of its exterior action. Regardless of the moral quality of the end of the agent, the exterior action and the circumstances, a human action is good if and because its consequences are good.

Despite their otherwise vast differences, all three moral theories agree that the ethical quality of a human action is dependent on the moral quality of only one of its elements. For all of them, there is only one source of morality. As they reduce the ethical quality of a human action to one element only, these theories can therefore also be called "single source" or "reductionist" moral theories. Figure 1 summarises the main tenets of the three reductionist moral theories that have been introduced so far:

But although possibly identifying as motivist, deontologist or consequentialist at first sight, not all moral philosophers advance their favoured moral theory in a pure form—in fact, that is the exception rather than the rule. Many of them advance moral theories that are somewhat of a combination of motivism and consequentialism, consequentialism and deontology, or motivism and deontology.[6]

There is, however, not only disagreement between but also within the respective classes of moral theories. For example, consequentialists not only argue about the question which consequences (e.g. money, physical harm or pain, bodily pleasure) to take into account but also whether only the consequences for the agent (egoism) or also for other persons are of relevance, how to weigh the consequences, and whether only direct or also indirect consequences should be counted. Deontologists, on the other hand, debate among themselves which moral duties the human person actually possesses. So, when zooming in on a group of moral theories, one could get the impression that what holds together the moral sub-theories gathered under its roof is their common rejection of the other classes of moral theories rather than unreserved agreement and unanimity as to their own position.

Knowing about these different kinds of moral theories we are now confronted with a tough choice: Which moral theory should we use to evaluate whether a human action is a good or evil? However, before deciding on one of the three moral theories we should address a more fundamental question. For choosing one from among the three moral theories introduced above means accepting the validity of their joint underlying premise. This premise states that the moral quality of a human action is

[6] An example of a combination of consequentialism and deontology is the moral theory named "rule consequentialism". According to this theory, "an act is morally wrong if and only if it is forbidden by rules justified by their consequences" (Hooker 2015). So, rule consequentialism stipulates that there is some set of rules that bind an agent and by doing so introduces a somewhat deontological element in an otherwise consequentialist moral theory.

An example of a combination of motivism and deontology is Immanuel Kant's moral theory. In his "Grundlegung zur Metaphysik der Sitten" ("Fundamental principles of the metaphysics of ethics") we read:

Denn bei dem, was moralisch gut sein soll, ist es nicht genug, daß es dem sittlichen Gesetze gemäß sei, sondern es muß auch um desselben willen geschehen.	For in order that an action should be morally good, it is not enough that it conform to the moral law, but it must also be done for the sake of the law.

(Kant 1785: 390)

So, according to Kant it is not only the lawfulness of an action that makes it good; rather the agent must also act with the right purpose in mind, namely fulfilling his duty because it is his duty.

determined by one of its elements only. But why should we accept this premise to begin with? The moral theory advanced by this book suggests that we should not do so.

For what all reductionist moral theories fail to appreciate is the nature of the human action as a voluntary action, that is, as an action that springs from the deliberate will of the agent. Why is this problematic? An action can only be subject to moral evaluation insofar as it is a human action, that is, insofar as it is deliberately willed. An action is deliberately willed to the extent that its elements are deliberately willed. As all elements of a human action can be deliberately willed, there is no reason to single out one or fewer than all elements of a human action and neglect the rest. Picking fewer than all elements as a source of morality arbitrarily ignores the multifaceted nature of human actions.

The point of departure for the evaluation of the moral quality of a human action should be what is deliberately willed by the agent when performing the action. In principle, each element of a human action can influence the moral quality of a human action because and insofar as the element is deliberately willed. Taking ethics seriously as the science of good and evil in human actions means that one cannot single out fewer than all elements of a human action as a source of morality and exclude the other elements as irrelevant. Rather, as all elements of a human action can play a role in its ethical evaluation because and insofar as they are deliberately willed, the ethical evaluation of a human action requires us to check all of its elements.

This book, therefore, advances a holistic or comprehensive moral theory according to which all elements of a human action—insofar as and because they are deliberately willed—have to be taken into consideration when evaluating the ethical quality of a human action:

> Insofar as and because they are deliberately willed, the moral quality of (a) the end of the agent (*finis operantis*), (b) the exterior action and its natural end (*finis operis*), (c) the circumstances of the exterior action as well as (d) the consequences of the exterior action impact—albeit to varying degrees—the moral quality of the overall human action.

In contrast to the single source and therefore insufficient moral theories that hinge the morality of a human action on one of its elements only, the moral theory advanced by this book follows a fourfold goodness principle. When evaluating the ethical quality of a human action, this principle takes

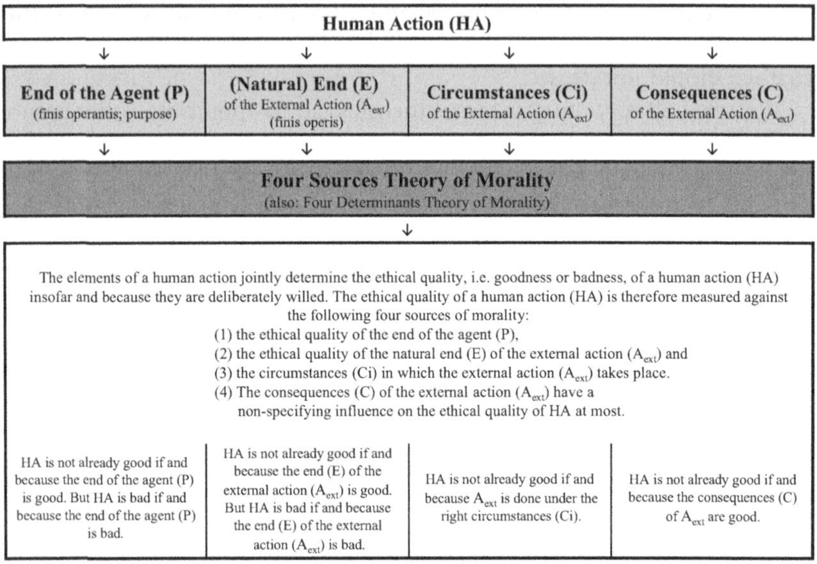

Fig. 2 The four sources theory of morality

into account—albeit to varying degrees—all four elements of a human action insofar as and because they are deliberately willed. In other words: All elements of a human action are a human action's sources of morality. Given this fact, this book's moral theory could—for want of a better term—be named the "four sources moral theory" or the "four sources theory of morality". The main tenets of this moral theory can be summarised as follows (cf. Fig. 2):

For a complete account of the four sources theory of morality the interested reader is referred to Erk (2019: 112–129). For the purposes of this book it is not necessary to repeat this theory in its entirety and show how each of the elements of a human action influences its ethical quality.

The important message of this chapter is that all elements of a human action can, in principle, be deliberately willed when performing a human action; moreover, some of them (namely the end of the agent as well as the end of the exterior action) must be deliberately willed. Therefore, there is no reason to categorically exclude any of the four elements as a source of morality. The only moral theory that does justice to the nature of human actions is the four sources theory of morality, according to which all elements of a human action are its sources of morality and must be taken into consideration insofar as they have been deliberately willed.

In order to establish the exact overall morality of a human action, we have to check the ethical quality of all its elements. This can be a complicated task. In this context, it is worth noting that the two most important sources of morality are the end of the agent (*finis operantis*) and the end of the human action's exterior action (*finis operis*). The evilness of either of those two elements—insofar as they have been deliberately willed—suffices to make the entire human action evil. The other elements of a human action do not have the same specifying weight: The consequences of the exterior action have a merely non-specifying influence that can make an otherwise evil human action more or less evil but not good or an otherwise good human action more or less good but not evil.[7] The circumstances of quantity and quality, time, place, mode or manner, agent and helps or influences can have a specifying influence on the ethical quality of a human action but only if they change the nature of the exterior action; otherwise they have a merely mitigating or aggravating influence.

This account can be decomplicated further. As mentioned in Chap. II, Sect. 5, there are two kinds of killing actions: essential killings and accidental killings.

- In essential killings the death of the killed human being results as an essential consequence of the action's exterior action. Put differently, in cases of an essential killing the death of the killed has been the natural end of the killing action's exterior action. Although the death of the killed is an actual consequence of the killing action, the source of morality that is relevant to the evaluation of the morality of an essential killing is not death qua being a consequence but death qua being the natural end of the exterior action. So, the key element

[7] This statement holds true for the actual accidental consequences of an exterior action, that is, those actual consequences that are not the end of the agent or the end of the exterior action. Insofar as they have been successfully realised, the end of the agent and the end of the exterior action can also be counted among the consequences. As they are already taken into consideration qua their nature as separate elements of the human action, they need not be considered a second time.

The actualised end of the exterior action specifies the ethical quality of a human action because it is the end of the exterior action, not because it is a consequence of the exterior action. In the same fashion, the actualised end of the agent specifies the ethical quality of a human action because it is the end of the agent, not because it is a consequence of the exterior action.

whose ethical quality to scrutinise in order to establish the impermissibility of an essential killing action is the end of its exterior action.
- In accidental killings the death of a human being is an accidental consequence of the killing action's exterior action. The death of the killed has come about although it is not the natural end of the exterior action. The sources of morality that are relevant to the evaluation of the morality of an accidental killing are both the exterior action that has accidentally resulted in the death of the killed and the death of the killed qua its nature as an accidental consequence.

Leaving aside the concrete circumstances of the exterior action, the two key elements whose ethical quality we must be able to scrutinise are the natural end of the exterior action of a killing action as well as its accidental consequences. Being able to do so is sufficient to establish at least the impermissibility of an essential and an accidental killing action. So, how do the moral quality of (a) the exterior action and (b) the exterior action's accidental consequences influence the ethical quality of the overall action?

1.1.1. The Exterior Action as a Source of Morality

Let's start with the exterior action and its natural end. How exactly does the ethical quality of the end of the exterior action influence the action's overall morality? Depending on its specific goodness or badness, the *finis operis* of an action has the following influence on the overall moral quality of the action:

- If the **exterior action** (as defined by its natural end) is **good**
  - The goodness of the exterior action is a necessary but not a sufficient condition for the overall goodness of the human action; it does not guarantee the goodness of the human action as a whole.
  - A good exterior action increases the goodness of an otherwise good human action but does not suffice to make the human action good in its entirety.
  - A good exterior action makes an otherwise bad human action only less bad but not good.
- If the **exterior action** (as defined by its natural end) is **indifferent/ neutral**
  - If the exterior action of the human action is indifferent/ neutral, this has no influence on the overall moral quality of the human action. The level of goodness or badness of a human action remains unchanged.

- If the **exterior action** (as defined by its natural end) is **bad**
  - If the exterior action of a human action is bad (and cannot be excused), the human action is bad in its entirety. The badness of its exterior action suffices to make a human action bad in its entirety.
  - A bad exterior action makes an otherwise bad human action worse.

1.1.2. Accidental Consequences as a Source of Morality

With respect to the influence of an exterior action's accidental consequences on the ethical quality of the overall action we can orientate ourselves by the possible degrees of voluntariness of these kinds of consequences as they have been distinguished earlier (cf. Chap. II, Sect. 3). In this chapter, it has also been argued that a consequence can only be an accidental consequence if it is not intended, that is, that there is no such thing as a directly voluntary accidental consequence. As an accidental consequence can only be unknowingly accepted or indirectly voluntary, we only need to concern ourselves with the influence of these two kinds of consequences on the ethical quality of the overall action.

- **Unknowingly accepted accidental consequences**
  Foreseeable but unforeseen accidental consequences do not add to the goodness or badness of the human action. The goodness or badness of such unknowingly accepted accidental consequences has no influence on the overall moral quality of a human action; it is morally irrelevant, so to speak.
- **Indirectly voluntary accidental consequences**
  An accidental consequence that has been foreseen (and that, by necessity, has been foreseeable) is a non-specifying consequence. Such an indirectly voluntary (also: knowingly accepted) accidental consequence can increase or decrease the goodness or badness of the human action but cannot change its overall moral species. It makes a good or an evil action better or worse, but it cannot make an otherwise evil action good or an otherwise good action evil.
- When it comes to determining the exact influence of a knowingly accepted accidental consequence on the ethical quality of an action, there are several principles to be kept in mind:

- The more avoidable an evil indirectly voluntary accidental consequence, the greater its influence on the ethical quality of an action, that is, the worse it makes the action.
- The more the agent was in a position to avoid an evil indirectly voluntary accidental consequence, the greater the consequence's influence on the moral quality of an action, that is, the worse it makes the action.
- The worse an evil indirectly voluntary accidental consequence, the greater its influence on the ethical quality of an action, that is, the worse or less good it makes the action. In the same fashion, the better a good indirectly voluntary accidental consequence, the better or less evil it makes the action.
- The greater the causal proximity of the exterior action and an indirectly voluntary accidental consequence, the greater the consequence's influence on the ethical quality of the action.
- The more immediate an indirectly voluntary accidental consequence, the greater its influence; the more remote the consequence (and therefore the less under the control of the agent and the more subject to external influences), the lesser its influence.

Helpful as these insights regarding the sources of morality of a human action may be, there is one big question that remains to be answered: What do the terms "good" and "evil" actually mean? Knowing that the goodness or evilness of all elements of a human action has—albeit to varying degrees—an influence on the action's overall ethical quality does not yet tell us how to actually determine their goodness or evilness. This issue is addressed by the following chapter.

1.2. The Standard of Morality: The Good[8]

In order to know whether the exterior action of a human action—or its end of the agent, its circumstances or its consequences, for that matter—is good or evil one has to first know what the concepts "good" and "evil" actually mean. Addressing this question, this chapter contains a treatise on the good that details its nature as well as content and also elaborates on the contradictory opposite of good, namely evil (also: bad). It familiarises

[8] The content of this chapter draws on, and in part reproduces verbatim, material published in Erk (2019: 130–159).

the reader with the main tenets of the essentialist standard of morality, that is, the standard of morality that takes recourse to the essence or nature of things and that is also at the heart of (traditional) natural law theory.

1.2.1. The Nature of the Good

As the science of good and evil in human actions, moral philosophy is necessarily concerned with elucidating which of the things that agents apprehend as good are just seemingly, that is, subjectively, good and which of them are truly, that is, objectively, good. But how do we know whether something that has been judged as good by our intellect is not just seemingly (subjectively) good but truly (objectively) good? What is the nature of objective goodness?

At its core, objective goodness involves perfection, that is, completion or fullness of being (also: flourishing, perfection). A being is good if it has achieved fullness of being, something it can achieve by actualising the capacities (also: potentialities; faculties; potencies; powers) granted to it by its nature.

Ratio autem boni in perfectione consistit.	The nature of the good consists in perfection.

(Thomas Aquinas, Contra Gentiles, lib. 1 cap. 39 n. 5)

The good is a mode of being. Something is good if it is the way it is supposed to be. How a being is meant to be is not, however, determined by the being itself. Rather, it is determined by the respective being's inherent essence or nature, that is, the nature it has been endowed with (by its creator[9]) and whereby the being is what it is.[10] The good is perfection relative to a nature.

A being is objectively good if it has realised its nature or essence, that is, "if it possesses the perfections proper to its nature" (Ross 2012: 61) and "is actually lacking in nothing that is due to its nature" (Coffey 1918: 172). Understood like this, the term "perfection" contained in the

[9] As it is used here, the term "creator" is not necessarily synonymous with "God". Rather, it is meant to give expression to the fact that every (contingent, i.e. non-necessary) being has an efficient cause. So, the creator of something is the being that has brought this something into existence by a conscious effort, as, for example, the carpenter who creates a chair.

[10] For the concept of nature or essence cf. Chap. V, Sect. 1.3.

definition of the good should lose its overburdening undertone. For what goodness amounts to is normality, not extraordinarity.

> It is of the essence of a triangle to be a closed plane figure with three straight sides, and anything with this essence must have a number of properties, such as having angles adding up to 180 degrees. These are straightforward objective facts, and remain so even though there are triangles which fail perfectly to match that description. A triangle drawn hastily on the cracked plastic seat of a moving bus might fail to have sides that are perfectly straight, and this its angles will add up to something other than 180 degrees. Even a triangle drawn slowly and carefully on art paper with a straight edge and a Rapidograph pen will contain subtle flaws. Still, the latter will more perfectly approximate the essence of triangularity than the former will. It will be a better triangle than the former one. Indeed, we would call the former a bad triangle and the latter a good one. (Feser 2009: 176, 2015: 381)

All beings carry the measure of their objective goodness within them, namely their inherent nature or essence. The statement "X is objectively good" means "X is the way X is meant to be (by its creator)" or "X has realised the nature X has been endowed with (by its creator)". Or as Lisska (1996: 102f) puts it: "To function well means to reach the developmental potential of one's essential properties".

1.2.2. The Absence of Good: Evil

One cannot deal with the concept "good" without also touching upon the concept "evil" (also: bad). What does the term evil signify? Evil is not an entity, that is, it is not something that is; it is something that is not. Evil is a privation (also: deprivation; Latin: *privatio*; Greek: στέρησις). What is absent or non-existent is the contradictory opposite of evil, namely good. Generally speaking, evil is the privative absence of good:

Malum est privatio boni	Evil is the privation of good.

(Thomas Aquinas, I^{a}-IIae q. 18 a. 8 ad 1; cf. I^{a} q. 48)

Privation does not, however, simply mean absence; rather, it is "absence plus a need" (Oderberg 2014: 72). Not every absence of good is (an) evil but only the privative absence of good, that is, the privation of (a) necessary or due good. A privative absence of something which is not due to a being is not (a) bad.

Malum enim, ut dictum est, nihil est aliud quam privatio eius quod quis natus est et debet habere: sic enim apud omnes est usus huius nominis malum.	As has been said, evil is nothing else than the privation of something which a subject is designed and bound to have. Such is the use of the word evil among all men.

(Thomas Aquinas, Contra Gentiles, lib. 3 cap. 7 n. 2)

But how do we know that a good is a due good? The measure or standard of whether something is due to a being is the respective being's nature or essence. If evil is the privative absence of a good due to a being's nature and if good is perfection of being as required by the nature of a being, this allows us to conclude that evil is the privative absence of a perfection which is due to a being in light of its nature. When something is deprived of the perfection which is proper to it, that is, if it is not as it should be or does not possess all the goods it is supposed to possess given its nature, it is said to be evil or bad.

So, if a stone is deprived of water this cannot be considered an evil as water is not a good that a stone needs as stone, that is, to be a full-fledged stone. But dehydration in plants or animals is a different story. Although they need water to varying degrees, without water no plant or animal can be the plant or animal it is supposed to be given its nature or essence. As water is a good that all plants or animals necessarily need to be the plant or animal they are supposed to be, dehydration is an evil. In the same fashion, an oak tree that does not have the ability to see cannot be said to be a bad oak tree. For it is not in the nature of an oak tree to be able to see. But a hawk that cannot see is a bad hawk for it is in the nature of a hawk to be able to see.

1.2.3. An Essential Content of the Good: The Basic Human Goods

So far, we have been concerned with the good in general. But it is one thing to know what the good is—acting on it is another. When acting we cannot simply will the realisation of the good or goodness in general. The good or goodness is too abstract to act on it. When acting, we must will the realisation of a particular good or particular goods: "No one can in acting simply pursue good—one has to pursue some particular good"

(Murphy 2019). To be able to act on it, the notion of the good and goodness must first be operationalised.

In order to make our understanding of the good action-guiding, we have to break the abstract good down into concrete goods. This requires us to first clarify the difference between "the good" or "goodness" on the one hand and "a good" on the other hand. Generally speaking, a good is something that is good for a being if and because it furthers its goodness. The objective goodness *of* a being, that is, its objective ontological or absolute goodness, is the yardstick by which we can determine whether something is objectively good *for* a being. A bad or an evil is something that detracts from its goodness, that is, is evil for the being.

The question we must therefore answer next is: What is good for a being? In case of human beings, this question has three answers: What is good for a human being is determined by

- the basic human goods (BHGs)
- the common good
- the virtues and
- the natural ends of his faculties

These concretisations of the good serve as the action-guiding standards by which we can establish the morality of a human action.

The following paragraphs shall elaborate on the first of these concretisations of the good. For an overview of the latter three concretisations the interested reader is referred to Erk (2019: 144–157; 203–206). Focusing only on the so-called basic human goods (BHGs) is sufficient for the purposes of this book. For as we will see, one of the BHGs is "life". So, as soon as an exterior action or an agent's intention is in violation of this BHG, it does not matter whether it is a virtuous action that does not pervert the natural ends of the agent's faculties and furthers the exterior common good—the resulting human action is already a bad action by virtue of its violation of the most fundamental BHG, namely life.

1.2.3.1. Nature and Natural Inclination

So, let us turn our attention to the basic human goods. As has been stated above, something is objectively good for a being if it furthers the respective being's objective goodness, that is, its being the way it is objectively meant to be. Everything that is perfective of a being's nature is objectively good and an objective good for that being.

In the case of human beings, this means that only those things that further the realisation of and are in accordance with his nature as a human being can be objectively good for a human being. This also holds true for human actions which are good only insofar as they are in accordance with the nature of man: An action that is in harmony with man's nature is good and an act that is out of harmony with man's nature is bad.

> We have it, then, that a morally good act is an act that makes towards the progress of human nature in him who does it, and which is freely done, Similarly, a morally evil act is a bar to progress, or a diversion of it from the right line, being also a free act. Now, that act only can make for the progress of human nature, which befits and suits human nature, and suits it in its best and most distinctive characteristic. (Rickaby 1918: 111)

So, a human being acts well if he acts in accordance with his predetermined and inherent nature as a human being. But how can we tell whether an action is in accordance with the nature of a human being? To be able to answer this important question, we need to first ask and answer a more fundamental question: How do we come to know the nature of a human being? What is the nature of man?

Unfortunately, the nature of a thing is not directly accessible to our senses. It can, however, be abstracted from sensory experience gained through observation: By observing its behaviour, that is, "by finding out how it behaves, operates, functions, changes (if at all), what powers it has, what similarities or dissimilarities it bears to other things, and so on" (Oderberg 2007: 56), we can gain important—albeit necessarily imperfect—indirect insights into the nature of a being. The reason why we can do so is that all beings have a "natural inclination"[11] towards the things that are in accord with their nature and help them become the way they are meant to be (by their creator).

In other words, all beings have a natural inclination to pursue the things that are good and avoid the things that are bad for them. So, whatever a

[11] A natural inclination (also: natural tendency, natural disposition, natural appetite; Latin: *inclinatio naturalis, appetitus naturalis*) is an intrinsic behavioural disposition towards activity and operation that aids in the realisation of a being's nature. As "the tendency to achieve its end which is imparted to a thing in virtue of its having a nature" (Pakaluk 2018: 19), a natural inclination inclines a being to pursue its own good, that is, the realisation of its nature, in its own way. For a more detailed description of this concept cf. Erk (2019: 167 (FN 17)).

being has the natural inclination to pursue, that is, whatever satisfies the natural inclination of a being, is objectively good and an objective good for this being as it helps it become good, that is, become how it is meant to be (by its creator). In contrast, whatever a being has the natural inclination to avoid is objectively bad and an objective bad for this being.

How do we learn about the natural inclinations of a being? The key is observation: "By observing behavior, we realize inclination; upon knowing inclination we naturally discover the good" (Jensen 2013: 171). When it comes to human beings, the observation of our natural inclinations is more complicated than with other beings: The human being is a special kind of being as he can consciously choose to ignore or act against his natural inclinations. Therefore, the statement that human beings have a natural inclination to pursue the things that are good and avoid the things that are bad for them does not entail an automatism. While this statement holds true in general, that is, for human beings on aggregate, it must be interpreted correctly with respect to individual human beings. It does not suggest that every single human being is always behaving in conformity with his nature. Possessing an inclination and acting on it are two different things. Human beings can—and, as experience unfortunately shows, do—act against their natural inclinations:

> In man the intrinsic tendency which in other bodies unifies, directs, and controls all their actions, is not of a compelling but only of an inviting nature and that, as a consequence, he is able to produce and choose his own aims and to determine his actions in the manner he pleases. [...] Man is not like the other beings driven by an intrinsic force against which there is no resistance possible, but only drawn by an extrinsic attraction against which there may be resistance. (Baschab 1937: 253f)

While this makes the task harder, it does not, however, make the discovery of the natural inclinations of human beings as well as their nature impossible.

> What is genuinely good for someone, accordingly, may in principle be something he or she does not consciously want, like children who refuse to eat their vegetables, or an addict convinced that it would be bad to stop taking drugs. [...] Knowing what is truly good for us requires taking an external, objective, 'third-person' point of view on ourselves rather than a subjective 'first-person' view; it is a matter of determining what fulfills our nature, not our contingent desires. (Feser 2015: 384)

Nature conveys a norm that allows for exceptions in individual cases. Accordingly, natural inclination is best understood by long-term and cross-generational observation. It is by long-term and cross-generational observation of how human beings behave under normal circumstances as well as the analysis of how their behaviour reflects on their development and the development of their fellow men and descendants that the natural inclinations of man are revealed and insights into the nature of man gained. The focus of the observation and measure is not an individual being, but the class of beings to which this individual belongs.

So, whatever a class of beings has the natural inclination to pursue is objectively good and an objective good for every individual representative of this class of beings as it helps it become good, that is, become how it is meant to be (by its creator). In contrast, whatever a class of beings has the natural inclination to avoid is objectively bad and an objective bad for every individual representative of this class of beings.

1.2.3.2. The Basic Human Goods

The things towards which human beings have a natural inclination and that are, in consequence, objectively good for and the goods of human beings are called "basic human goods" (BHGs). These basic human goods (also: intrinsic human goods) are the ends that a human being "must realize in order to flourish as the kind of organism it is" (Feser 2015: 382). As such, these goods entail a standard of goodness. So, the question we have to address next is the following: What are the basic human goods, that is, the goods that human beings are naturally inclined to pursue?

Throughout history, many philosophers have undertaken the task of "formulating propositionally, and in as illuminating a way as possible, what items need be affirmed as intrinsically good in order to make sense out of our inclinations" (Murphy 2019). If one compares the different compilations of BHGs that can be found in the literature, one will see—somewhat unsurprisingly—that there is no single set or list of BHGs that everybody agrees on.[12] But although it is unrealistic to expect that there will ever be perfect agreement on a finite list of basic human goods, there are, however, several basic human goods that are accepted by virtually everybody. For the purposes of this book it is sufficient to concentrate on these mostly agreed upon basic human goods and not dwell on the candidates that are deemed controversial and questionable.

[12] For an overview of different compilations of basic human goods cf. Erk (2019: 313–315).

The following items are part of this agreed upon common core of distinct basic goods that the human person is naturally inclined to pursue (cf. Oderberg 2000: 41ff):

1. **Life**
   Life is the most fundamental of the human goods. Without being alive, we cannot pursue any goods. The good of life, however, does not only encompass our biological existence. We do not simply naturally tend towards existence but a healthy and integrated existence. Therefore, the good of life "encompasses such sub-goods as proper diet, exercise, fresh air and rest" (Oderberg 2000: 41) but also all goods that further the numerous physiological processes and operations that maintain and restore our bodies to optimal functioning.
2. **Acquisition of knowledge and pursuit of truth**
   Human beings are "enquiring creatures, naturally curious, always looking for new and better ways to solve problems, to understand themselves and the world around them" (Oderberg 2000: 41). From the moment we can use our rational capacities we use them in the pursuit of knowledge: "We want to know as much as we can within the limits imposed on us by ability and environment, including all the other matters that require our daily attention" (Oderberg 2000: 42).
   We do not, however, just want to gather information regardless of its truth value; we want to gain true knowledge. We also do not just want to know that something is; we want to know about the true causes of things and achieve true understanding of why things are as they are. In this sense, the pursuit, acquisition and possession of true knowledge and understanding is a basic good for man.[13]
3. **Friendship**
   Every one of us, "even the most reclusive of hermits needs others and human flourishing requires at least some level of wanted interaction with others" (Oderberg 2000: 42). The good of friendship encompasses not only those "wanted interactions that involve fellow-feeling, sympathy, kindness, love and care" (Oderberg 2000: 43) but also "the family, which provides a natural locus for the promotion of other

[13] One of the objects of knowledge is the good itself: "Without knowledge of the good, the good life as a whole could not even begin to be lived. [...] In short, just as life and health are what perfect our bodies (and fulfil us thereby), so knowledge and understanding perfect our minds" (Oderberg 2000: 42).

goods, such as life (via procreation and physical and economic sustenance) and knowledge (via education of children)" (Oderberg 2000: 42), as well as "social living, in particular living in a self-governing community, or perhaps state whose sole purpose is to promote the peaceful and harmonious coexistence of its members" (Oderberg 2000: 43).

4. **Work and play**
   Work and play (in the sense of leisure and relaxation) are important human goods as they "occupy and exercise our minds and bodies, engage us in enjoyable endeavours, and bring their own special satisfactions" (Oderberg 2000: 43)—"with work at its best a form of play and vice versa" (Oderberg 2000: 43). Although being distinct human goods and as other basic human goods as well, both work and play also serve in the promotion of other basic human goods such as life, knowledge and friendship.
5. **Appreciation of beauty** (both in nature and through art)
   The appreciation of beauty, "both in nature and in the products of human activity" (Oderberg 2000: 43), is another distinct good to which the human person has a natural inclination. Men of all times and places have appreciated beauty as it "brings its own special pleasures and satisfactions and satisfies the highest appetites of the mind" (Oderberg 2000: 44). As in the case of work and play, this good also contributes to the promotion of other basic human goods and sometimes overlaps them.
6. **(True) religious belief and practice**
   The overwhelming majority of men of all cultures and times have been and are religious believers and practitioners. Religion, that is, the belief in and worship of God (cf. II^{a}-IIae q. 81), has been and is "practised by the overwhelming number of people in all places and at all times" (Oderberg 2000: 44). Even many who would call themselves non-believers, materialists or atheists "nevertheless readily testify to an appreciation of some sublime principle of the cosmos that surpasses understanding, but is real, and dwells behind everything that happens" (Oderberg 2000: 44). Religious belief and practice reflect a natural inclination of the human person that is brought to light, for example, by the experience of the many mysterious and irreducibly transcendent aspects of reality or the awe-inspiring complexity of the universe.

> Combining this religious tendency with our quest for truth mentioned above, "we can see that what perfects human nature is not mere religious belief and practice of one sort or another, but only true belief and the practice which best expresses it" (Oderberg 2000: 44). True belief is belief in the true God and not whatever might be misapprehended as god.

This list of distinct basic human goods is, in principle, open to additions—as long as the good to be added is truly representative of a natural inclination of man and not already contained in one of the above goods.

The basic human goods contained in the list of basic goods are, however, not on par. Rather, they are hierarchically ordered with the higher goods presupposing the lower ones. Whether a good ranks higher or lower in the list depends on the question of whether this basic human good is a good (a) that man has in common with all living beings (e.g. preservation of life), (b) that man has in common with all animals (e.g. procreation and education of children) or (c) that is exclusive to man as a rational being.

> The lower goods are subordinate to the higher ones in the sense that they exist for the sake of the higher ones. The point of fulfilling the vegetative and sensory aspects of our nature is, ultimately, to allow us to fulfill the defining rational aspect of our nature. (Feser 2015: 384; cf. I[a]-IIae q. 94 a. 2)

Although much more could be said about the natural inclinations of man and the resulting basic human goods, these remarks should be sufficient for the purposes of this book. The exact content of the list of basic human goods does not have to concern us any further. We can work with the core of basic human goods that is commonly agreed upon. For the BHG that is in the spotlight of this book is one that all authors agree upon and that is the most fundamental BHG, namely life.

1.3. Summary

The contents of the previous two chapters provide us with a general understanding of how to determine the objective ethical quality of a human action. Being able to do so is the necessary first step when thinking about the (im)permissibility of the killing of human beings in general as well as selected killings such as suicide, lethal self-defence, abortion (including infanticide) and euthanasia. Let us recapitulate what we have learned so far:

- **The sources of morality of human actions**
  Insofar as and because they are deliberately willed, the moral quality of (a) the end of the agent (*finis operantis*), (b) the exterior action and its natural end (*finis operis*), (c) the circumstances of the exterior action as well as (d) the consequences of the exterior action impact—albeit to varying degrees—the moral quality of the overall human action.
  Although all elements of a human action can play a role in the evaluation of the exact ethical quality of a human action, the ethical species of a human action—that is, its being either a good or evil action as a whole—is entirely determined by the ethical quality of its end of the agent (*finis operantis*) and the natural end of its exterior action (*finis operis*). Only these two elements have the power to specify the ethical quality of a human action in and of themselves. A human action is evil in its entirety if either its exterior action or its end of the agent is evil. Its unknowingly accepted accidental consequences do not add to the goodness or badness of a human action. The directly voluntary (also: knowingly accepted) accidental consequences of an exterior action can increase or decrease the goodness or badness of the human action but cannot change its overall moral species.
- **The standard of morality of human actions**
  The general standard of morality of human actions is the good. The term "good" means perfection of being according to nature (also: essence). A being is good if it is lacking in nothing that is due to its nature. This objective goodness *of* a being is the yardstick by which we can determine whether something is objectively good *for* a being. Amongst other things, what is good for a human being is the possession of the so-called basic human goods (BHGs). Besides the common good, the virtues as well as the natural ends of man's faculties, these basic human goods are an important part of the concrete standard of morality of human actions.
  An action that violates the standard of morality is an objectively evil action. Accordingly, an action that violates at least one basic human good is an objectively evil action.

With respect to the evilness of a human action, the following somewhat simplifying rule applies:

An action is an objectively evil action if its end of its agent and/or the end of its exterior agent violate at least one basic human good.

The exact extent of such action's evilness can, however, only be determined by taking into consideration the moral quality of all other elements of the action, too.

When it comes to the opposite end of the spectrum of morality, things get slightly more complicated. It is much easier to determine whether an action is evil than to determine whether it is fully good. For there is, unfortunately, no single factor that guarantees the goodness of an action. An action does not automatically become good if it furthers the realisation of at least one basic human good. An action can only be good if it is good in every respect. This has already been pointed out by Thomas Aquinas[14]:

Non tamen est actio bona simpliciter, nisi omnes bonitates concurrant, quia quilibet singularis defectus causat malum, bonum autem causatur ex integra causa [...].	However, an action is not good simply, unless it is good in all elements; for, evil results from any single defect, but good from the complete cause [...].

(Thomas Aquinas, Iª-IIae q. 18 a. 4 ad 3; cf. Iª-IIae q. 19 a. 6 ad 1; Iª-IIae q. 19 a. 7 ad 3; Iª-IIae q. 20 a. 2 co.)

Therefore, establishing the goodness of a particular action requires a somewhat elaborate study of the ethical quality of its elements and the composition of the results to form an overall picture. A human action can only be good if it (a) furthers the realisation of at least one basic human

[14] Because here Thomas Aquinas cites Dionysius Areopagita as the source of this principle, it is also called the "Dionysian Principle". In Dionysius Areopagita's original words the principle reads as follows:

Τὸ ἀγαθὸν ἐκ μιᾶς καὶ τῆς ὅλης αἰτίας, τὸ δὲ κακὸν ἐκ πολλῶν καὶ μερικῶν ἐλλείψεων.	The good comes from the one and whole cause, but the evil is from many and partial defects.

(Dionysius Areopagita, De Divinis Nominibus, Chapter IV, § 30)

As Rickaby (1918: 31) points out, "whoever knows this principle, does not thereby know the right and wrong of every action, but he knows how to go about the enquiry. It is a rule of diagnosis".

good, (b) furthers the realisation of the common good, (c) is a virtuous action, that is, is in line with at least one of the virtues, and (d) is in conformity with and does not frustrate the natural end of the faculties that a human being uses to perform the action. The exact mechanisms of the determination of the moral quality of a human action do not have to concern us here. The interested reader can look these mechanisms up in Erk (2019: 157–163).

Unless one is already ethically competent (in the sense of an unconsciously held competence) to do so, assessing one's past, current and future action might take some time. However, the more often we actually assess our actions and by doing so train our ethical understanding, the easier it gets. Practice makes perfect—in the true sense of the word.

2. The Permissibility of Human Actions

The previous chapter has given us the basic tools with which to determine the ethical quality of an action, in particular its evilness. However, this has only been a first, albeit important, step. For what this book is actually about is not so much the ethical quality of selected kinds of killings but their moral permissibility. So, what does it mean to ask whether suicide, lethal self-defence, abortion, euthanasia, organ transplantation and cooperation in suicide are ethically permissible? How do we get from an action's goodness or evilness to its permissibility or impermissibility?

In general, an action is said to be impermissible if it is prohibited to perform this action. An action is morally impermissible if there is a moral duty to not perform it. So, in order to be able to learn whether the killing actions listed in the previous paragraph are impermissible or permissible, we must answer the question whether there is a moral duty to refrain from them. Doing so requires us to familiarise ourselves with the concepts of "duty" and "right". The following section is tasked to do so. In the next but one section the moral quality of an action will be tied to its (im)permissibility. As will be argued, an action is morally impermissible insofar as it is an evil action. Another section will then establish the general moral duty to realise the basic human goods, especially the basic human good "life".

Taking these steps will enable us to actually and concretely address the problem of the (im)permissibility of the killing actions suicide, lethal self-defence, abortion, euthanasia, organ transplantation and cooperation in suicide.

2.1. A Primer on Duties and Rights[15]

Generally speaking, every action available to us is (a) either commanded or not commanded and (b) either prohibited or not prohibited. An action that is commanded and not prohibited is a *command* (also: *command-duty*). An action that is not commanded but prohibited is a *prohibition* (also: *prohibition-duty*). An action that is neither prohibited nor commanded, that is, that is permitted and facultative, is called a *liberty* (also: *liberty-right*, privilege, licence, permission). Therefore, for every action we do there either is a command-duty that commands us to do it, a prohibition-duty that forbids us to do it, or a liberty that allows us either to do or not do it; there is no fourth option.

Having subdivided the realm of actions, let us shed further light on the concepts of duty and liberty.

- In general, a duty denotes a behavioural constraint, that is, a constraint on the duty-bearer's behaviour which makes him unfree to either commit (prohibition) or omit (command) a certain action. In other words: A duty is "a necessity laid on the will" (Cronin 1930: 212). Stating that somebody bears a duty to φ therefore amounts to stating that he (or to be more precise: his will) is under a necessity to will φ, that is, that it is necessary for his will to will φ. The presence of a duty to φ is always equivalent to the non-presence or absence of a liberty-right to φ. In short: A duty is a non-liberty.
  The group of duties can be further divided into the subgroups of commands and prohibitions. The former duties command the duty-bearer to either commit or omit an action; the latter forbid him to either commit or omit an action.
- While a duty corresponds to a behavioural constraint, a liberty-right corresponds to a behavioural freedom. It designates an action to which its holder does not have a duty, that is, which he has the choice to perform or not, without being prevented by a contrary command or prohibition. A liberty-right to φ is the absence of or freedom from both the command to φ and the prohibition to φ.

[15] The content of this chapter draws on, and in part reproduces verbatim, material published in Erk (2019: 188–192).

Basic Distinctions	Right	Duty
Active	Liberty (Liberty-Right)	Duty (Prohibition or Command)
Passive	Claim-Right	Logically impossible

Fig. 3 Duties and rights: Basic distinctions

Rights that—as is the case with all liberty-rights—enable their holder to carry out a certain action, that is, whose content is an action of the holder of this right, are referred to as *active rights*. In addition to these active rights there are also rights that do not grant their holder the right to an action of their own, but rather a right to an action of another person; such rights therefore grant a claim that somebody other than the holder of the right performs a certain action. These rights are called *passive rights* or *claim-rights*, whereby the content φ of a claim-right is that action or behaviour that the bearer of the claim-right is entitled to see performed.

These initial distinctions yield the following categories or instances of rights and duties (cf. Fig. 3):

Rights, that is, both liberty-rights and claim-rights, as well as duties can be further specified by means of a variety of parameters: their (a) subject, (b) content, (c) counterpart, (d) addressee as well as (e) justification.

(a) With regard to the *subject* of the right or duty, that is, the right-holder or duty-bearer, it is possible to distinguish between three types of rights and duties:

- *Particular rights or duties* are assigned to individual, specified persons.
- *Local rights or duties* are only assigned to the individual members of a certain (temporally, locally and/ or topically) limited group of persons.
- *Universal rights or duties* are held by all persons.

(b) The *content* of a duty or right can either be positive or negative as well as perfect or imperfect.

- An action can either consist in a commission, that is, something that is done, or an omission, that is, something that is not done. In both cases the agent is active, but in a different manner: When somebody commits something, he performs a positive action; when he omits something, he performs a negative action. Depending on whether its content is a commission or omission, a *duty or right* is called either *positive or negative.* While duties and claim-rights are either positive or negative, liberties are always both positive and negative.
  As far as duties are concerned, both duties that command a commission and duties that prohibit an omission are positive duties, while both duties that prohibit a commission and duties that command an omission are negative duties. And to add a bit more complexity: Both a positive command-duty and a negative prohibition-duty have the same content as do a negative command-duty and a negative prohibition-duty. While the former command a commission and therefore prohibit an omission, the latter command an omission and therefore prohibit a commission.
- A *perfect duty or right* has a strictly defined content; it can only be satisfied by performing a narrowly specified action in a specified fashion. An *imperfect duty or right* allows for some degree of freedom in deciding how and/or when to fulfil it and sometimes also towards whom to perform it. It either contains some general behaviour that can be satisfied in various ways or it consists in the indication of some state of affairs or situation that can or must be brought about without specifying the actions that serve as means to that end.

(c) In contrast to liberty-rights, duties and claim-rights have a *counterpart*, that is, somebody against whom the duty or claim-right is held. With regard to their kind of counterpart, duties and claim-rights can be differentiated into horizontal duties/claim-rights and vertical duties/claim-rights:

- *Horizontal claim-rights and duties* are called horizontal because they exist between two or more human persons: A horizontal claim-right is a claim-right whose corresponding duty rests with one other, several other or all other human persons. In the same fashion, a horizontal duty is a duty whose corresponding claim-right rests with one, several or all other human persons.
- Depending on its counterpart, we can distinguish between the following kinds of horizontal duties and claim-rights:

- Duties/claim-rights in personam (*special duties/claim-rights*), that is, duties/claim-rights held against a single (i.e. one other) human person
- Duties/claim-rights in *rem publicam* (*communal duties/ claim-rights*), that is, duties/claim-rights held against some (i.e. more than one but fewer than all) other human persons
- Duties/claim-rights in *rem* (*general duties/claim-rights*), that is, duties/claim-rights held against all other human persons

• In contrast to horizontal duties whose subject and counterpart are human persons, the counterpart of a *vertical duty* is not a human person: A vertical duty is a duty whose corresponding claim-right does not rest with any other human person(s) but ultimately with God. This is why vertical duties are sometimes also described as having no counterpart.

(d) A right or a duty is not just made up of a subject and counterpart. There is also somebody who benefits from or is affected by a right or duty. Rights and duties usually have an *addressee* who can be a single person, some (i.e. more than one but not all) persons or all persons. The addressee can differ from but also coincide with the subject or counterpart of the right or duty. For example, in case of a contract made in favour of a third party, a person A concludes a contract with a counterpart B, in which A obligates himself vis-à-vis B to do something for the benefit of C.

(e) Rights and duties don't just fall out of the sky, rather they require a *justification*. It is not just who has a right/duty to whom, with which content benefitting/affecting whom, that needs to be clear. It must also be clear why this right/duty exists. With regard to their justification, it is possible to distinguish three different types of rights and duties:

• *Legal rights and duties* owe their existence to legislation. They are set and codified by an entity authorised to do so in an orderly procedure intended for this purpose. In other words: Legal rights/duties are rights/duties written down and contained in a collection of laws.
• *Conventional rights and duties* are grounded in and the product of custom and/or social practice. They are not dependent on codification or inclusion in a legal system. Even though they can of course be translated into legal rights and duties, they are by their nature pre-positive (i.e. exist and are valid, despite not being codified and contained in a legal system) and to some extent also stand above

positive rights/duties (i.e. exist and are valid, even when they contradict legal rights/duties). With respect to their validity in terms of time and place, both conventional and legal rights and duties are relative.

- What distinguishes ***moral rights and duties*** from both legal and conventional rights and duties is the fact that the latter two arise from human will, while the former are a given for the human will. As they are predetermined for the human mind, human beings therefore cannot invent or create, but only discover, recognise and acknowledge moral rights and duties. Moral rights and duties are found by means of moral philosophical reflexion. They exist independently of customs, legislative processes or public opinion.
  Moral duties and rights are universal, that is, are possessed by all people of all times at all times, and have validity independently of, and even in contrast to, all legal and conventional rights and duties. They are the framework which legal as well as conventional rights and duties must not transgress and within which legal and conventional law has to operate.

Equipped with this vocabulary and these fundamental distinctions we can now turn to the question of how the existence of moral rights can actually be justified. As this is a book about moral philosophy, the kind of impermissibility that interests us is not legal or conventional impermissibility. Asking about the impermissibility of certain actions from an ethical perspective means asking about their moral impermissibility. In order to know whether certain actions—such as selected killing actions—are morally impermissible or permissible, one needs to first understand how moral duties and rights are justified to begin with. Why are there and where do moral duties and rights come from? It is answering this question that the next chapter is devoted to.

2.2. The Good and the Moral Duty to Pursue It[16]

As stated in the previous chapter, a duty is a necessity laid on the duty-bearer's will, that is, a necessity binding his will. He who has a duty to φ is not free not to will the realisation of φ. Given its nature of being borne by

[16] The content of this chapter draws on, and in part reproduces verbatim, material published in Erk (2019: 192–200).

all men of all times at all times, a moral duty, therefore, is a necessity laid on the will of all men of all times at all times. As shall be argued in this chapter, the most fundamental of these necessities is the necessity to do and pursue what is good and to avoid what is evil.

In order to prove that every human person possesses the moral duty to do and pursue what is good and to avoid what is evil (D_{Good}), we have to establish that it is necessary for the will of man to will doing as well as pursuing what is good and avoiding what is evil. In other words: We have to prove that the will of man is not free not to do and pursue what is good, and not to avoid what is evil.

The first step we have to take in this respect is to find out what kind of necessity is implied in the statement that the moral duty to do and pursue what is good and to avoid what is evil is a necessity laid on the will of all men of all times at all times. In general, something is necessary if it is not possible for it not to be; in contrast, what is contingent (also: non-necessary) can or cannot be. While necessary beings contain no potentiality of not-being, contingent beings contain potentiality of non-being ("potentia ad non esse" (Contra Gentiles, lib. 2 cap. 30 n. 2)).

Necesse est enim quod non potest non esse.	That is necessary which must be.

(I[a] q. 82 a. 1 co.; cf. Contra Gentiles, lib. 2 cap. 83 n. 3; Super Sent., lib. 3 d. 16 q. 1 a. 2 co.; Metaphysics, 1015a 28)

Contingens est, quod potest esse et non esse.	That is contingent which can be or cannot be.

(I[a] q. 86 a. 3 co.)

So, if we say that a moral duty is a necessity laid on the will of all man of all times and at all times, this means that a moral duty makes it somehow impossible for the will to refrain from willing the content of this duty—the will cannot not will but has to will the content of a moral duty:

Im Verhältnis zum Willen heisst dasjenige notwendig, was der Wille nicht nichtwollen kann.	In relation to the will, that which the will cannot not want is called necessary.

(Commer 1884: 232)

How can we imagine this necessitating? Here it helps to gain an overview of the kinds of necessity that actually exist. Typically, we can distinguish between the following three senses or classes of necessity:

1. **Necessity of nature**
   The necessity of nature—which is also called natural necessity (Latin: *necessitas naturalis*), absolute necessity (Latin: *necessitas absoluta*) or necessity of natural inclination (Latin: *necessitas naturalis inclinationis*)—arises from features intrinsic to a being, that is, its essence or nature as determined by its formal and/or material cause: "For, given a certain nature, certain properties, relations, and acts must follow, and with absolute necessity" (Cronin 1930: 215). Something that is the case by necessity of nature must be the case because it is a necessary consequence of a being's nature.
   For example, it is absolutely necessary that the interior angles of a triangle, that is, a closed plane figure with three straight sides, equal two right angles, that is, 180 degrees. This is part of the very definition essence or nature of a triangle and cannot not be the case.
2. **Necessity of efficient cause**
   The fact that something must be can also result from an external principle, namely an efficient cause or agent outside a being. Something that is the case by necessity of efficient cause (Latin: *necessitas ex cause efficiente*) must be the case because it is the necessary consequence of an external force or agent impacting a being in a way that this being cannot resist. This is why this kind of necessity is also called necessity of compulsion, force, coercion or constraint (Latin: *necessitas coactionis*).
   For example, if a man is pushed hard enough so that he is unable to resist the force, there is no option for him but to fall to the ground; it is necessary for him to fall. Or to give another example: That day and night alternate is a necessity of efficient cause which is grounded in the rotation of the Earth.
3. **Necessity of end**
   The third kind of necessity is also called final necessity or utility. According to the necessity of end (Latin: *necessitas finis; necessitas ex (suppositione) fine*), the reason that something must be arises from a being's end or final cause. The necessity referred to in this case, however, does not pertain to the end itself; it is not the end that is necessary but the means to it. The means to an end are to be willed by necessity on the condition or supposition that the end is truly willed: He who wills an end must also will the means necessary to achieve or realise the willed end. This includes both doing and pursuing what furthers as well as avoiding what hinders the realisation of the willed end. The necessity of end is a conditional necessity.

For example, it is necessary for a young boy to train if he truly wills to become a professional football player.

With these kinds of necessity in mind, we can now continue to ask which of them is actually capable of impacting the will of man in such a way that—as a result—the will becomes unfree not to will to do and pursue what is good and avoid what is evil. Which kind of necessity can necessitate the will of man of all times and at all times in this fashion?

As should be rather obvious, it is not the second kind of necessity. For the necessity of efficient cause has no necessitating impact on (the freedom of) the will; it is foreign to the will: "the will cannot be violently compelled to an act by any agent outside itself" (Cronin 1930: 215). No external force or agent can compel the will to will a certain thing. By pushing us, an external agent can make us fall to the ground against our will, but he cannot force our will to will that we fall to the ground. The will of man cannot be compelled by some external force in such a way that it could not act otherwise; it cannot be forced into willing something if it does not have an inclination to it, that is, if this something is not apprehended by reason as good. The will simply eludes, that is, is not subject to, external force:

Non potest contingere ut voluntas aliquid coacte vel violente velit.	It is impossible for the will to will anything by force or violently.

(Thomas Aquinas, De Veritate, q. 22 a. 5 co.)

What is done by necessity of efficient cause is—by definition—contrary to the will: "coactum enim est quod est voluntati contrarium" (Contra Gentiles, lib. 3 cap. 138 n. 2). And because it cannot be laid on the will of man, necessity of efficient cause cannot ground any duty, let alone the moral duty that is the focus of our investigation.

So, what about the other two kinds of necessity? Can they be laid upon and, accordingly, bind our will? The necessity of end binds our will if we truly and seriously will an end E that can only be realised by means of φ-ing: If we really will the realisation of E, we are not free to not also will to φ; rather, φ-ing becomes a necessity to our will. And as a duty is a necessity laid on our will limiting its freedom, this is tantamount to saying that whenever we truly and seriously will (the realisation of) an end that can only be realised by φ-ing, we are under a duty to φ.

However, a duty arising from necessity of end does not automatically constitute a genuine *moral* duty. For it is characteristic of moral rights and duties that they are universal, that is, that they apply to all men of all times at all times. But this is not automatically true of duties grounded in necessity of end. For a duty grounded in necessity of end is usually a hypothetical duty, that is, a conditional duty depending on an "if". A hypothetical duty to φ is a merely relative duty laid on the will of only those who actually will the end that can only be realised by means of φ-ing. So, while necessity of end imposes the duty to will φ on all those who will an end, it cannot impose the duty to will the end on everybody's will. Therefore, a hypothetical duty cannot be a genuinely moral duty. Only a categorical duty, that is, an absolute and unconditional duty that depends on no "if", can be a genuinely moral duty necessitating the will of man.

A duty grounded in necessity of end can, however, become a categorical duty if it can be shown that the end which the duty is meant to realise is a necessary end: If the will as such is not free to not will a certain end, then all human persons are under the genuinely moral duty to take the means to realise that end. So, what we have to do in order to establish the duty to do and pursue what is good and avoid what is evil as a moral duty is to show that this duty serves the realisation of an end that the will of man is not free not to will. And this is where the third kind of necessity, natural necessity, comes into play.

The will is an appetitive power. The will is the rational appetite of man that desires whatever has been apprehended as good by reason. The will can do so because it naturally, that is, by its very nature or natural inclination, wills the good in general: "voluntas vult naturaliter bonum" (De Veritate, q. 22 a. 6 ad 5; also cf. Contra Gentiles, lib. 3 cap. 109 n. 6).[17] The will is fixed on desiring the good in general (though not any particular good) by an irresistible law of nature; it casts itself towards what is absolutely and perfectly good in the same way as iron is attracted to a magnet:

> Will is rational appetite; an appetite which embraces nothing of necessity, except what is apprehended as absolutely good and desirable in every respect. The Rational Will can be irresistibly drawn only by that which rea-

[17] To be more precise: Because the will is the kind of thing it is (i.e. as it is its very nature) the will cannot but will the highest good which is *beatitudo* (English: beatitude, happiness; German: (Glück-)Seligkeit) and what is included in it: "ex necessitate volumus esse beati" (I^{a} q. 82 a. 2 co.; also cf. I^{a} q. 82 a. 1 co.; De Veritate, q. 22 a. 5 co.; De Malo, q. 6).

> son proposes as so universally attractive that it contains no lack of good, no dissatisfactory feature. As long as the thought of an object reveals any lack of good, any disagreeable aspect, the Will has not that which it is naturally longing for, perfect good; and it is able to reject this object. [...] We, however, though necessitated to seek after the good in general, are physically free to reject any particular form of it presented to us. (Shallo 1916: 279f; cf. 276; cf. Commer 1884: 233)

Because it is what it is, the will is naturally incapable of not willing the good in general and, accordingly, also naturally incapable of willing the evil in general. In other words, the will is under the natural necessity to will the good in general and to not will the evil in general; it is in its nature not to be able to do otherwise. The basic act of the will, therefore, is not one of free choice: The will cannot will evil in general and must will good in general, that is, the perfect and therefore highest good.

Free choice only enters (a) if the will is required to choose whether to pursue a particular good, that is, a concrete good that has been presented as good to the will by reason, or (b) if the will is required to choose between two such particular goods. While the good in general is perfectly good, that is, good in every respect, particular goods are not good in every respect. Virtually everything we do in this life has both positive and negative aspects, even if the former by far outweigh the latter. So, while the will, on the one hand, must will particular goods because they are good in some respect (e.g. exercising is good for one's health), the will must at the same time also not will them because they are bad in other respects (e.g. exercising takes time and possibly other resources, such as money for equipment, and implies physical exertion and sometimes pain). And this is where the freedom of the will enters: The will is free because it is not forced to choose in a certain fashion when it comes to particular goods or when choosing between goods; we are free to choose whether to go to the gym to exercise or not.[18]

By now we know that the will wills the good by natural necessity. But to establish that the duty to do and pursue what is good and avoid what is evil is not a hypothetical but a categorical duty requires us to show that the good that the will wills by natural necessity has the nature of an end. For if it has, then the good that the will wills by natural necessity is a necessary

[18] For an overview of philosophical proofs of the freedom of the will cf. Lehmen (1911: 484ff).

end. And if it is a necessary end, that is, an end that cannot not be willed, then the duty to actually pursue that end is a categorical duty.

So, does the good have the nature of an end? As can be easily shown, it does. For what is an end? An end is both a starting point and a terminus: It is that for the sake of which something happens (cf. Sententia Ethic., lib. 1 l. 9 n. 3) and that in which the appetite of an agent finds its rest (cf. Contra Gentiles, lib. 3 cap. 3 n. 3). Accordingly, the good has the nature of an end if an appetite strives for and finds rest in its attainment. As should have become clear, there is such appetite, namely the rational appetite of man, that is, the will: The will strives for and finds its rest in (the realisation of) the good in general. We can, therefore, conclude that the good in general has the nature of an end.[19]

Bonum habet rationem finis, malum autem rationem contrarii.	Good has the nature of an end, but evil the nature of a contrary.

(Thomas Aquinas, I^{a}-IIae q. 94 a. 2 co.; cf. I^{a} q. 5 a. 4 co.; I^{a}-IIae q. 1 a. 4)

Where does this leave us? We now know that (a) the good has the nature of an end and (b) the will wills the good by natural necessity. Combining these two premises allows us to conclude that the good is an end of the will by natural necessity: Qua its nature, the will cannot not have the good as its end. Therefore, good in general is the necessary end of the will.

Let us sum up: Some paragraphs ago it was established that he who wills (the realisation of) an end must—by necessity of end—also will the means to realise that end. And in the preceding paragraph it was established that—by natural necessity—the will must will the good in general as its necessary end. From these two premises we can conclude that as a being endowed with a will man bears the categorical duty to take the means necessary to realise good in general. This duty has both a positive and a negative aspect: Its bearer is required (a) to do and pursue what furthers the realisation of the good and (b) not to do and pursue what frustrates the realisation of the good, especially what furthers the realisation of the contradictory opposite of good in general, namely evil in general.

[19] This is why Thomas Aquinas can state that the object of the will is the good and the end: “Obiectum autem voluntatis est bonum et finis” (I^{a}-IIae q. 1 a. 3 co.).

Fagothey (1963: 79f) expresses this as follows: “Every end is a good and every good is an end. An end would not be sought unless it were somehow good for the seeker, and the good by being sought is the end or purpose of the seeker’s striving”.

Weil das Gute die Bedeutung des Zweckes hat, das Uebel aber die Bedeutung des Gegentheils, so kommt es, dass der Verstand alle Dinge, wozu der Mensch eine natürliche Neigung hat, als Güter natürlich wahrnimmt und folgerecht als solche, die man bei der Handlung verfolgen muss, und das Gegentheil von diesen Dingen als Uebel, die man vermeiden muss. Die Ord-nung der Vorschriften des Natur-gesetzes richtet sich daher nach der Ordnung der natürlichen Neigungen.	Because the good has the nature of an end, but the evil has the nature of the contrary, it follows that the mind naturally perceives all things, to which man has a natural inclination, as goods and consequently as something which must be pursued when acting, and the contrary of these things as evil which must be avoided. The order of the precepts of natural law therefore follows the order of natural tendencies.

(Commer 1886: 119)

Given this conclusion, we have, therefore, proven what we have set out to prove, namely that there is a genuinely moral duty laid on the will of man to take the means necessary to realise the good, that is, to do and pursue good in general and avoid evil in general. Every being endowed with a will and, therefore, every human person is under the following moral duty D_{Good}:

D**$_{Good}$:**

Every human person possesses the universal vertical moral duty to do and pursue good in general. This duty comprises (a) the command-duty to do and pursue what furthers the realisation of the good as such and (b) the prohibition-duty not to do and pursue what hinders the realisation of the good as such, especially what furthers the realisation of the evil as such.[20]

[20] De Wulf (1922: 108f) expresses this duty as follows: "In the first place we are bound to will our end, i.e., our well-being, and to seek it where it is to be found [...] and not to look for it exclusively in those secondary goods which cease to be good when not controlled by reason. In the second place we are morally bound to will whatever is indispensable in order to reach this end, and to avoid that which must of necessity turn us away from it. [...] Moral obligation consists in the necessity of willing our supreme good, combined with the liberty of choosing the concrete objects wherein it is in fact realized".

When it comes to human persons the moral duty D_{Good} is compulsory in a moral sense. This means that the human person is morally bound to realise D_{Good} but is physically free to not do so: "The main difference between the law of nature as it applies to the life of man and to the actions of the rest of the universe is not a difference in the nature or degree of goodness, but simply a difference in the manner of subjection to the same law: both are bound by the same law, but in man its fulfillment must proceed from self-determination and its course must be self-directed, i.e. it is the moral law of his nature; whereas, in the rest of the universe, its fulfillment is by intrinsic necessity, i.e., it is the physical law of its nature. One of the consequences of this difference is that man may fail to attain his natural end, whereas the rest of the universe cannot fail" (Baschab 1937: 256).

However, having gotten this far we are faced with a practical problem. For how can the moral duty D_{Good} actually be met? How can good in general be done and pursued? What is it that we have to do and pursue to realise the good and avoid its frustration?

As we cannot simply pursue good but have to pursue some particular good, this fundamental moral duty has to be concretised in order to be actionable. The term "the good" in D_{Good} has to be substituted with a particular good. This can be done by means of the concretions of the good in general as they have been outlined in Chap. III, Sect. 1.2, that is, the basic human goods (BHGs), the common good, the virtues and the natural ends of the faculties of human beings. Accordingly, the general duty to do and pursue what is good and to avoid what is evil (D_{Good}) is substantiated by and gives rise to several specific and moral duties of the human person:

- The moral duty to realise and not violate the basic human goods
- The moral duty to contribute to the realisation of and not violate the common good
- The moral duty to develop and not frustrate the virtues
- The moral duty to not frustrate the natural ends of one's faculties

Since we have concentrated only on the basic human goods when explicating the standard of morality and since this suffices for the purposes of this book, the following section only deals with the moral duty to realise the basic human goods as well as the moral duties and rights that directly derive from this duty. For a detailed explication of the rest of the abovementioned moral rights and duties, the interested reader is referred to Erk (2019: 203ff).

2.3. The Moral Duty to Realise the Basic Human Goods[21]

The first and foremost good that is to be pursued by every human person is the good of her own goodness. Therefore, the duty to do and pursue what is good and avoid what is evil primarily pertains to the human person's own goodness: Every human person has the fundamental (moral) duty to realise her own goodness, that is, do and pursue what is necessary

[21] The content of this chapter draws on, and in part reproduces verbatim, material published in Erk (2019: 200–203).

to become good and avoid what hinders her becoming good or what makes her bad.

What does this exactly imply? From Chap. III Sect. 1.2 we know that the goodness of the human person consists in her flourishing, that is, the perfection of her nature by having become how she is meant to be (by her creator). Combining these insights warrants the conclusion that every human person has the fundamental (moral) duty to perfect her nature, that is, to become how she is meant to be as predetermined by her inherent nature. Every human person has the (moral) duty to use her capacities only in ways that are consistent with her nature as a human person, that is, that further the realisation and perfection of her nature as a human person:

Die persönliche Vollendung zu suchen, ist nicht zunächst ein Recht, sondern eine Pflicht.	To seek personal perfection is, first of all, not a right but a duty.

(Utz & Hieronimi 1969: 41)

Das Sittengesetz fordert vom Menschen nur, sein wesenhaftes, in seiner Natur vorgezeichnetes Selbst zu sein. Daher ist jede Pflicht des Menschen im Grunde eine Pflicht der Treue gegenüber seinem wahren Selbst.	The only thing moral law demands of man is to be himself as determined by his nature. For this reason, every duty of man is in essence a duty of loyalty against his true self.

(Messner 1955: 53)

So, the first and foremost moral duty of every human person is that of self-perfection; every human person possesses this duty simply because she is what she is. However, because she cannot achieve self-perfection directly or pursue goodness *in abstracto*, the human person has to fulfil this duty indirectly by pursuing those goods that she is naturally inclined to, namely the basic human goods (BHGs), and to avoid what is detrimental to them, frustrates them or destroys them. In other words, the human person possesses the moral duty D_{BHG}:

D_{BHG}:
Every human person has the universal vertical moral duty to perfect herself, that is, to realise the basic human goods for herself. This duty comprises (a) the command-duty to do what furthers the realisation of the basic human goods and (b) the prohibition-duty to not do what hinders their realisation or violates them.

Having presented this fundamental moral duty of every human person in detail is, however, not the end. Having a closer look at this duty reveals that it can be unfolded further and that it directly grounds a moral right which in return corresponds to another duty. Let us start with the right: As is the case with every duty, D_{BHG} grounds a claim-right on the part of the duty-holder. By virtue of possessing a duty, the duty-holder automatically possesses the negative claim-right to not be hindered in performing this duty; it would just be nonsensical to possess a duty and not have the negative claim-right to fulfil it. D_{BHG}, therefore, grounds the negative moral claim-right (CR_{BHG}) to not be hindered in fulfilling D_{BHG}. This claim-right is, however, not only held by all human persons (because all human persons possess D_{BHG}) but also against all human persons; it is, in other words, not only a universal but also a general claim-right.

As has been stated above, someone can only have a claim if somebody else has a duty to fulfil this claim; if there were no such corresponding duty, there would be no claim. The duty that CR_{BHG} corresponds with however, is not D_{BHG}. From a logical perspective, the only duty that CR_{BHG} can correspond with is the negative moral command-duty ($D_{Non\text{-}Hindrance}$) to not hinder the holder of CR_{BHG} in performing D_{BHG}. As CR_{BHG} is held by all and against all human persons, its corresponding duty $D_{Non\text{-}Hindrance\ BHG}$ cannot but be held by all and against all human persons as well. Therefore, $D_{Non\text{-}Hindrance\ BHG}$ is a universal general negative moral command-duty.

Summing up what has been stated so far allows us to note that every human person has the fundamental vertical moral duty D_{BHG} and that this duty grounds the moral claim-right CR_{BHG} which corresponds with the horizontal moral duty $D_{Non\text{-}Hindrance\ BHG}$:

D_{BHG}:
Every human person has the universal vertical moral duty to perfect herself, that is, to realise the basic human goods for herself. This duty comprises (a) the command-duty to do what furthers the realisation of the basic human goods and (b) the prohibition-duty to not do what hinders their realisation or violates them.

↓ ↓ ↓ ↓

This duty grounds the following claim-right:

↓ ↓ ↓ ↓

CR_{BHG}:
Every holder of D_{BHG} has the universal general negative horizontal moral claim-right to not be hindered in fulfilling D_{BHG}.

↑ ↑ ↑ ↑

This claim-right corresponds with the following duty:

↓ ↓ ↓ ↓

$D_{\textit{Non-Hindrance BHG}}$:
Every human person possesses the universal general horizontal negative moral command-duty to not hinder the holder of CR_{BHG} in fulfilling D_{BHG}.

There is, however, more to D_{BHG} than has been uncovered so far. For upon closer inspection, the vertical duty D_{BHG} is not an atomic duty but a molecular duty (also: cluster-duty), that is, a duty that consists of several duties. If we hold that every human person has the duty to perfect herself by pursuing and realising the basic human goods and avoiding what is detrimental to her, then this duty applies to all basic human goods listed in Chap. III, Sect. 1.2.3.

Therefore, D_{BHG} contains the duty to pursue the basic human good "life" and avoid what is detrimental to it ($D_{BHG\text{-}Life}$), the duty to pursue the basic human good "friendship" and avoid what is detrimental to it ($D_{BHG\text{-}Friendship}$), the duty to pursue the basic human good "work and play" and avoid what is detrimental to it ($D_{BHG\text{-}Work\&Play}$), and so forth. D_{BHG} is not one but (at least) six duties. And there is more: Each of the six sub-duties of D_{BHG} comprises both the command-duty to pursue the respective basic human good and the prohibition-duty to avoid what hinders its realisation or violates it. Given their nature as duties, each of these sub-duties furthermore grounds a claim-right on the part of the duty-holder to not be hindered in fulfilling the respective sub-duty. And each of these claim-rights corresponds with a horizontal duty to not hinder the holder of the claim-right in fulfilling the sub-duty which grounded the claim-right.

2.4. Summary

As the preceding sections have shown, good and evil are not just neutral concepts. Rather, they possess normative force. As the yardstick of goodness, the nature of human beings grounds their most fundamental duty, namely the moral duty D_{Good}. This duty is a molecular duty that comprises (a) the universal positive vertical moral command-duty to do and pursue what furthers the realisation of good as such and (b) the universal positive vertical moral prohibition-duty not to do and pursue what furthers the realisation of evil as such. These sub-duties basically amount to the command-duty to only perform good actions (a) and the prohibition-duty to not perform evil actions (b). As an action is morally impermissible insofar as there is a prohibition-duty to not perform it, this means that evil actions are morally impermissible.

With respect to an action's evilness, we have learned that an action is an objectively evil action if its end of its agent and/or the end of its exterior agent violate at least one basic human good. Therefore, the general moral duty D_{Good} finds a concrete expression in the moral duty D_{BHG}. This duty is composed of (a) the universal positive vertical moral command-duty to do what furthers the realisation of the basic human goods and (b) the universal positive vertical moral prohibition-duty to not do what violates them. Put differently, the latter aspect of D_{Good} implies that every action that violates one of the basic human goods is morally impermissible. *As life is one of the basic human goods, this necessitates the conclusion that every action that violates the basic human good "life" is morally impermissible.*

The moral duty D_{BHG} is accompanied by the duty-bearer's universal general negative horizontal claim-right CR_{BHG} to not be hindered in fulfilling this duty. The moral claim-right CR_{BHG} in turn corresponds with the universal general horizontal negative moral duty $D_{Non\text{-}Hindrance\ BHG}$ to not hinder the holder of CR_{BHG} in fulfilling his duty to further and not violate the basic human goods.

Bibliography

Baschab, Charles Reinhard. *A Manual of Neo-Scholastic Philosophy. Fourth revised Edition.* St. Louis & London: B. Herder Book Co., 1937.

Coffey, Peter. *Ontology or the Theory of Being: An Introduction to General Metaphysics.* London, New York, Bombay, Calcutta & Madras: Longmans, Green and Co., 1918.

Commer, Ernst. *System der Philosophie. Zweite Abtheilung: Naturphilosophie und Psychologie.* Münster: Nasse'sche Verlagsbuchhandlung, 1884.

Commer, Ernst. *System der Philosophie. Vierte (Schluss-) Abtheilung: Ethik.* Münster & Paderborn: Ferdinand Schöningh, 1886.

Cronin, Michael. *The Science of Ethics. Volume I: General Ethics.* Third Impression. Dublin: M.H. Gill and Son, 1930.

de Wulf, Maurice. *Mediaeval Philosophy Illustrated from the System of Thomas Aquinas.* Cambridge, MA: Harvard University Press, 1922.

Erk, Christian. *Moral Philosophy. Fundamental Concepts.* Second, revised and expanded Edition. Heusenstamm: editiones scholasticae, 2019.

Fagothey, Austin. *Right and Reason. Ethics in Theory and Practice.* Third edition. Saint Louis: The C.V. Mosby Company, 1963.

Feser, Edward. *Aquinas.* London: Oneworld Publications, 2009.

Feser, Edward. "In Defense of the Perverted Faculty Argument." In: Feser, Edward. *Neo-Scholastic Essays.* South Bend, Indiana: St. Augustine's Press, 2015. pp. 378–415.

Feser, Edward & Joseph Bessette. *By Man Shall His Blood Be Shed: A Catholic Defense of Capital Punishment.* San Francisco: Ignatius Press, 2017.

Grisez, Germain Gabriel. "The First Principle of Practical Reason: A Commentary on the Summa Theologiae, 1–2 Question 94, Article 2." *Natural Law Forum* 10 (1965): 168–201.

Hittinger, Russell. *A Critique of the New Natural Law Theory.* Notre Dame, IN: University of Notre Dame Press, 1987.

Hooker, Brad. "Rule Consequentialism." *The Stanford Encyclopedia of Philosophy.* November 18, 2015. Available at: https://plato.stanford.edu/entries/consequentialism-rule Accessed: March 28, 2022.

Jensen, Steven J.. *Living the Good Life. A Beginner's Thomistic Ethics.* Washington, D.C.: The Catholic University of America Press, 2013.

Kant, Immanuel. "Grundlegung zur Metaphysik der Sitten (1785)." In: Kant, Immanuel. *Akademieausgabe von Immanuel Kants Gesammelten Werken. Band 4 (Kritik der reinen Vernunft, Prolegomena, Grundlegung zur Metaphysik der Sitten, Metaphysische Anfangsgründe der Naturwissenschaft).* pp. 385–464. Available at: http://www.korpora.org/Kant/verzeichnisse-gesamt.html and http://www.korpora.org/Kant/aa04 Accessed: March 28, 2022.

Lee, Patrick. "The New Natural Law Theory." In: Angier, Tom (Ed.). *The Cambridge Companion to Natural Law Ethics.* Cambridge: Cambridge University Press, 2019. pp. 73–91.

Lehmen, Alfons. *Lehrbuch der Philosophie auf aristotelisch-scholastischer Grundlage zum Gebrauch an höheren Lehranstalten und zum Selbstunterricht. Zweiter Band: Kosmologie und Psychologie.* Dritte, verbesserte und vermehrte Auflage. Herausgegeben von Peter Beck. Freiburg im Breisgau: Herdersche Verlagsbuchhandlung, 1911.

Lisska, Anthony J.. *Aquinas's Theory of Natural Law: An Analytic Reconstruction.* Oxford: Clarendon Press, 1996.

Long, Steven. "Fundamental Errors of the New Natural Law Theory." *The National Catholic Bioethics Quarterly* 13.1 (2013): 105–131.

Messner, Johannes. *Ethik. Kompendium der Gesamtethik.* Innsbruck, Wien & München: Tyrolia Verlag, 1955.

Murphy, Mark C. "The Natural Law Tradition in Ethics." *The Stanford Encyclopedia of Philosophy.* May 26, 2019. Available at: http://plato.stanford.edu/entries/natural-law-ethics Accessed: March 28, 2022.

Oderberg, David S.. *Applied Ethics. A Non-Consequentialist Approach.* Oxford: Blackwell, 2000.

Oderberg, David S.. *Real Essentialism.* London: Routledge, 2007.

Oderberg, David S. "The Metaphysics of Privation." In: Hüntelmann, Rafael & Johannes Hattler (Eds.). *New Scholasticism Meets Analytic Philosophy.* Heusenstamm: editiones scholasticae, 2014. pp. 63–88.

Pakaluk, Michael. "Natural Inclinations in Aquinas' Account of Natural Law." In: Wolfe, Christopher & Steven Brust (Ed.) *Natural Law Today. The Present State of the Perennial Philosophy.* Lanham: Lexington Books, 2018. pp. 19–32.

Rickaby, Joseph. *Moral Philosophy. Ethics, Deontology and Natural Law.* Fourth Edition. London: Longman, Green and Co., 1918.

Ross, John Elliot. *Ethics. From the Standpoint of Scholastic Philosophy.* Heusenstamm: editiones scholasticae, 2012. (Reprint of the third edition, New York: The Devin-Adair Company, 1938.)

Shallo, Michael W.. *Lessons in Scholastic Philosophy.* London: Hirschfeld Brothers Limited, 1916.

Tollefsen, Christopher. "The New Natural Law Theory." *Lyceum* 10.1 (2008): 1–8. Available at: http://lyceumphilosophy.com/?q=node/97 Accessed: March 28, 2022.

Utz, Arthur F. & Gerda Hieronimi. *Grundsätze der Sozialpolitik. Solidarität und Subsidiarität in der Altersversicherung.* Stuttgart: Seewald Verlag, 1969. (Veröffentlichungen des Instituts für Gesellschaftswissenschaften Walberberg, Band II)

IV: Interlude I: Life and Death

So far, we have been speaking about life and death as if these two concepts were the most common in the world. And in a sense, they are. We all live, and we all die. Life and, whether we like it or not, death, too, are part of our existence. But, once we start to ask ourselves what these two taken-for-granted phenomena actually mean, we may begin to realise that we do not really have a clear grasp of them: What does it mean to live? What does it mean to be dead? What is life? What is death? For a killing action to have occurred, a human being must have lost his life. It would be somewhat superficial to assess the ethical quality of an action without knowing what it is that is lost when a human being is killed.

Therefore, this chapter is devoted to an explication of two of the most fundamental concepts of the ethics of killing, namely life and death.[1] Killing a human being means causing the death of this human being. Every killing necessarily requires that somebody has died; if there is no killed, then there has been no killing. So, in order to establish that some action is actually a killing and not something else (e.g. an assault), we need to be able to diagnose death. The diagnosis of death requires two things: (a) a set of operational criteria for determining that death has occurred (e.g. complete and irreversible loss, i.e. absence of previously present brain function) and (b) specific medical tests showing whether or not the criteria

[1] The content of this chapter draws on, and in part reproduces verbatim, material published in Erk (2011, 24–42, 2014a: 124–133, 2014b: 106, 2015: 163–194, and 2016: 24–29).

C. Erk, *The Ethics of Killing*,
https://doi.org/10.1007/978-3-031-07183-6_4

have been met (e.g. electroencephalography; cerebral blood flow scan; testing of cranial nerve reflexes).

However, the criteria and tests for the diagnosis of death do not just fall from the sky. They logically derive from, that is, are mere operationalisations of, an underlying definition of death. Recording the electrical activity of the brain by means of an electroencephalogram (EEG), performing a cerebral blood flow scan and testing cranial nerve reflexes are tests specifically designed to ascertain whether a human being has lost all brain activity. Making the complete and irreversible loss of brain activity the criterion of death only makes sense if death is defined as a state in which a human being's brain has completely and permanently lost its activity. In other words, the diagnosis of death presupposes a definition of death—and is only as good as its underlying definition of death.

In order to come up with the best possible criteria and medical tests for the diagnosis of death we should therefore be concerned with defining death as exactly as possible. And this is what we are setting out to do next. So, what is death?

1. What is Death?

One of the first things we (should) note when addressing the question "What is death?" is that we are somewhat barred from gaining any substantial insights into the occurrence we call death. Finding a definition of death is somewhat foredoomed by the fact that we cannot make any direct or positive substantial statement about death. We know that death marks the end of a process we call dying and that death is—as Augustine of Hippo remarked—the only thing on earth that is certain:

Cetera nostra et bona et mala incerta sunt; sola mors certa est. [...] Numquid potes dicere: Forte moritur, forte non moritur?	Everything else of ours, both good and evil, is uncertain; death alone is certain. [...] Can you actually say: Perhaps he will die, perhaps he will not die?

(Augustine of Hippo, Sermones 97, 3)

Natus es; certum est quia morieris: et in hoc ipso quia mors ipsa certa est, dies mortis incertus est.	You are born; it is certain that you will die; and in this certainty of death itself, the day of your death is uncertain.

(Augustine of Hippo, In Psalmos 38, 19)

But soon after these rather superficial insights our inquisitive efforts hit a (natural) roadblock. Death is not something we can observe and define in the same fashion as we can observe and define, for example, an observable material object or even an emotion. For there is only one way to experience death: by actually dying. The problem is that those of us who do experience and observe death in this way can usually not tell the rest of us about it: "But nobody can have any direct experience of the 'real nature' of death—except, possibly, the dying person himself. And it is inherent in the nature of this experience that it is not communicable" (Pieper 2000: 7f). It is not in the natural order of things that we survive our death to tell the tale. As an object of investigation death is drenched in a kind of darkness that is impenetrable to the eye of the living; it eludes our attempts to get a direct or positive (scientific or philosophical) grasp of it.

This does not mean, however, that there is nothing substantial to say about death. For although we cannot make any direct or positive statement about the essence of death (and the essence is what is supposed to be captured by a definition), there is something substantial to be said about death in an indirect or negative fashion: Death is the absence of something (X) that has been present before and that is lacking after death has occurred. Regardless of its actual content, the nature of death as an absence or negation is something that is at least implicitly and unconsciously recognised by all definitions of death—be they brain, cardiorespiratory or otherwise based. So, the only way we can actually answer the question "What is death?" is by saying: "Death is the absence of some previously present X."

But what is this ominous X that is supposed to be present before and absent in death? The most straightforward approach to answering this question is to look for death's contradictory opposite, that is, that which is incommensurate with death. Although not part of the world of the living, death has a contradictory correlate in the world of the living, namely life itself. Life and death are mutually exclusive categories: Something that lives cannot be dead; something that is dead cannot be alive; and what has not been alive cannot die. Death is the absence of life in a previously living being, and being dead simply means having lost one's life. Or to put it in philosophically more precise terms: "Mors est privatio vitae" (Suppl. IIIae q. 75 a. 3 s.c.), that is, death is privation of life. In death, "the vital principle 'leaves' the body which up to this moment has been animated by it" (Pieper 2000: 24). So, ultimately, the first and most fundamental answer to the question "What is death?" and thus the only general definition of death we can give is the following: *Death is the completed*[2] *loss of life.*[3]

Accepting this insight has certain implications for our endeavours. For it requires us to resist the temptation to take the second step without having taken the first. If death is the absence of previously present life, that is, if every definition of death is just the negation of a definition of life, every definition of death must necessarily be preceded by a definition of life. Just as the definition of death enjoys logical priority over the criteria and tests of death, so does the definition of life over the definition of death. Any complete formulation of death must start with and emanate from a definition of life. Death can only be fully understood if looked at through the lens of life. So, let us direct our attention to the phenomenon called "life" and find an answer to the following question: What is life?

2. What is Life?

It would not be an exaggeration to call the question of life the mother of, if not all questions, then at least all bioethical questions. This chapter aims at probing into the essence of life and by doing so answer the question "What is life?" The answer is an important step towards understanding what is actually taken when somebody's life is taken and when it can be said to have been taken.

[2] To make things not unnecessarily complicated and avoid confusion, this definition of death only includes the completeness but not the permanency (also: irreversibility) of the loss of life. In principle, I agree with Bernat (2006: 37) that death is irreversible, that is, that "if the event of death were reversible it would not be death but rather part of the process of dying that was interrupted and reversed". However, I only agree to the extent that the irreversibility in question is understood as natural irreversibility. While death is irreversible from a human perspective, death—as attested by, for example, the Apostles' Creed as well as the Niceno-Constantinopolitan Creed of the Catholic Church and their professed doctrine of a bodily resurrection—is reversible and not permanent from a supernatural, that is, God's perspective.

[3] Some might object that the same that is true for life (i.e. that it is present before and absent after death) also applies to, for example, brain activity, cardio-respiratory activity, integrated functioning or the capacity to smile. But, although all of this is indeed present before and absent after death, death can be defined as the loss of these (or some other) activities or capacities only if and insofar as they are taken to be constitutive of life.

2.1. The Signs of Life

Usually, the competence to answer the question "What is life?" is attributed to the natural sciences in general and to biology in particular. The latter seems to be an obvious choice since biology is the natural science that has the term "life" (Greek: βίος; read: bíos) in its name: "In its broadest sense, biology is the study of living things—the science of life" (Raven et al. 2017: 2).

So, what does biology tell us about the phenomenon called "life"? In general, biology takes the term "life" to denote either a particular being that is alive or everything that lives. From the point of view of biology, instead of life (e.g. in the expression "all life on earth") one could also simply speak of living beings. All living beings on earth have three characteristics in common: (a) living beings consist of at least one cell: "All living things are composed of one or more cells" (Mason et al. 2017: 3);[4] (b) these cells are made up of the same types of non-living molecules—"mainly carbohydrates, fatty acids, nucleic acids, amino acids" (Sadava et al. 2016: 2)—and bounded by a membrane; (c) these cells furthermore contain a genetic code—the so-called genome—in which the information for the respective living being's structure and organisation is stored in the form of long-chain nucleic acids.

But, even if we know that a being is composed of cells which are made up of non-living molecules and contain a genetic code, it is not yet certain that the respective being actually lives. Meeting these criteria is just a necessary but not a sufficient condition for the actual possession of life. While such a being is apt to possess life, it does not necessarily have to possess life. For even a corpse—at least up to a certain stage of decomposition—is composed of cells which are made up of molecules and contain a genetic code. So, how do we know that a being is actually a living being, that is, a being that actually possesses life? What sets living beings apart from non-living beings?

Generally speaking, a being can be considered to be alive if certain activities can be observed on it. Or as Urry et al. (2017: 3) and Campbell et al. (2017: 50) have put it: "We recognize life by what living things do". So, what are the activities typically associated with life and shared by living

[4] By implication, this allows for the conclusion that the cell is the basic structural and functional unit of all known life. There is no such thing as sub-cellular life. Living beings composed of one cell are called unicellular; living beings consisting of more than one cell multicellular.

in contrast to non-living beings? What are the empirically observable signs or marks of life? If we consult textbooks of biology on this question, we can find that the signs of life usually include the following activities or processes:[5]

- Metabolism (catabolism and anabolism), that is ingestion and processing/utilisation of energy and nutrients
- Growth and development (morphogenesis)
- Regeneration
- Regulation, that is, maintenance of homeostasis (internal biological balance) and order (low entropy)
- Reception and processing of information; response to stimuli from and interaction with environment (sensitivity/irritability)
- Reproduction/propagation
- (Evolutionary, i.e. cross-generational) adaptation

So, according to biology, possessing life means exhibiting a more or less broadly defined range of activities or processes. Living beings differ from non-living beings in that the former show the signs of life while the latter do not. However, it is important to note that living beings do not have to show all signs of life simultaneously or continuously. In principle, a being is alive if it shows at least one of these signs of life.

However, while the signs of life help explain what it means to be alive, they are not a definition of life. Giving criteria by which to recognise that something is alive is not tantamount to defining life. Why this is so might become more obvious if we replace the term "life" in the question "What is life?" with the term "wind". If we know how to ascertain that the wind is blowing, this does not yet tell us anything about what is blowing.

[5]The following list has been compiled based on Hillis et al. (2019: 2), Mader & Windelspecht (2019: 2ff), Morris et al. (2019: 8), Shuster et al. (2018: 24), Phelan (2018: 25), Campbell et al. (2017: 51), Mason et al. (2017: 3), Raven et al. (2017: 2f), Urry et al. (2017: 3), Sadava et al. (2016: 2f), Klubertanz (2014: 39–46 (§21–24)), Koshland (2002), Bedau (1996: 334–337) and Jessop (1988).

This compilation is not to be regarded as exhaustive. There is, for example, even a working definition of life that has been developed by the National Aeronautics and Space Administration (NASA) and is used by astrobiologists (cf. Brenner 2010).

The list does, however, include the essential features of life found in the biological textbook literature. The exact scope of the list as well as the exact terminology—that can vary from author to author—are not relevant for our purposes.

Answering the question "What is wind?" by "Wind is the rustling of the leaves in a tree" is not a satisfactory answer. The problem with this answer is that it confuses cause and symptom. It points to a sign of (the presence of) wind but does not touch upon its essence. For what is it that makes the leaves rustle? Wind is not the rustling of the leaves but what causes this observable phenomenon. The same holds true for the phenomenon we call "life": Life is not the activities of living beings but what causes them. Life expresses itself in a range of characteristic activities. But rather than being these activities, life is what makes them possible.

In order to find an answer to the question "What is life?" we must therefore take a philosophical look beyond the empirical observable signs of life.

2.2. Life as Capacity for Endogenous Activity

This search for the essence of life starts from the self-evident philosophical axiom that one cannot give what one does not have (***nullus potest dare quod non habet***). This axiom does not only apply to the particular activity of giving but to activity in general: Just as no one can give what he does not have, so no being can perform an action if it does not have the ability (also: faculty; power; capacity; capability) to do so. Or as Coffey (1918: 301) puts it: "Action is the actuality of operative power or faculty".

> It is true, Aristotle concedes, that you can only know that a faculty exists when it is active and that its activity allows us to determine its nature. But this does not imply that the faculty only exists when it is being used. Previous to this activity there must be an ability to act, just as previous to change there must be ability to become a new thing. Without this ability there would neither be any activity nor change. (Elders 1993: 160f)

As it is valid for actions in general, the above-mentioned axiom also applies to the special kind of actions that concern us here, namely the activities exhibited by living beings: For a living being to display a sign of life, it must have the ability to do so. Exhibiting a sign of life in an empirically observable fashion is the expression of the underlying capacity to exhibit this sign.

A being who possesses life possesses the capacity to show signs of life. It is the possession of the something we call "life" that grants this capacity. Expanding on what the biology textbooks tell us, life is not (or only in a figurative sense) that which lives, but that which en-livens everything that lives and by doing so enables it to display signs of life. Life is the power that is at the root of all activities of living beings and signs of life. Life is not function; it is capacity for function. In other words: Life is that which grants beings who possess it the capacity to display signs of life; living is being capable of displaying signs of life; and to be alive is to exhibit signs of life.

However, this statement can and must be further clarified. For if we take another look at the list containing the signs of life, we can see that all signs have two things in common:

- Whenever a living being shows a sign of life, it exhibits some form of activity. It does something—for example, metabolising, growing, reproducing or regenerating—in an empirically observable fashion. What the signs of life have in common is that they are different kinds of activities, namely the most fundamental activities of all living beings, be they plants, animals or human beings. As all other activities exhibited by living beings (e.g. motion) are based on and necessitate at least one sign of life-activity, we could also state that life is not only the capacity to display signs of life but the capacity of activity in general.
- The activity exhibited by living beings is furthermore characterised by the fact that it has its cause not outside of but within the respective being. The signs of life are induced—be it deliberately willed or not—by the living beings themselves and not from the outside. As they have the principle of their activity within them, living beings can and do become active by their own power. In other words: The efficient cause (Latin: *causa efficiens; causa movens*) of the activity that is characteristic of living beings lies within them. Living beings are pictures that can paint themselves.
- The activity that is characteristic of living beings is also characterised by the fact that it serves the being's self-perfection: "Living things act for themselves in order to perfect themselves, where by perfection I mean that the entity acts so as to produce, conserve, and repair its proper functioning as the kind of thing it is—not to reach a state of absolute perfection, which is of course impossible for any finite being" (Oderberg 2007: 180). Living beings are capable not only of "transient causation" but also of "immanent causation": While in transient causation, "the activity terminates in something distinct

> from the agent" (Oderberg 2020: 86), immanent causation is "causation that originates with an agent and terminates in that agent for the sake of its self-perfection" (Oderberg 2020: 86). An example of immanent causation would be the digestion of food; an example of transient causation would be the act that is at the heart of this book, namely killing.
>
> Hence, "in immanent causation the agent is both the cause and the effect of the action, and the cause itself is directed at the effect as perfective of the agent" (Oderberg 2020: 86).[6] Living beings can and do act upon themselves for their self-perfection; inanimate things may change other objects (e.g. a boulder that rolls down a slope and hits another boulder, which hits and dents a parked car) but they do not act upon themselves. The activity caused by immanent causation does not, however, entail or require the consciousness of the agent engaging in this activity. Furthermore, that living beings act for their self-perfection has to be understood "as a ceteris paribus law, where the exceptions can be spelled out in a principled way, the main—perhaps only ones—being: (a) damage to the organism may prevent its doing so; (b) hostile environmental conditions may prevent its doing so" (Oderberg 2020: 88).

An example from everyday life helps to illustrate these points: In common parlance, not only living beings but also buildings can and do grow. However, there is an important difference between the growth of a building and the growth of a living being: Unlike living beings, buildings cannot grow actively, that is, out of themselves, but only passively, that is, by means of external causes (architect, engineer, craftsman, construction worker, etc.). A building can only grow to the extent that external forces act on it. In contrast, a living being can grow by itself. So, if we speak of life as the capacity to display signs of life, this capacity is the living being's capacity for endogenous activity, that is, to become active from within and by its own power. While inanimate objects (e.g. a stone or molecule) are acted upon, a living being acts from within in a fashion that is self-initiated and directed to its preservation and perfection.

[6] For further explications on the distinction between immanent and transient causation cf. Oderberg (2007: 180, 195 and 2013: 213), Feser (2014a: 99, and 2014b: 18) as well as Klubertanz (2014: 47f (§ 26)) and (2014: 55 (§ 28)).

These insights allow us to concretise the statement made above, according to which life is the ability to display signs of life. As should have become clear, in the end it is not important for the understanding of life which activities a living being exactly shows or can show. It is decisive how it does this, namely out of itself:

Propria autem ratio vitae est ex hoc, quod aliquid est natum movere seipsum, large accipiendo motum, prout etiam intellectualis operatio motus quidam dicitur. Ea enim sine vita esse dicimus, quae ab exteriori tantum principio moveri possunt.	But life is essentially that by which anything has power to move itself, taking movement in its wide sense so as to include the 'movement' or activity of the intellect. For we say that those things are without life which can be moved by an exterior principle only.

(Thomas Aquinas, Sentencia de Anima, lib. 2 l. 1 n. 9; cf. I^{a} q. 18 a. 1 co.)

A living being that possesses life possesses the capacity to become active, that is, to cause its activity—and therefore also the activities characteristic of life—by itself. So, *life is the capacity for endogenous activity*, that is, activity that originates with and is caused by the living being itself.

It should be emphasised that the endogenous activity to which this statement refers does not have to be of a certain kind. Among the endogenous activities exhibited by living beings there are more important and less important activities, and indeed, there are vital and non-vital activities. But in the end, any kind of endogenous activity is endogenous activity. It is therefore inadmissible to restrict the endogenous activity that is characteristic of life to any one or even several selected activities only—even if it were the most critical activity. Life is the capacity for any kind of endogenous activity, that is, for endogenous activity in general.

2.3. *The Soul as the Principle of Life*

So far, we have established that life is the capacity for endogenous activity. But where does this capacity come from? It should be a logical matter of course that the capacity called life must originate from somewhere; it needs a source or principle.[7]

Living beings possess the capacity for endogenous activity we call life because and insofar as they possess something that grants them this

[7] According to Thomas Aquinas, a principle is that "a quo aliquid procedit quocumque modo" (I^{a} q. 33 a. 1 co.), that is, that whence something proceeds in any way.

capacity. This life-principle, the source of all capacities and activities of a living being, is called *soul* (Latin: *anima*; Greek: ψυχή (read: psychḗ)):

Anima dicitur esse primum principium vitae in his, quae apud nos vivunt; animata enim viventia dicimus, res vero inanimatas vita carentes.	The soul ('anima') is defined as the first principle of life of those things which live; for we call living things 'animated', and those things which have no life 'inanimated'.

(Thomas Aquinas, I^a q. 75 a. 1 co.)

It is the soul as a life-giving principle that animates, that is, enlivens, a being and thus makes it a living being.[8] The soul is the life spark. A being possesses the capacity for endogenous activity called life if and only if it possesses a soul:

[8] Accordingly, Thomas Aquinas also states that "animam ingredi corpus est vivificare corpus" (Sentencia de Anima, lib. 1 l. 6 n. 14), that is, that it is the soul's presence that gives life to the body, and that "certa est [...] quod anima vivificet" (Sentencia de Anima, lib. 1 l. 1 n. 6), that is, that it is certain that the soul gives life.

This insight can also be found in Plato's Phaedo (105d), where he asserts that "if the soul takes possession of anything it always brings life to it". However, Plato is wrong in assuming that the soul is united to the body as a helmsman is united to a ship (cf. Phaedo, 115c 2–115e 4). For Plato the soul is man (cf. Alcibiades I, 130c 5; 129e 11) or—as Thomas Aquinas has summed up Plato's position—man is a soul using a body (cf. I^a q. 75 a. 4 co.). Plato could therefore be said to be an early precursor of contemporary proponents of body–self dualism (also: mind–body dualism) who hold that a person is her thoughts, beliefs, desires, memories and other mental processes and uses a physical body just as a craftsman uses a tool or as a helmsman steers a ship.

But contrary to what Plato and body–self dualists think, "the soul [...] is not an entire man, and my soul is not I" (Super I Cor., cap. 15 l. 2); man "non est anima tantum" (I^a q. 75 a. 4 co.), that is, is not the soul alone, but something composed of body and soul. Man is a physical being; it belongs to his nature to have a body (cf. III^a q. 5 a. 1 co.). Accordingly, "the soul does not have the perfection of its own nature, except in union with the body" (De spiritualibus creaturis, a. 2 ad 5; also cf. Chap. VIII, Sect. 9).

The soul is not only the moving cause of the body. As its form, the soul is united to the body as an imprint is united to a coin: "The soul is connected to with the body not like the workman with his tool, nor does it 'dwell' in him like the sailor in the boat or the prisoner in his cell; rather it is 'in' him the way the stamped form (a coat of arms, a portrait, an eagle) unites with a blank of liquefied silver to make a coin. The form impressing itself on the body from within—that is the true definition of the soul: anima forma corporis" (Pieper 2000: 31; cf. I^a q. 90 a. 4 co.; I^a q. 76 a. 7 s. c.; De unitate intellectus, cap. 1). So, soul and body are not two independent substances that interact; rather, soul and body are one substance. And as they are two principles of the same single substance, that is, the human being, the question of efficient–causal interaction between the incorporeal soul and the corporeal body does not arise to begin with.

Corpus autem, vivificatum per animam, ex seipso movetur: sed corpus mortuum vel immobile manet, vel ab exteriori tantum movetur.	A body that is made alive by a soul moves by itself; but a dead body either remains without movement or is only moved from outside.

(Thomas Aquinas, Contra Gentiles, lib. 3 cap. 139 n. 17)

> The soul is that by which the living thing ultimately has the operations of life. (Klubertanz 2014: 50 (§ 27))

The soul lends its capacity for endogenous activity to the being it enlivens and animates. A living being's capacity for endogenous activity is therefore derivative of its soul. However, possessing a soul grants a living being not only life and the capacity for endogenous activity but also existence as this particular living bodily being.

ἡ ψυχὴ τοῦ ζῶντος σώματος αἰτία καὶ ἀρχή.	The soul is principle and cause of the living body.

(Aristotle, De Anima, 415b8)

Ab anima est, et quod sit, et quod sit corpus vivum.	The living body gets both its being and its life from the soul.

(Thomas Aquinas, Sentencia De Anima, lib. 2 l. 1 n. 15; cf. lib. 2 l. 7 n. 10)

> Hence, the soul is that real and really distinct principle of a living thing in virtue of which the thing is living. (Klubertanz 2014: 51 (§ 27))

A living being exists only as this particular living being, if and as long as it lives. Socrates is only Socrates as long as Socrates lives; Socrates' corpse was but no longer is Socrates. By informing a body, that is, a certain quantity of matter, the soul causes the existence of a distinct living being. Living beings are therefore compounds of an immaterial soul and matter. If the soul stops informing the matter a living being is made of, this being ceases to live and to exist.

At this point, it makes sense to summarise what we have established so far about the phenomenon we call "life". When asked "What is life?" we can give the following answer:

- Life is the capacity for endogenous activity, that is, activity that originates with and is caused by the living being itself. The empirically observable signs of life are actualisations of this capacity.

- A living being is capable of endogenous activity, that is, possesses life, if and insofar as it possesses a soul. It is the immaterial life-principle named "soul" that grants living beings their capacity for endogenous activity. As such, the soul is not only the principle of life but also the principle and efficient cause of all activity shown by a living being. Therefore, living beings carry the principle and efficient cause of their activity within them.
- The soul is not only the principle of a living being's life but also of its existence. If an immaterial soul joins with a material body, it causes a living being to exist and grants it the capacity for endogenous activity.

Now that we have a sufficiently robust understanding of life and—by implication—also of its contradictory opposite, namely death, we can turn our attention to an almost inevitable follow-up question: When is death, that is, when has a previously living being actually died?

3. When is Death?

Posing the question "When is death?" means asking how to determine when a living being has crossed the line that separates life from death. How can we reliably determine that a previously living being has lost its life and has therefore died? On a general level, the answer to this question is rather straightforward and can be drawn from the insights that have been established so far. These insights read as follows:

- Death is the completed loss of life.
- Life is the capacity for endogenous activity.
- A living being possesses the capacity for endogenous activity called life if and insofar as it possesses a soul.

These premises allow for the conclusion that death is the completed loss of soul. Given this definition of death, a living being has died if its soul has completely left its body: "Death, put simply, is the separation of the soul from the body" (Oderberg 2019: 362; cf. Contra Gentiles, lib. 2 cap. 57 n. 12). In death, the connection between the body and the soul is abolished and their unity dissolved.[9]

[9] It has to be stressed that is not man's body or his soul which dies; what suffers death is man himself, that is, the whole formed by body and soul (cf. Pieper 2000: 27f). However, when body and soul separate, they do not simply part ways as two unrelated things; given the nature of the soul as the form of the body (cf. the previous footnote), death is rather the separation "of something that belongs together by nature" (Pieper 2000: 34).

So, in order to be able to establish that and when a living being has died, we need to know that and when a living being no longer possesses a soul. The real question we have to answer is not "When is death?" but rather "How long is life?" Answering this question is somewhat challenging as the separation of the soul from the body eludes our observation and requires us to give empirically observable criteria for an unobservable event. So, let's tackle this challenge.

From the fact that a being shows empirically observable signs of life, we can safely conclude that the being in question possesses a soul and life. Otherwise, it would not be capable of showing any signs of life. However, from the fact that a being does not show some particular sign of life, we cannot conclude without further ado that this being no longer possesses a soul. When we observe that somebody speaks, we know for sure that he has the capacity to speak. When we encounter somebody who does not speak during our encounter, we cannot, however, say for sure whether he is simply keeping quiet but is able to speak or whether he is a mute and thus has no capacity to speak.

The only way to assert with certainty that a being has lost its soul and life is to demonstrate that it is impossible for the being to show any signs of life and any activity. This is the case only if the being's physical-material infrastructure, that is, its body, is in a state that is incompatible with the exhibition of signs of life and makes the exhibition of any sign of life or activity impossible.[10] In this state, its body has become "unfit to be informed by the soul, necessitating the departure of the soul" (Spencer 2010: 854). Not observing signs of life or activity is a first indication for the loss of soul and life, but it is not enough to ascertain a being's death with certainty. A being's death can only be asserted with certainty if it is impossible for this being to show any signs of life or activity. So, a being is not dead when it does not show any signs of life or activity; it is dead if it cannot show any signs of life or activity.

In order to determine that death has occurred we therefore need criteria by means of which we can determine with certainty that a body cannot

[10] This insight has already been expressed by Thomas Aquinas when he states that "cessantibus propriis dispositionibus forma in materia remanere non potest" (Q. d. de anima, a. 9 ad 16; cf. a. 14 ad 20), that is, that a form cannot remain in matter when the proper dispositions cease to exist. When the proper dispositions are removed, that is, when the matter does not allow for an exhibition of the signs of life, the union of soul and body is destroyed, and the soul departs the body. The question what this disposition of the body actually consists of is touched upon in Chap. VIII, Sect. 1.

show any signs of life or activity. Luckily, medical science can help us here. Medicine, in particular forensic pathology, has developed a range of "signs of death" (Latin: *signa mortis*) by which we can determine whether a human being is dead or alive.[11] These signs can be divided into the early certain signs of death, the late certain signs of death as well as the uncertain signs of death, and accordingly form what can be called three "stages of death":

- Stage 1: Uncertain signs of death
  - Paleness (*pallor mortis*)
  - Reduction of body temperature (*algor mortis*)[12]
  - Respiratory arrest (no breathing)
  - Cardiac arrest (no pulse)
  - No neuronal activity
  - No reflexes (areflexia)
  - Loss of consciousness
  - Corneal opacity
  - Fixed pupils, dilated pupils
  - Muscle atony/flaccidity
- Stage 2: Early certain signs of death (early postmortem changes)
  - Postmortem lividity (*livores mortis*)
  - Postmortem rigidity (*rigor mortis*)

[11] The list containing the signs of death is compiled based on Buris (1993: 11ff), Payne-James et al. (2011: 42ff), Prahlow & Byard (2012: 145ff), Dettmeyer et al. (2014: 36ff), Madea et al. (2014, 2015, 60ff), Vij (2014: 78ff), Madea & Kernbach-Wighton (2017) and Vanezis (2019).

[12] To be more precise, we should better speak of a change in body temperature that occurs postmortem. The temperature of a dead body changes until it matches the ambient temperature. While the change is usually a steady decline, a dead body's temperature can also rise, for example if somebody has died in a desert, or stay the same, for example if somebody has suddenly died in a room the temperature of which matches the temperature of a healthy human being.

- Stage 3: Late certain signs of death (late postmortem changes)[13]
  - Autolysis (self-digestion, e.g. of stomach by gastric acids)
  - Putrefaction (mainly anaerobic bacterial reduction process, accompanied by bloating and ammoniacal odours)
  - Decomposition (aerobic microbiological process, usually accompanied by pungent, rotten odours)
  - Adipoceration, that is, formation of corpse wax (adipocere) (occurs in cold, wet environment)
  - Colonisation with small animals (e.g. fly eggs)
  - Skeletonisation
  - Mummification (occurs in hot, dry environments)
  - Fossilisation

As the name already indicates, the uncertain signs of death do not suffice to determine a human being's death with certainty. The problem with these signs of death is that they not only occur after death but can also occur in living human beings (e.g. in case of poisoning). A body showing the uncertain signs of death may well be only apparently dead; so, rather than having lost its life, this body might be in the states of "vita minima" and "vita reducta", that is, in the process of increasing devitalisation (cf. Madea 2014: 51; Dettmeyer et al. 2014: 35; Madea et al. 2015: 46).

Only if a human body shows the certain signs of death can it be safely assumed that the body can no longer show any signs of life or endogenous

[13] The extent and onset of the late certain signs of death are influenced by a range of intrinsic and extrinsic factors. As Madea et al. (2014: 92) point out, the onset and extent of postmortem changes are accelerated if death occurs in a hot, moist environment or under high ambient temperatures, if the body surface is insulated by warm clothing or other covering, if the subject was overweight or had a high fat content, if the subject suffered from an underlying infection or sepsis, if the subject was intoxicated (e.g. with illicit drugs such as heroin) or if the subject suffered or died from open wounds or during a surgical procedure. The onset and extent of postmortem changes are decelerated if death occurs in a cold, dry environment or under low ambient temperatures, if the subject was scantily dressed, naked or undressed shortly after death, or if the subject was stored in a cooling device (fridge, freezer) shortly after death.

activity whatsoever.[14] For only then is the body as the soul's physical-material infrastructure in such a state that it is no longer capable of showing any endogenous activity. And diagnosing this state is the only way to safely assert that a body is no longer capable of showing endogenous activity. However, it is not necessary to wait for the late certain signs of death to occur; the early certain signs of death suffice for the certain determination of death.[15] And since the three stages are passed through one after the other, a corpse that shows the early certain signs of death (stage 2) necessarily also shows the uncertain signs of death (stage 1).

Thus, as a result of our considerations, we can state the following with respect to the diagnosis of death—or to be more precise, the diagnosis of the absence of previously present life:

> A previously living being is dead, that is, has completely lost its life and therefore can no longer show any signs of life or endogenous activity, if it shows no empirically observable signs of life and if it shows the early certain signs of death.

This criterion is an attempt to make the empirically unobservable occurrence we call death empirically observable. If the conditions listed in the above-mentioned criterion of death are met, we know that a previously living being's soul has left its body. As the soul is immaterial and therefore unobservable, the criterion does not allow for a conclusion as to the exact moment when the soul has left the body—it only allows for the conclusion that the soul can no longer be in the body and must therefore have left the body.

[14] As the formulation "safely assumed" indicates, this judgement is a judgement made with moral certainty. As McHugh & Callan (1958: §643) state, a judgement is morally certain "when error is impossible according to what is customary among mankind, the opposite of what is held by the mind being so unlikely that it would be imprudent to be moved by it". In contrast, something is metaphysically certain "when error is absolutely impossible", and something is physically certain "when error is impossible according to the laws of nature" (McHugh & Callan 1958: §643; Prümmer 1961: 212). When confronted with the decision to engage or not engage in a certain action, "certitude in the wide sense, which excludes all fear of error and every serious or prudent doubt, but not one or other slight and imprudent doubt" (McHugh & Callan 1958: §645), is sufficient for acting with a safe conscience. This degree of certitude is present in relying on the certain signs of death for the determination of death.

[15] This position is slightly less strict than the position of Oderberg (2019: 374), who holds that "only putrefaction—the physical decomposition of the body—can be a certain sign that death has occurred" Oderberg's position seems to be in line with Aquinas' assertion that "when the soul departs, the body is dissolved" (Contra Gentiles, lib. 2 cap. 58 n. 8). The early certain signs of death are sufficient for the determination of death because—like the late certain signs of death—they are already certain signs, too.

4. Human Life and Human Death

So far, we have been concerned with life and death in general. In the course of our deliberations, we have introduced a definition of life and established a reliable criterion for the diagnosis of death. However, as this is a book about the ethics of killing human beings, we should not continue without having asked ourselves what distinguishes human life from the life of other living beings. So, what is distinctive and special about human life—and, in consequence, human death?

4.1. Human Life

If we look at the nature surrounding us, we can distinguish three irreducibly different types or degrees of life. What sets these kinds of life apart from each other is the range of activities they can display:

> First there is vegetative or plant life, which is characterized by the capacities for nutrition, growth, and reproduction. Second, there is animal life, which like vegetative life involves nutrition, growth, and reproduction but adds to these its own distinctive capacities, namely sensation, appetite, and locomotion. Finally, there is human life, which includes all the basic capacities of vegetative and animal life—nutrition, growth, reproduction, sensation, appetite, and locomotion—and adds to them the distinctive human capacities for intellectual activity and volitional action. (Feser 2014b: 19; cf. Seifert 1997: 24; Frey 2019: 211f)

As can be seen from this classification, there are certain kinds of activities that are displayed (a) by plants, animals and humans, (b) by animals and humans only and (c) by humans only. In other words, humans possess not only the capacities that plants and animals also possess but are also capable of displaying uniquely human, namely rational, activities. The vegetative life that enlivens plants grants them the capacity for the vegetative activities; the animal life that enlivens animals grants them the capacity for both the vegetative and sensitive activities; the life that animates human beings grants them the capacity for the vegetative, sensitive and rational activities.

The different kinds of life are usually named after the qualitatively highest capacity they grant.[16] Therefore, plant life is also called vegetative life

[16] This is a classic case of a so-called *pars pro toto* where something is named after a part or after a limited characteristic that is itself not necessarily representative of the whole.

or nutritive life, animal life is also called sensitive life, and human life is also called rational life or intellectual life. But although human life can also be referred to as rational life, it would not be correct to state that humans possess the three kinds of life in an additive fashion. They do not possess vegetative life plus sensitive life plus rational life. Living human beings can and do possess one kind of life only, namely human life. Accordingly, living human beings also can and do possess only one kind of soul, namely a human soul. *Possessing a human—or as is it also called "rational" or "intellectual"*[17]*—soul means being a living human being.*

It is characteristic of the human soul that it grants the capacity for vegetative, sensitive as well as rational activity. A human being's rational soul is not only the cause of this human being's rational activity but of all empirically observable signs of life. The rational soul is not only the cause and principle of his intellectual activity; it is also the cause and principle of the typically vegetative signs of life, such as metabolism, regeneration or growth and development (cf. Nguyen 2018: 417). As such, the rational or human soul is just the most powerful type of soul. But still it is only one soul:

Sic ergo dicendum quod eadem numero est anima in homine sensitiva et intellectiva et nutritive. [...] Sic igitur anima intellectiva continet in sua virtute quidquid habet anima sensitiva brutorum, et nutritiva plantarum. [...] Ita nec per aliam animam Socrates est homo, et per aliam animal, sed per unam et eandem.	We must therefore conclude that in man the sensitive soul, the intellectual soul, and the nutritive soul are numerically one soul. [...] Thus, the intellectual soul contains virtually whatever belongs to the sensitive soul of brute animals, and to the nutritive souls of plants. [...] So Socrates is not man by one soul, and animal by another, but by one and the same.

(Thomas Aquinas, I^{a} q. 76 a. 3 co.; cf. I^{a} q. 76 a. 4 co.; I^{a} q. 118 a. 2 ad 2; Q. d. de anima, a. 11 co.)

It is not the case that human beings live by a vegetative soul and are made human by a rational soul. As "the soul is the substantial form of man" (I^{a} q. 76 a. 4 s. c.), that is, that which makes a human being a human being, no human being can—by definition—have more than one

[17] Similar to the alternative labelling of the three kinds of life, the different kinds of soul that enliven the three classes of living beings are sometimes also named after the qualitatively highest capacity they grant. Accordingly, the plant soul, that is, the type of soul that enlivens plants, is called "vegetative soul" or "nutritive soul"; the animal soul, that is, the type of soul that enlivens animals, is also called "sensitive soul"; and the human soul, that is, the type of soul that enlivens humans, is also called "rational soul", "intellective soul" or "intellectual soul".

soul.[18] A human being in which there were more than one soul could not be one:

Nihil enim est simpliciter unum nisi per formam unam, per quam habet res esse [...]. Si igitur homo ab alia forma haberet quod sit vivum, scilicet ab anima vegetabili; et ab alia forma quod sit animal, scilicet ab anima sensibili; et ab alia quod sit homo, scilicet ab anima rationali; sequeretur quod homo non esset unum simpliciter.	For nothing is absolutely one except by one form, by which a thing has existence [...]. If, therefore, man were 'living' by one form, the vegetative soul, and 'animal' by another form, the sensitive soul, and 'man' by another form, the intellectual soul, it would follow that man is not absolutely one.

(Thomas Aquinas, I^a q. 76 a. 3 co.; cf. Contra Gentiles, lib. 2 cap. 58 n. 8)

As human beings are human because and insofar as they possess a human, that is, rational, soul, the only type of soul human beings can possess is a rational soul. If they were to possess another type of soul, they would and could not be human beings. The vegetative, sensitive and rational souls are categorically different types of soul that cannot and do not develop into each other by means of some evolutionary upgrading or downgrading process (cf. Heaney 1992: 26f). The types of souls are not in potentiality to one another. A rational soul has not developed into a rational soul; it has always been and will always remain a rational soul; it has never been and will never be a vegetative or sensitive soul. And neither will a vegetative or sensitive soul ever be or become anything else than a vegetative or sensitive soul.

4.2. Human Death

What does the insight regarding human life mean for the death that human beings die, that is, human death? If death is the complete loss of life, then human death is the complete loss of human life. The life of a human being

[18] According to the metaphysical theory named "hylomorphism" (also: "hylemorphism"), every concrete material substance, that is, every particular existing material thing, is a composite of matter and form: "The form of a substance is its organizational structure; the matter is that which is given organizational structure by the form. [...] *Substantial* form is that specific aspect of a substance's organizational structure by virtue of which it is the kind of substance it is. (A seat's roundness isn't part of the substantial form of a chair—a chair could have a square seat instead, for instance, and still be a chair—but having some kind of seat would be)" (Feser 2006: 220).

has ended if he has lost the life that is characteristic for him as a human being, that is, human life. As we know, human life is not the result of an addition of vegetative, sensitive and rational life; human life is rational life, that is, that distinct type of life that grants the capacity for vegetative, sensitive and rational activity. And this rational life is granted by the possession of a human, that is, rational, soul—a soul which "virtually contains the sensitive and nutritive soul" (Thomas Aquinas, I^a^ q. 76 a. 3 co.) but which nevertheless alone does whatever these other types of soul do in other things (cf. I^a^ q. 76 a. 4 co.).

If human beings do not possess three types of life and if they possess not three but only one life-principle, then they cannot die gradually. If there is only one life and one soul, then the logic of human death is a binary one. A human being is either dead or not. Given the nature of human life and the human soul, human beings cannot lose their life one bit at a time; they cannot lose their rational life while keeping their vegetative life. When a human being loses his rational life, he necessarily also loses the capacity to exhibit any signs of life or any endogenous activity. The nature of the soul as that which makes some matter an individual human being does not allow for gradations. If the rational soul is gone, all life and all capacities are gone.

In conclusion, this means that a human being has died if he has lost all capacity for endogenous activity. This loss can only be reliably diagnosed if a human being's body, that is, the physical-material infrastructure that is necessary to exhibit endogenous activity, is in such a state that the only conclusion that is left to draw is that the human being is not capable of exhibiting such activity. This is the case at the earliest when the body shows the early certain signs of death at least. Accordingly, the criterion of human death can be formulated as follows:

> A previously living human being is dead, that is, has completely lost his life and therefore can no longer show any signs of life or endogenous activity, if he shows no empirically observable signs of life and if he shows the early certain signs of death.

At this point, we must put an unfortunately widespread medical myth to rest, namely that growth processes can still be observed in (even buried) corpses. This claim is found, for example, in Bernat (1998: 17), who writes that "hair and nails continue to grow for days after death". Unfortunately, this is not the case. Appearances are deceptive here: The supposed growth of hair and nails does not come about through growth processes (which would require the existence of life in the corpse); it is the result of decomposition processes. After death, dehydration causes the skin and other soft tissues to shrink. The tissue surrounding the hair and nails retracts and makes them appear longer. While the hair and nails remain the same length, the change in the surrounding tissue creates the optical illusion of growth:

> Shortly, finger pads and acra dry up, with their consistency becomes more compact and their colour changing to a reddish-brownish tone. The drying process causes contractions of the finger pads and toe pads, thus it seems as if nails are still growing. The 'growing' of a beard after death is also caused by the fluid loss of the facial skin. (Madea et al. 2014: 90; also cf. Vreeman & Carroll 2007; Henßge & Madea 2004: 119)

5. Summary

Before turning our attention to the question whether circulatory death or brain death can constitute human death, it is helpful to briefly sum up this book's position on (human) life and (human) death as it has been developed so far, starting with its three most fundamental premises:

- (P1) Death is the absence, that is, the completed loss, of life.
- (P2) Life is the capacity for endogenous activity.
- (P3) A living being possesses the capacity for endogenous activity called life if and insofar as it possesses a soul.

These three premises logically result in the following conclusion containing this book's general definition of death:

- (C1) Death is the completed loss of soul, that is, the complete separation of the soul from the body.

The next step has been to develop this general definition of death into the definition of human death. This has been done by introducing the following two additional premises:

- (P4) Human death is the completed loss of human life.

- (P5) Human life is granted by the possession of a human (i.e. rational) soul.

Combining (C1) with (P4) and (P5) results in conclusion (C2):

- (C2) Human death is the completed loss of rational soul, that is, the complete separation of rational soul and body.

Having defined human death, the next step has been to propose a criterion of death. The question of when human beings—as opposed to other living beings—die has been answered by reference to the nature of the human soul as expressed in the premise below:

- (P6) The human soul grants the capacity for vegetative, sensitive as well as rational endogenous activity but is numerically one.

Combining (C2) with (P6) results in the following definition of human death:

- (C3) A human being has died if he has lost all capacity for endogenous activity, be it vegetative, sensitive or rational.

This means that as long as a human being shows any vegetative, sensitive or rational endogenous activity, that is, any signs of life, his rational soul cannot yet have separated from his body. Conclusion (C3) has then been developed further by means of the following premises:

- (P7) The only way to assert with certainty that a human being has lost all capacity for vegetative, sensitive or rational endogenous activity—and therefore his rational soul—is to demonstrate that it is impossible for him to show any such activity.
- (P8) It is impossible for a human being to show any vegetative, sensitive or rational endogenous activity if his physical-material infrastructure, that is, his body, makes the exhibition of any such activity impossible.
- (P9) Only if a human body shows the early certain signs of death can it be safely assumed that the body can no longer exhibit endogenous activity whatsoever.

Combining (P7), (P8) and (P9) with (C3) results in the criterion for the diagnosis of human death—or to be more precise, the diagnosis of the absence of previously present human life—we have been looking for:

- (C4) A previously living human being is dead, that is, has completely lost his life and therefore can no longer show any signs of life or endogenous activity, if he shows no empirically observable signs of life and if he shows the early certain signs of death.

6. Brain Death as Human Death?

If one takes the above criterion for the diagnosis of death as a yardstick, it becomes clear that neither clinical criterion of death, that is, neither circulatory death nor brain death, can be used to diagnose death reliably.

- Brain death is diagnosed by the permanent[19] loss of all or a certain kind of brain activity: Whole-brain death is defined as the complete loss of all brain activity, that is activity of the cerebrum, cerebellum and brainstem. Partial-brain death definitions (e.g. brainstem death, cortex death) require the loss of activity in some part of the brain only.
- Circulatory death is diagnosed by the complete and permanent loss of all cardiopulmonary function, that is, function in the heart and lungs.[20]

Neither the complete and irreversible cessation of blood circulation and breathing nor the complete and irreversible cessation of all (let alone partial) brain function as such are sufficient conditions for the reliable diagnosis of death. These are necessary conditions, but they are not suitable to serve as sufficient conditions.

[19] This condition is meant to rule out a merely transient loss of brain function. According to Bernat (2004: 162) the permanent (also: irreversible) loss of brain function signifies a prolonged loss "that cannot be restored using contemporary technologies". The permanency of the loss is established by sequential neurological examinations separated by a time interval that varies as a function of, for example, age and etiology.

[20] Instead of "circulatory death" one could also speak of "cardiopulmonary death", "cardio-respiratory death" or "circular-respiratory death". These terms are synonymous and refer to the complete and irreversible cessation of circulatory and respiratory functions.

This becomes especially clear when looking at the brain death criterion of death.[21] Given the nature of human life and the rational soul, a human being's rational soul cannot have left his body yet, as long as it shows any form of endogenous activity. However, brain-dead human beings show a wide range of vegetative activities:

> The organism of patients who are correctly diagnosed as 'whole brain dead' continues to function as an integrated whole: patients digest food, regulate salt and water homeostasis, maintain their temperature, grow hair, heal wounds, fight infections, react to stress, grow in length, go through puberty, and even gestate fetuses. (Monteverde & Rid 2012)[22]

Given the notion of life and soul established above, the endogenous activity present in brain-dead human beings leaves only one conclusion to be drawn: Brain-dead human beings still possess the life that is distinctive of them as human beings, namely rational life—along with their rational soul. What they have lost, however, is the functioning infrastructure that a

[21] As Bishop (2019: 421) points, out the brain death criterion is an "artifact of advancing technology": "Prior to the first cardiac transplant in 1967, physicians did not speak of 'brain death.' The closest thing to brain death was coma dépassé ('beyond coma'), which was theorized not as death but as a deep coma, more akin to sleeping than to end of life. No one thought of people in this state as 'dead.' […] This suggests that before transplantation became a possibility, people treated patients in coma dépassé as regular patients; there was no question as to whether they were alive or not".

Giacomini (1997) has studied the internal files of the ad hoc committee of the Harvard Medical School that redefined death as brain death in 1968. Her work allows for the conclusion that the primary purpose of many members of the committee was to find a way to legitimise transplantation of vital organic material and increase the number of potential donors. With the dead donor rule applicable, the only way to do so was to redefine death as brain death and convert irreversible coma to brain death.

[22] For more examples of such activities cf. Shewmon (2001: 467f), who offers an almost two-page overview of non-brain-mediated integrative functions that are possessed by at least some brain-dead persons. With respect to the successful gestation of a foetus, Spike (1999) describes the case of a 20-year-old pregnant woman, Ellen Higgins, who was declared brain-dead when her foetus was estimated to be 16.5 weeks of gestation. But even in the state of brain death, Higgins—being kept alive by a ventilator and other life-support measures—continued her pregnancy for months and even delivered a living baby: "An elective C-section was done at 31 weeks, after 100 days in the ICU [Intensive Care Unit], with Ellen's family present, delivering a healthy baby boy, 1440 grams […]. But Ellen's internal bleeding after the C-section couldn't be stopped, and so her family chose to stop the ventilator 20 minutes after the C-section, and her heart stopped minutes later. […] The baby spent 8 weeks in the NICU [Neonatal Intensive Care Unit], with no unusual complications" (Spike 1999: 62).

rational soul needs to exhibit its specifically rational capacities. But just because a rational soul does not have the physical-material means necessary to exhibit some of its capacities, this does not mean that it is no longer present. If the soul were to leave the body in the moment of brain death, then—given the nature of the soul—the body would not be able to exhibit any signs of life whatsoever. As this is obviously not the case, brain death cannot be death. The diagnosis of brain death may imply a prognosis of death, but stating that a person is going to die at some point in the future is not the same as stating that she has died.[23]

Neither brain death nor circulatory death is a reasonable criterion of death. The only sound criterion that is grounded in a meaningful definition of life is a soul-based criterion of death. According to this criterion, a previously living human being is dead, that is, has completely lost his life

[23] This book's account of death is incompatible with brain death for reasons other than those, for example, put forward by Shewmon. As Shewmon shows in his arguments against the integrative-unity rationale for brain death, the brain is not the "master integrator organ" (Shewmon 2010: 257) or "critical system" (Shewmon 1998: 126) that "confers integrative unity upon the body, transforming it from a mere collection of organs and tissues to an 'organism as a whole'" (Shewmon 2001: 457). Rather, "the integration is clearly a nonlocalized emergent phenomenon involving the mutual interaction among all the parts" (Shewmon 2007: 309). If the integrative-unity rationale for brain death were true, embryos could not be alive as they have no central integrating organ. As a side note: Beside this biological challenge to the brain's actual role as master integrator, ascribing the brain this role is philosophically unsound, too. For as Accad (2015: 221) rightly points out, a part cannot be an integrative cause of an organism, "for this role is exclusively assumed by the substantial form of the body, i.e., the soul". Or as Spencer puts (2010: 865) it: "On a hylomorphic theory no organ constitutes the unity of the organs, for this is done by the soul".

Shewmon's argument aims at proponents of brain death such as Bernat (2006: 38) who define death as "the cessation of functioning of the organism as a whole" and assume that the brain is the primary integrator that guarantees this functioning of the organism as a whole. Accordingly, brain death is taken to be the criterion of death of such an organismic definition of death. However, although it is usually at the centre of the controversy over brain death, the question whether the brain is the master integrator organ of the human body is only tangentially relevant to this book's argument. For its argument is based on the premise that a definition of death can only be derived from a logically prior definition of life. Whether a definition of death makes sense is not primarily dependent on the definition itself but its underlying definition of life.

Unfortunately, such a definition of life seems to be entirely missing from the treatises of the proponents of brain death. They stipulate that death consists in the loss of the integrated functioning of the organism (which is supposedly guaranteed by the brain) but do not explicate the notion of life that is logically preceding and inherent to this stipulation. As long as this explication is not made, there is not really anything to engage with.

and therefore can no longer show any signs of life or endogenous activity, if he shows no empirically observable signs of life and if he shows the early certain signs of death.

This position does not doubt that—without medical intervention—her soul will leave a brain-dead person rather sooner than later. It does, however, doubt the assertion made by proponents of brain death that a person's soul has already left her body at the time she is declared brain-dead. For this would be contrary to the nature of the soul as the principle of all endogenous activity. Life is the capacity for any kind of endogenous activity, that is, for endogenous activity in general.

Bibliography

Accad, Michel. "Of wholes and parts: A Thomistic refutation of 'brain death'." *Linacre Quarterly* 82.3 (2015): 217–234.

Bedau, Mark A.. "The Nature of Life." In: Boden, Margaret A. (Ed.). *The Philosophy of Artificial Life*. Oxford: Oxford University Press, 1996. pp. 332–357.

Bernat, James L.. "A Defense of the Whole-Brain Concept of Death." *The Hastings Center Report* 28.2 (1998): 14–23.

Bernat, James L. "On Irreversibility as a Prerequisite for Brain Death Determination." In: Machado, Calixto & D. Alan Shewmon. *Brain Death and Disorders of Consciousness*. New York: Springer, 2004. pp. 161–167. (Advances in Experimental Medicine and Biology, Volume 550)

Bernat, James L.. "The Whole-Brain Concept of Death Remains Optimum Public Policy." *Journal of Law, Medicine & Ethics* 34.1 (2006): 35–43.

Bishop, Jeffrey P.. "When is somebody just some body? Ethics as first philosophy and the brain death debate." *Theoretical Medicine and Bioethics* 40.5 (2019):419–436.

Brenner, Steven A.. "Defining Life." *Astrobiology* 10.10 (2010): 1021–1030.

Buris, László. *Forensic Medicine*. Budapest et al.: Springer-Verlag, 1993.

Campbell, Neil A., Lisa A. Urry, Michael L. Cain, Steven A. Wasserman, Peter V. Minorsky & Jane B. Reece. *Biology: A Global Approach*. Eleventh Edition. New York: Pearson, 2017.

Coffey, Peter. *Ontology or the Theory of Being: An Introduction to General Metaphysics*. London, New York, Bombay, Calcutta & Madras: Longmans, Green and Co., 1918.

Dettmeyer, Reinhard B., Harald F. Schütz & Marcel A. Verhoff. *Forensic Medicine—Fundamentals and Perspectives*. Berlin & Heidelberg: Springer-Verlag, 2014.

Elders, Leo. *The Metaphysics of Being of St. Thomas Aquinas in a Historical Perspective*. Leiden, New York & Köln: E. J. Brill, 1993.

Erk, Christian. *Health, Rights and Dignity. Philosophical Reflections on an Alleged Human Right*. Frankfurt, Paris, Lancaster & New Brunswick: ontos Verlag, 2011.

Erk, Christian. "Das Eigentliche des Todes. Ein Beitrag zur Be-Lebung der Debatte über Hirntod und Transplantation." *Ethik in der Medizin* 26.2 (2014a): 121–135.

Erk, Christian. "The Diagnosis of the Absence of Life." *Bioethica Forum* 7.3 (2014b): 105–108.

Erk, Christian. *Rationierung im Gesundheitswesen. Eine wirtschafts- und sozialethische Analyse der Rationierung nach Selbstverschulden*. Berlin & Boston, MA: de Gruyter, 2015.

Erk, Christian. "(Brain) Death, (Brain) Life and the Value of Life." *International Journal of Bioethics and Health Policy* 1.1 (2016): 27–34.

Feser, Edward. *Philosophy of Mind. A Beginner's Guide*. London: Oneworld Publications, 2006.

Feser, Edward. *Scholastic Metaphysics. A Contemporary Introduction*. Heusenstamm: editiones scholasticae, 2014a.

Feser, Edward. "*Remarks on God, Reason and Reality*." Paper presented at the Symposium "God, Reason and Reality." November 8, 2014b. Dominican School of Philosophy and Theology, Berkeley, CA. Available via: https://web.archive.org/web/20161010002912/https://www.dspt.edu/files/FeseronGodReasonReality.pdf Accessed: March 28, 2022.

Frey, Jennifer A.. "Happiness as the constitutive principle of action in Thomas Aquinas." *Philosophical Explorations* 22.2 (2019): 208–221.

Giacomini, Mita. "A change of heart and a change of mind? Technology and the redefinition of death in 1968." *Social Science and Medicine* 44.10 (1997): 1465–1482.

Heaney, Stephen J.. "Aquinas and the Presence of the human rational Soul in the early Embryo." *The Thomist* 56.1 (1992): 19–48.

Henßge, Claus & Burkhard Madea. "Leichenerscheinungen und Todeszeitbestimmung." In: Brinkmann, Bernd & Burkhard Madea (Hrsg.). *Handbuch Gerichtliche Medizin, Band I*. Berlin & Heidelberg: Springer, 2004. pp. 79–225.

Hillis, David M., Mary V. Price, Richard W. Hill, David W. Hall & Marta J. Laskowski. *Principles of Life*. Third Edition. New York: Sinauer Associates, 2019.

Jessop, Nancy M. *Theory and Problems of Zoology*. New York: McGraw-Hill, 1988. (Schaum's Outline Series)

Klubertanz, George Peter. *The Philosophy of Human Nature*. Heusenstamm: editiones scholasticae, 2014. (Reprint of the second edition, 1951)

Koshland, Daniel E. Jr.. "The Seven Pillars of Life." *Science* 295.5563 (2002): 2215–2216.

Madea, Burkhard. "Nature and Definition of Death." In: Madea, Burkhard (Ed.). *Handbook of Forensic Medicine*. Hoboken, NJ: Wiley Blackwell, 2014. pp. 49–56.

Madea, Burkhard & Gerhard Kernbach-Wighton. "Early and Late Postmortem Changes." In: Houck, Max M. (Ed.). *Forensic Pathology*. London: Academic Press, 2017. pp. 41–54.

Madea, Burkhard, Clauss Henssge, Saskia Reibe, Michael Tsokos & Gerhard Kernbach-Wighton. "Postmortem Changes and Time since Death." In: Madea, Burkhard (Ed.). *Handbook of Forensic Medicine*. Hoboken, NJ: Wiley Blackwell, 2014. pp. 75–133.

Madea, Burkhard, Katja Jachau, Saskia Reibe, Peter Schmidt, Gerhard Kernbach-Wighton, Oliver Peschel, Véronique Henn, Christoph Meissner, Manfred Oehmichen, Michael Thali, Rüdiger Lessig, Stefan Pollak & Ulrich Zollinger. "Thanatologie." In: Madea, Burkhard (Hrsg.). *Rechtsmedizin: Befunderhebung, Rekonstruktion, Begutachtung*. Berlin & Heidelberg: Springer-Verlag, 2015. pp. 33–170.

Mader, Sylvia S. & Michael Windelspecht. *Biology*. Thirteenth Edition. New York: McGraw-Hill Education, 2019.

Mason, Kenneth A., Tod Duncan, George B. Johnson, Jonathan B. Losos & Susan R. Singer. *Biology*. Eleventh Edition. New York: McGraw-Hill, 2017.

McHugh, John A. & Charles J. Callan. *Moral Theology: A Complete Course Based on St. Thomas Aquinas and the Best Modern Authorities*. Revised and enlarged by Edward P. Farrell. New York City: Joseph F. Wagner; London: B. Herder, 1958. Available at: http://www.gutenberg.org/ebooks/35354 Accessed: March 28, 2022.

Monteverde, Settimio & Annette Rid. "Controversies in the Determination of Death." *Swiss Medical Weekly* 142 (2012): w13667.

Morris, James, Daniel Hartl, Andrew Knoll, Robert Lue, Melissa Michael, Andrew Berry, Andrew Biewener, Brian Farrell, N. Michele Holbrook, Jean Heitz, Mark Hens, Elena R. Lozovsky, John Merrill, Randall Phillis & Debra Pires. *Biology. How Life Works*. Third Edition. New York: W. H. Freeman, 2019.

Nguyen, Doyen. "Why the Thomistic Defense of 'Brain Death' Is Not Thomistic: An Analysis from the Perspectives of Classical Philosophy and Contemporary Biophilosophy." *The Thomist: A Speculative Quarterly Review* 82.3 (2018): 407–446.

Oderberg, David S.. *Real Essentialism*. London: Routledge, 2007.

Oderberg, David S. "Synthetic Life and the Bruteness of Immanent Causation." In: Feser, Edward (Ed.). *Aristotle on Method and Metaphysics*. Basingstoke: Palgrave Macmillan, 2013. pp. 206–235.

Oderberg, David S.. "Death, Unity, and the Brain." *Theoretical Medicine and Bioethics* 40.5 (2019): 359–379.
Oderberg, David S.. *The Metaphysics of Good and Evil.* New York & London: Routledge, 2020.
Payne-James, Jason, Richard Jones, Steven Karch & John Manlove. *Simpson's Forensic Medicine.* 13th Edition. London: Hodder Arnold, 2011.
Phelan, Jay. *What is Life? A Guide to Biology.* Fourth Edition. New York: W. H. Freeman, 2018.
Pieper, Josef. *Death and Immortality.* Translation by Richard and Clara Winston. South Bend, IN: St. Augustine's Press, 2000. Originally published in German as: *Tod und Unsterblichkeit.* München: Kösel-Verlag, 1968.
Prahlow, Joseph A. & Roger W. Byard. *Atlas of Forensic Pathology.* New York: Springer, 2012.
Prümmer, Dominicus M.. *Manuale Theologiae Moralis Secundum Principia S. Thomae Aquinatis in Usum Scholarum.* Tomus I. Editio Quinta Decima. Barcinone, Friburgi Brisgoviae & Romae: Herder, 1961.
Raven, Peter H., George B. Johnson, Kenneth A. Mason, Jonathan B. Losos & Susan R. Singer. *Biology.* Eleventh Edition. New York: McGraw-Hill, 2017.
Sadava, David E., David M. Hillis, H. Craig Heller & Sally D. Hacker. *Life: The Science of Biology.* Eleventh Edition. Sunderland, MA: Sinauer, 2016.
Seifert, Josef. *What is Life? The Originality, Irreducibility and Value of Life.* Amsterdam and Atlanta, GA: Rodopi, 1997. (Value Inquiry Book Series (VIBS), Volume 51)
Shewmon, D. Alan. "'Brainstem Death,' 'Brain Death' and Death: A Critical Re-Evaluation of the Purported Equivalence." *Issues in Law & Medicine* 14.2 (1998): 125–145.
Shewmon, D. Alan. "The Brain and Somatic Integration: Insights into the Standard Biological Rationale for Equating "Brain Death" With Death." *Journal of Medicine and Philosophy* 26.5 (2001): 457–478.
Shewmon, D. Alan. "Mental Disconnect: 'Physiological Decapitation' as a Heuristic for Understanding Brain Death." In: Sánchez Sorondo, Marcelo (Ed.). *Working Group on The Signs of Death. The Proceedings of the Working Group 11–12 September 2006.* Vatican City: Pontificia Academia Scientiarum, 2007. pp. 292–333. (Pontificiae Academiae Scientiarum Scripta Varia 110)
Shewmon, D. Alan. "Constructing the Death Elephant: A Synthetic Paradigm Shift for the Definition, Criteria, and Tests for Death." *Journal of Medicine and Philosophy* 35.3 (2010): 256–298.
Shuster, Michèle, Janet Vigna & Matthew Tontonoz. *Biology for a Changing World.* Third Edition. New York: W. H. Freeman, 2018.
Spencer, Mark K.. "A Reexamination of the Hylomorphic Theory of Death." *The Review of Metaphysics* 63.4 (2010): 843–870.

Spike, Jeffrey. "Brain Death, Pregnancy, and Posthumous Motherhood." *Journal of Clinical Ethics* 10.1 (1999): 57–65.

Urry, Lisa A., Michael L. Cain, Steven A. Wasserman, Peter V. Minorsky & Jane B. Reece. *Campbell Biology*. Eleventh Edition. New York: Pearson, 2017.

Vanezis, Peter. "Diagnosing Death and Changes after Death." In: Vanezis, Peter & Philip Beh (Eds.) *Essential Forensic Medicine*. Hoboken, NJ: Wiley, 2019. Chapter 16.

Vij, Krishan. *Textbook of Forensic Medicine and Toxicology: Principles and Practice*. Fifth Edition. London: Elsevier, 2014.

Vreeman, Rachel C. & Aaron E. Carroll. "Medical Myths." *British Medical Journal* 335 (2007): 1288–1289.

V: Interlude II: Moral Status, Personhood and Dignity

In the last but one chapter we have established that there is a moral duty to further the realisation of the basic human goods and not violate them. As life is one of the basic human goods, this general moral duty comprises the specific moral duty to not violate the basic human good "life". It is therefore morally impermissible to violate the basic human good "life". So far, so good. But do all human beings actually possess this duty? And do all human beings possess the corresponding claim-right to not have their basic human good "life" violated? For some, this conclusion would be premature. In their mind, the subject of moral rights and duties is not human beings as such but only insofar as they are persons.

In order to be able to fully assess the moral (im)permissibility of killings of human beings by human beings, we have to therefore concern ourselves with the concept of personhood. Assuming that it is persons and not human beings as such who can be the bearers of moral duties and holders of moral rights, do the concepts "human being" and "person" have the same extension? Are all human beings persons and are all persons human beings? If a being is not under the moral duty to not kill a human being, then killing a human being cannot be morally impermissible for this being. And if a human being does not possess a moral claim-right not be killed, then it is not morally impermissible to kill it.

C. Erk, *The Ethics of Killing*,
https://doi.org/10.1007/978-3-031-07183-6_5

In general, a being can possess moral rights and duties insofar as it has moral status (also: moral standing).[1] Moral status, in return, is granted by virtue of being a person. This is one line of argumentation that can be found in the ethical literature. Another, shorter, line of argumentation manages without the mediation of the concept of moral status and connects personhood with the possession of moral rights and duties directly. Regardless of which line of argumentation is actually chosen, the endpoint of both arguments is the same: A being B can possess moral rights and duties only if B is a person.[2] This statement is commonly accepted among moral philosophers and can be taken to be virtually uncontested.[3] We can, therefore, formulate the foundational premise (FP) of this chapter as follows:

(FP) A being B can possess moral rights and duties only if B is a person.

This chapter is mainly meant to investigate whether and, if so, to what extent human beings fulfil this condition. Its main task is to find out whether all human beings are persons and whether all persons are human beings. By doing so we can establish whether all human beings are subject to the moral duty not to kill a human being and whether all human beings have the moral right not to be killed by another human being.

A second objective that this chapter aims to achieve is to familiarise the reader with the dignity of the human person. The second part of this chapter therefore explicates the special kind of value that is possessed by persons only. It then continues to show how this value grounds a general moral claim-right that is relevant to the topic of this book as it gives persons the moral claim-right not to be killed.

[1] Warren (1997: 21) describes the concept "moral status" as follows: "To have moral status is to be morally considerable, or to have moral standing. It is to be an entity towards which moral agents have, or can have, moral obligations. If an entity has moral status, then we may not treat it in just any way we please".

[2] The observant reader might have realised that the chapter on moral rights and duties did not speak of human beings but human persons. The chapters preceding this chapter did not use the term "person". This choice of words was intended to emphasise that human beings are subject to rights and duties insofar as they are persons.

[3] Of course, there are those who advance a different position, for example Marquis (1989). In order to avoid a presumed standoff between conflicting interpretations of what it means to be a person, Marquis tries to ground the right to life of foetuses by appealing to their having a future-like-ours. Interesting as it may be to see whether his argument succeeds, this would be beyond the scope of this book.

1. Categorising the Definitions of Personhood

The answer to the question who a person is depends on the answer to the question what a person is. So, what is a person? What is it that grants a being personhood? At first glance, the answers to this question are almost as numerous as the authors who have tried to address it and by doing so define personhood. If one analyses and compares the individual definitions, it becomes apparent that they can be grouped into three ultimately incompatible and contradictory response clusters or definitional approaches[4]:

- A first class of definitions consider the possession of a certain capability (or a set of certain capabilities) to be the decisive characteristic of being a person. While they agree that capabilities are at the heart of personhood, these definitions differ on the question of whether a being only needs to *possess* or also *exercise* the capability (or: capabilities) to be a person. Other definitions within this definitional cluster—while also emphasising the importance of capabilities—stipulate that it is not the being itself who has to possess the capability but the kind or species to which it belongs.
- For a second group of definitions, personhood depends on whether a being is recognised as a person.
- A third class of definitions does not consider the possession of a certain capability or capabilities as decisive for personhood, but the possession of a certain, namely a rational, nature.

The analytical and comparative look at the literature as well as the history of the concept shows that virtually every answer to the question "What is a person?" can be assigned to one of the following three definitional approaches:

- Capability-Based Definitions of Personhood
- Relational Definitions of Personhood
- Ontological Definitions of Personhood

[4] This statement is based on an extensive comparison of almost 80 definitions of personhood from antiquity to the present published in Erk (2015: chapter 3 of the appendix; also cf. chapter V.1-V.4).

These definitional approaches will be described and analysed in more detail below.

1.1. Capability-Based Definitions of Personhood

The majority of the contemporary definitions of personhood are capability-based definitions of personhood. What the definitions belonging to this definitional approach have in common despite all differences is that they are based on the following argumentation pattern (AP), which understands personhood in essence as an empirically observable quality:

(AP1) A being B is a person P only if B possesses the capabilities $C_1, \ldots, C_n$.

This line of argumentation assumes that the possession of a more or less large number of (empirically observable) capabilities determines whether a being is a person or not. However, the definitions that follow this argumentation pattern do not form a homogeneous block. Rather, they differ with respect to the number as well as the type of capabilities that are deemed necessary for personhood, and they also differ with respect to the question in what form exactly the capability needs to be possessed.

- Among the capability-based definitions of personhood there are definitions that postulate only one capability as constitutive of being a person ($n = 1$) and there are definitions which make personhood dependent on the possession of more than one capability ($n > 1$).
  Approaches that build on one capability consider the possession of this capability as a necessary and sufficient condition for being a person. In definitions that require several capabilities, each capability is only a necessary condition for personhood, while the sufficient condition for being a person is the simultaneous possession of all required capabilities.
- With regard to the kind of the presupposed capabilities, there are definitions which require the possession of intellectual, physical or social capabilities—or a combination of them.

Interesting as they may be from a theoretical perspective, these distinctions are not overly relevant in practice. For virtually all capability-based definitions link personhood to one or more intellectual capabilities. These mental capabilities can come in various shapes and forms, such as self-awareness, self-consciousness, self-distance, reasoning, reflection, valuation of existence, valuation of the future, communication, agency, memory, second-level preferences and autonomy. As diverse as these capabilities may seem, they—and along with them social capabilities, the capability to enter into relations with other beings and even the capability to love—are all variants and directly derivative of one general capability, namely rationality.[5] So, although they highlight different aspects of it, what virtually all proponents of capability-based definitions of personhood stipulate is that personhood is grounded in the possession of the capability named rationality.

A far more important distinction within the class of capability-based definitions of personhood can be made based on the question how exactly a being has to possess the required capability in order to be a person. When it comes to the possession of a capability, it would be too simple to distinguish only between the capability's presence and absence. For as Aristotle pointed out in his treatise "On the Soul", a being B can be said to possess a capability C, such as the capability to read, in more than one way. So, when we say that B actually possesses the capability to read, we can mean two things:

[5] In principle, rationality can be defined as the capacity of abstraction (cf. Hüntelmann 2017: 25). Only beings possessing rationality can mentally grasp abstract entities or abstract from the concrete and material factors of, for example, a triangle or a human being and recognise the essence and nature of a triangle or a human being. Accordingly, only rational beings are able to understand the laws of their being and existence. Besides this capacity to form concepts or ideas, Oderberg (2020: 102) takes rationality to consist of two further intellectual operations, namely the combination of concepts into judgements and the combination of judgements into discursive reasoning in order to reach conclusions: "The symptoms of, and evidence for, these fundamental powers are all of the things we familiarly take to be constitutive of rationality: language, culture, invention, political organization, law, religion, and so on". Another aspect of rationality is affectivity, that is, the capability to experience mental suffering as well as happiness.

As Oderberg (2020: 102) points out, the capacity we call self-consciousness, that is, the capacity "to reflect on and make judgments about our own states and activities, ponder ourselves as distinct individuals with our own histories and futures, and the like", is a variety of rationality: "It is plausible that even self-consciousness is made possible by rationality. [...] Self-consciousness seems to require the ability to form a concept of the self and to make judgments about the self; without such powers our self-awareness would not be distinctive" (Oderberg 2020: 102). This full-fledged self-consciousness is more than and qualitatively distinct from whatever self-awareness animals possess that grants them the ability to distinguish self from other and to be aware of their own sensations.

- B possesses the capability to read and is currently reading.
- B possesses the capability to read but is currently not reading.

What is the difference between these two cases? The difference is not the possession of the capability, because in both cases B actually possesses the capability in question; in both cases, B can read. However, in one case B is currently making use of the capability; in the other he is not. So, the actual difference is that in the first case the capability to read is fully actualised while in the second case it is not. By not only possessing but also using the capability, B has realised all the potential that comes with the possession of the capability; in the second case, this potential is not fully realised. While in the first case the capability is possessed in actuality only, that is, in fully actualised form, in the second case the capability is possessed both in actuality and in potentiality; in the second case, there exists an unrealised potential that is absent in the first case.

In moral philosophy, the difference between these two kinds of existence of a capability is covered by the distinction between *actus primus* (first act) and *actus secundus* (second act). If B has a capability, that is, the capability to read, in the sense of *actus primus*, then this means that B can read, but does not (currently) use this ability—either because he chooses not to or because he is temporarily or permanently hindered. In contrast, to possess a capability in the sense of *actus secundus* means to not only possess the capability but to also make actual use of the potential for exercising the capability that is granted by its possession. Or as Thomas Aquinas puts it: "The first act is said to be in potentiality to the second act, which is operation" (I^a q. 76 a. 4 ad 1). Compared to the second act, the first act contains a certain amount of currently unactualised but actualisable potential. This is why the first act is also called *potentia secunda* (second potentiality).

So far, we have learned that a capability can be possessed in actuality only, that is, in a fully actualised form, or in both actuality and potentiality. But this is not the end of the story. For a being B can be said to possess a capability in a third way, namely in potentiality only. In case of our guiding example, namely the capability to read, this would mean the following:

- B does not currently possess the capability to read but can acquire it.

In this case, B does not actually possess the capability to read. Nevertheless, he still possesses the capability in a potential fashion as he can learn to read. In moral philosophy, this kind of existence of a capability is termed *potentia prima* (first potentiality).

If we summarise what has been said so far, we can distinguish between three cases of possession of a capability. A capacity C can exist in a being B in either one of the following forms:

- **C exists in B in first potency**: Being B possesses capability C in potentiality only, that is, does not currently possess C but can acquire C.
- **C exists in B in first act (= second potentiality)**: Being B possesses capability C both in potentiality and actuality, that is, is currently possessing C but does not make use of C.
- **C exists in B in second act**: Being B possesses capability C in actuality only, that is, is currently possessing C and making use of C.

These insights are also illustrated in Fig. 1:

Capability C exists…	**in actuality** (also: actually; in actu)	**not in actuality**
in potentiality (also: in potency; potentially; in potentia)	C is (actually) possessed and not actualised (i.e. exercised), but can be (potentially) actualised C exists in **second potentiality (= first actuality)** (i.e. in potentia secunda = in actu primo) **Capability as Potential** (unused capability) **Example:** P possesses the capability for rational activity, i.e. can exhibit rational activity, but does not currently do so.	C is (actually) not possessed, but can be (potentially) possessed C exists in **first potentiality** (i.e. in potentia prima) **Potential for a Capability** (learnable capability; relative incapability) **Example:** P does not possess the capability for rational activity, i.e. cannot exhibit rational activity, but can acquire the capability for rational activity
not in potentiality	C is (actually) possessed and actualised (i.e. exercised) C exists in **second actuality** (i.e. in actu secundo) **Actualised Capability** (used capability) **Example:** P possesses the capability for rational activity, i.e. can exhibit rational activity, and currently exhibits rational activity.	C is (actually) not possessed and can also not be (potentially) possessed **C does not and cannot exist** **Absolute Incapability** **Example:** P does not possess the capability for rational activity, i.e. cannot exhibit rational activity, and can not acquire the capability for rational activity.

Fig. 1 The distinction between first potentiality, second potentiality/first act and second act

The application of the important distinction between first potentiality, second potentiality/first act and second act to the example of speaking a language might help to make them even clearer:

> Having the power of speech is [...] itself a first actuality relative to the actual exercise of that power, which is a second actuality relative to the mere having of the power. Similar distinctions can be made vis-a-vis potentiality. Someone who knows no English has the potential to speak it insofar as he might learn English. That is a 'first potentiality' for speaking English. Once he does learn the language he has a kind of standing ability to speak it on particular occasions if he wishes to. That is a 'second potentiality' for speaking English. (This second potentiality is in turn a first actuality insofar as it is a power that can be distinguished from the exercise of the power. The actual exercise of the power to speak English would, accordingly, be in turn a second actuality.) (Feser 2014: 41)

In the light of the distinctions just introduced it becomes obvious that the above argumentation pattern (AP1) needs to be broken down into the following three variants. By doing so, we can specify what kind of possession a capacity-based definition of personhood is actually talking about:

(AP1c) A being B is a person P only if B possesses the capabilities $C_1, \ldots, C_n$ in actuality only, that is, is possessing and currently making use of them in an empirically observable fashion.
(AP1b) A being B is a person P if B possesses the capabilities $C_1, \ldots, C_n$ both in actuality and potentiality, that is, is possessing and—although not currently doing so—can potentially make use of them.
(AP1a) A being B is a person P only if B possesses the capabilities $C_1, \ldots, C_n$ in potentiality only, that is, does not currently possess but can acquire them.

Excursus
In order to fill these somewhat theoretical argumentation patterns with life, the following list contains examples for each of the three argumentation patterns. Although many definitions are somewhat imprecise in their wording and need to be closely scrutinised as to what they actually want to express, all capability-based definitions of personhood can be assigned to one of these three argumentation patterns.

- **Examples of capability-based definitions of personhood that follow argumentation pattern (AP1c)**: A definition can be

(*continued*)

(continued)

placed in argumentation pattern (AP1c) if it makes personhood dependent on the actual exercise of a capacity. One way or another, those definitions usually state that in order to be a person a being must be doing something or exhibit some kind of activity.

- We must consider what person stands for;—which, I think, is a thinking intelligent being, that has reason and reflection, and can consider itself as itself, the same thinking thing, in different times and places; which it does only by that consciousness which is inseparable from thinking, and, as it seems to me, essential to it: it being impossible for any one to perceive without perceiving that he does perceive. (Locke 1690: Book II, Chapter 27, Section 9)
- What properties must something have in order to be a person, i.e., to have a serious right to life? The claim I wish to defend is this: An organism possesses a serious right to life only if it possesses the concept of a self as a continuing subject of experiences and other mental states, and believes that it is itself such a continuing entity. (Tooley 1972: 44)
- That is, to be a person, or a fully developed person anyway, is to have a conception of oneself, to see oneself as an individual with rights, to have plans and ambitions for oneself, whether or not oneself alone. (Solomon 1983: 214)
- To be a person, a being must be self-conscious, aware of its identity and its continued existence over time. (Parfit 1984: 202)
- Some human beings, for instance fetuses and newborn infants, are not rational and self-conscious, and therefore not persons, while some nonhuman animals, for instance chimpanzees, are rational and self-conscious and should be considered persons. (Singer & Kuhse 1994: 133)
- s I use the term, a 'person' is a being with a rich and complex mental life, a mental life of a high order of sophistication. (McMahan 2002: 45)

• **Examples of capability-based definitions of personhood that follow argumentation pattern (AP1b)**: A definition can be placed in argumentation pattern (AP1b) if it contents itself with

(*continued*)

(continued)

requiring the possession of a capability without making reference to its actual exercise.

- A person is a being that is conscious, in the sense of having the capacity for conscious thought and experiences, but not only that: it must have the capacity for reflective consciousness and self-consciousness. (Lockwood 1985: 10)
- Persons are beings capable of valuing their own lives. (Harris 1985: 16)
- All that is required for something to be a person is for it to have at least an active potential to perform self-conscious rational operations. The actual performance of such operations is accidental to a person's existence. (Eberl & Brown 2011: 51)
- We take 'person' to mean an individual who is capable of attributing to her own existence some (at least) basic value such that being deprived of this existence represents a loss to her. (Giubilini & Minerva 2013: 262)

- **Examples of a capability-based definition of personhood that follow argumentation pattern (AP1d)**: Definitions that belong to argumentation pattern (AP1d) usually stipulate that in order to be a person a being must belong to a biological species whose typical members actualise the required capabilities in the natural course of their development.

  - A person is any animal the physical make-up of whose species constitutes [that is, makes] the species' typical members thinking intelligent beings, with reason and reflection, and typically enables them to consider themselves the same thinking things, in different times and places. (Wiggins 1980: 188)
  - Although the following passage does not explicitly speak of persons but rather states towards which beings we have moral duties, it can be taken as an example of a capability-based definition of personhood that follows argumentation pattern (AP1d). For this class of beings is usually referred to as persons: "(T)he class of beings whom it is possible to wrong will include at least all those beings who are of a kind that is normally capable of judgment-sensitive attitudes" (Scanlon 1998: 186).

The argumentation patterns (AP1a), (AP1b) and (AP1c) make a statement about the minimum threshold that has to be crossed in order for a being to be a person. The application of argumentation pattern (AP1c) should be rather straightforward: A being is a person as soon and as long it actualises the required capabilities. However, the application of argumentation patterns (A1a) and (A1b) presents us with certain problems:

- In the case of (AP1a), how do we know that B can potentially possess the capabilities $C_1, \ldots, C_n$? How can we know whether and that B can acquire the capability to read if B does not possess it at the moment?
- In the case of (AP1b), how do we know that B possesses the capabilities $C_1, \ldots, C_n$? Moreover, even if we have observed that B has exercised the capabilities $C_1, \ldots, C_n$ at least once, how do we know that B still possesses them?

Proponents of (AP1a) usually solve this problem by referring to the species to which the being in question belongs. According to this solution, the conditions laid down in (AP1a) are considered to be fulfilled if it can be shown that B belongs to a biological species whose representatives usually actualise the capabilities $C_1, \ldots, C_n$ in the course of their natural development. Modifying (AP1a) in this fashion yields (AP1d):

(AP1d) A being B is a person P only if B belongs to a biological species whose typical members naturally actualise the capabilities $C_1, \ldots, C_n$ in an empirically observable fashion.

When it comes to (AP1b) things are a bit more complicated. How can one know and prove that a being B (still) possesses capability if B does not currently exercise it? There are two ways to solve this conundrum:

- The first way is to modify argumentation pattern (AP1b) by holding its condition fulfilled if B has made use of the required capabilities at least once. But, while this proves that B has acquired the required capabilities at some point in the past, it does not prove that B still possesses them. The only way to make sure that B still possesses the required capabilities is to observe their exercise. However, this would make (AP1b) basically collapse to (AP1c).
- The second way is to slightly relax the condition and hold it fulfilled if it can be shown that B belongs to a biological species whose representatives naturally possess the required capabilities in an actualised fashion. However, this would make (AP1b) basically collapse to (AP1d).

So, when it comes to the application of (AP1b) we are somehow caught between Scylla and Charybdis—or, to use a more modern proverb, between a rock and a hard place. As (AP1b) either collapsed to (AP1d) or (AP1c), it cannot be maintained as a separate argumentation pattern. So, our initial list of three argumentation patterns has shrunk to the two argumentation patterns (AP1d) and (AP1c).

Having categorised the capability-based definitions of personhood, there is one last step to be taken. As has already been mentioned above, the capability whose possession is decisive for and ultimately constitutive of personhood is rationality. Given the unanimity of this position, we can adapt our two argumentation patterns (AP1d) and (AP1c) by substituting the placeholder "the capabilities C_1, …, C_n" with the concrete capability rationality. This results in the following versions of (AP1d) and (AP1c):

> **(AP1d)** A being B is a person P only if B belongs to a biological species whose typical members naturally actualise the capability rationality in an empirically observable fashion.
> **(AP1c)** A being B is a person P only if B possesses the capability rationality and is currently making use of it in an empirically observable fashion.

Exercising the capability rationality means exhibiting rational behaviour, that is, performing rational acts such as reasoning, reflection and valuation of existence. As one cannot do what one does not have the capability to do, exhibiting rational behaviour logically presupposes the possession of the capability rationality. The argumentation pattern (AP1c) therefore does not require special reference to the possession of the capability rationality and can thus be further simplified as follows:

> **(AP1c)** A being B is a person P only if B is displaying rational behaviour in an empirically observable fashion.

1.2. Relational Definitions of Personhood

The second class of definitions of personhood can be named "relational". While capability-based definitions ground personhood in the possession of the capability rationality, relational definitions take personhood to be a direct function of certain forms of recognition or relationships.

According to the relational view of personhood, a being B can only be a person if B is recognised as a person by one or more human beings. Put simply, a being is a person if at least one human being says it is. Being a person means standing in an appreciative relationship with at least one human being that finds its expression in, for example, love, respect, esteem and/or solidarity. Personhood is something that is accorded to a being by at least one (relevant)[6] human being; it is constituted by the practical confirmation of its existence by some (relevant) third party.

Excursus

Below are some examples of relational definitions of personhood:

> But it seems to me useful to pose a preliminary question: How is one constituted a human person? Is it by a merely biological act? It seems to me astonishing that a spiritual being be constituted by a solely biological act. Does not the fact that the parents perceive the fetus as a human person make any difference in its institution as a human being, as a spiritual being? Is it not necessary that there be established a relation of person to person, a relation of generators with the fetus, for it to become a human person? (de Locht 1968: 155 (cited in Grisez 1989: 28))

> My position is that the only way a fetus can become a member of the human community, and therefore a person, prior to birth, is if the woman in whose body it exists welcomes it into the human community by her consent to the pregnancy. It is the consent of the woman to continuing the pregnancy that marks the beginning of personhood, no matter how unwillingly that consent is given. It is when she treats the biological reality within her as a Thou, that it becomes a Thou rather than an It. And it is when she treats it as a Thou, that society also has an obligation to treat it as a Thou. (Maguire 1989: 15)

> Thus we arrive at a definition of personhood, as I shall use the term in this book. It is a standing or a status that is bestowed on one human being, by another in the context of relationship and social being. It implies recognition, respect and trust. (Kitwood 1997: 8)

(*continued*)

[6] Usually, the recognition cannot simply come from any other human being. For example, in case of the unborn the relevant human being is usually taken to be the pregnant woman. According to this line of thinking, the unborn is a person if and because the pregnant woman recognises the human being inside her as a person; if she does not do so, then the unborn is not a person—and can also not become a person by acts of recognition by other human beings (e.g. the father).

(continued)

> To be recognized as a person constitutes one's being an actual person. No one, with whatever capacities, can be more than a potential person, if not taken to be a person by others. (Laitinen 2002: 474, cf. 2007)
>
> Personhood is rooted in relationships with others and, therefore, requires the presence of another human being. (Penrod et al. 2007: 63)

Generally speaking, the relational approach to defining personhood is based on the following argumentation pattern:

> **(AP2)** A being B is a person P only if B is recognised as a person by at least one other (relevant) human being.

What are we to make of this argumentation pattern? Attractive as it may sound at first sight, the argumentation pattern faces a grave logical problem. As commonly agreed, personhood is not merely a descriptive but a prescriptive concept. If, insofar as and because a being B is a person, B is owed to be treated in a certain kind of way. And this is where the relational-definition-of-personhood-cat bites its tail. (AP2) basically says that being a person is not the reason for being treated in a certain way but rather its consequence. According to this position, a being does not have moral status if, insofar as and because it is a person, but it is a person if, insofar as and because one deals with it in a certain way. It is a crucial difference whether the personhood of a being B is acknowledged and practically confirmed through corresponding acts of recognition (i.e. by behaving in line with B's moral rights and duties) or whether B is granted personhood by certain acts of recognition.

In the first case, B is already a person and is simply acknowledged as such; in the second case, B is not a person but made a person by the acknowledging behaviour of others. A purely relational definition of personhood as expressed in (AP2) thus reverses the relationship between personhood and recognition and empties the concept of personhood of any normative content. For what is it that is recognised by the act of recognition or practically confirmed by acts of love, respect, esteem, solidarity and so on? If there is not something to which these acts are a reaction, that is, if these acts are not already caused by being a person or the existence of some personhood-conferring property, then personhood becomes an arbitrary concept. But this would make the concept of personhood useless.

Moreover, there would be no logically compelling reason why personhood should only be attributable to human beings. If it is relationship that counts, personhood could be conferred on just about anything we can have a relationship with: We could recognise animals, trees, houses, cars and even our favourite coffee mug as persons—and somebody who breaks, for example, such a coffee mug would be guilty of murder. Given this grave deficiency, the argumentation pattern (A2) can safely be set aside in our further considerations of the concept of personhood.

1.3. Ontological Definitions of Personhood

The third class of personhood definitions includes all definitions that hold that personhood has something to do with "nature", "essence", "existence", "being", "mode of being" or "substance". For them, personhood is grounded in nature and being rather than capabilities and functioning. Regardless of what it is actually capable of doing or what it is doing at a certain point in time, a being B is a person if it is a certain kind of being, that is, if it possesses a certain nature.

Excursus

Below are some examples of ontological definitions of personhood:

> Persona est naturae rationabilis individua substantia. (Boëthius 1918: chapter 3)
>
> Persona est rationalis naturae individua substantia. (Thomas Aquinas, I[a] q. 34 a. 3 ad 1)
>
> What makes us persons is the kind of beings we are, the kind of nature we possess, and not a passing state or stage of that kind of being. (...) They are not human organisms first and persons only subsequently. (Iglesias 1984: 35)
>
> In speaking of the human person we are not speaking of any kind of capacity nor of any kind of attribute. [...] Personality discloses personhood; it does not constitute it. Personal attributes develop as consciousness develops; but persons do not develop, for they are not in the category of quality but of substance. (O'Donovan 1985: 129)
>
> But I would suggest that a person is not an individual with a developed capacity for reasoning, willing, desiring, and relating to others.

(*continued*)

(continued)

A person is an individual with a natural capacity for these activities and relationships, whether this natural capacity is ever developed or not. Individuals of a rational, volitional, self-conscious nature may never attain or may lose the functional capacity for fulfilling this nature to any appreciable extent. But this inability to fulfill their nature does not negate or destroy the nature itself, even though it may, for us, render that nature more difficult to appreciate or love. But that difficulty would seem to be a challenge for us as persons more than it is for them. (Joyce 1988: 201)

I think that all whole, bodily, substantial individuals of any species having a rational nature are persons. (Grisez 1989: 28)

To be a person is to belong to a kind of being characterized by rational (self-conscious, intelligent) nature. To have a particular nature is to be so constituted, dynamically integrated, as to have certain capacities (for example self-awareness and reasoning). (Finnis 1995: 48)

The genuine criterion for having a right to life is being a person, that is, a distinct substance of a rational nature. (Lee 2007: 97)

These goods contribute to the flourishing of the human person because he is an individual substance of a rational nature ordered towards these goods. (Beckwith 2011: 75).

If one analyses the definitions that can be attributed to the ontological definitional approach, one can see that they do not simply speak of any but rather a specific nature. For ontological definitions of personhood, the nature whose possession grounds personhood is a rational nature (N_R). So, albeit in a different fashion, rationality does not only play a central role in capability-based but also in ontological definitions of personhood. So, generally speaking, the ontological approach to defining personhood follows the following pattern of argumentation (AP3):

(AP3) A being B is a person only if B possesses a rational nature.

This requirement might seem straightforward at first sight. Upon closer scrutiny, however, we will see that the argumentation pattern requires some commenting. It is relatively easy to imagine what it means to possess a capability. But what is actually possessed when we say that a being possesses a rational nature? Imagining this might cause some difficulties. Let us first shed some light on the concept of nature. A being's nature—or as it is also called, essence—is that which makes the being the kind of being it is; it is that which is expressed in its definition.

> The essence of a thing is its nature, that whereby it is what it is. It is what we grasp intellectually when we identify a thing's genus and specific difference. (Feser 2014: 211; cf. Oderberg 2020: 13; Hüntelmann 2017: 13f)

To possess a certain nature means to be a certain kind of being. In a narrow sense, the nature of a being is given by its substantial form; in a broader sense, the nature of a material being is given by both its form and matter. Natures or essences do not exist as such; as universals they only exist in the beings whose nature they are and can only be recognised in individual beings by means of abstraction.

The nature of a being is not, however, a property of that being; rather, it is "that from which a thing's properties flow, that which explains its properties" (Feser 2014: 230; cf. Oderberg 2007: 23). As such, nature is also "the principle of action" (I[a] q. 39 a. 2 ad 3), that is, that whence the behaviour and actions of a being proceed. Nature is not only that whereby a being is what it is but also that whereby a being can do what it does. The possession of a certain nature grants the possession of the capabilities associated with that nature. Nature is both capability-granting and capability-limiting: Qua being a certain kind of being, a being possesses the capabilities associated with its nature but does not possess the capabilities not associated with its nature. So, while capability-based definitions ground personhood in the possession of the capability rationality, ontological definitions base personhood on the source and principle of this capability. For the ontological approach of defining personhood, a being is a person not because it possesses the capability rationality but because it is a kind of being whose nature it is to possess this capability. What is important for personhood is not the capability as such but that which underlies and grants the capability.

So far so good. But how can we know that a being possesses a rational nature? One way to ascertain this is to observe that a being displays rational acts, such as acts of self-awareness, self-consciousness, self-distance, reasoning, reflection, valuation of existence, valuation of the future, communication, agency, memory, second-level preferences and autonomy. If a being B does what only rational beings can do, B must necessarily possess a rational nature. In this case, B possesses a rational nature in actuality only. These considerations allow us to formulate a first modified version of argumentation pattern (AP3), namely (AP3c):

(AP3c) A being B is a person if B possesses a rational nature N_R in actuality only, that is, if B is displaying rational behaviour in an empirically observable fashion.

However, while sound in theory, this argumentation pattern is not actually advanced by scholars who subscribe to an ontological definition of personhood. The reason for this is most likely that argumentation pattern (AP3c) is indistinguishable from the capability-based argumentation pattern (AP1c). It is true that—from the perspective of ontological definitions—a being who exhibits the rational behaviour that is associated with the possession of a rational nature necessarily possesses a rational nature and, therefore, is a person. But it is not true that a being can only be said to possess a rational nature if it displays rational behaviour. Given the fact that (AP3c) does not really feature as an argumentation pattern of ontological definitions of personhood, we no longer have to worry about it and can skip it.

So, let us continue our enquiry. As we have learned in the chapter on the capability-based definitions of personhood, something cannot only be possessed in actuality only but also in potentiality only as well as both in actuality and potentiality. This distinction can also be applied to the possession of a rational nature: A being B can possess such a nature not only in actuality only but also both in actuality and potentiality or in potentiality only. The first form of possession of a rational nature is the basis for (AP3c). Taking the latter two forms of existence into account, we can formulate two additional modified versions of argumentation pattern (AP3):

(AP3b) A being B is a person if B possesses a rational nature N_R both in actuality and potentiality, that is, if B possesses N_R and—although currently not doing so—can potentially display the rational behaviour that is associated with the possession of N_R in an empirically observable fashion.
(AP3a) A being B is a person if B possesses a rational nature N_R in potentiality only, that is, if B does not currently possess but can acquire N_R.

While it is rather simple to prove that a being possesses a rational nature in the case of (AP3c), this proof proves to be somewhat difficult in the cases of (AP3a) and (AP3b). How can we assess whether a being B possesses a rational nature either in potentiality only or both in actuality and potentiality? Is there a way to prove that B possesses a rational nature even if B does not exhibit any observable rational behaviour (e.g. reasoning, reflection, valuation of existence)? The observant reader may notice that we have dealt with a similar problem in the chapter on the capability-based definitions of personhood. The same solution that has been proposed there can also be applied here, namely referring to the biological species to which the being in question belongs. According to this solution, the conditions laid down in (AP3a) and (AP3b) are met if it can be shown that B belongs to a biological species whose representatives usually display the rational behaviour that is associated with the possession of a rational nature in the course of their natural development. Modifying the argumentation patterns (AP3a) and (AP3b) in this fashion results in (AP3d):

> **(AP3d)** A being B is a person only if B belongs to a biological species whose typical members naturally display rational behaviour in an empirically observable fashion.

So, as a closer look reveals, (AP3d) is the soundest formulation of ontological definitions of personhood. This argumentation pattern is also advanced by, for example, Spaemann (2006: 297): "There is only one admissible criterion for human personhood: belonging biologically to the human family".

2. Personhood as Possessing Rational Life

Let us recapitulate what we have learned so far. We have learned that there are three main approaches to defining personhood, namely capability-based, relational and ontological definitions. Most of the personhood definitions found in contemporary ethical literature are capability-based; the second largest class is ontological definitions; relational definitions have the smallest share.

By design, *relational definitions* are somewhat inconsistent from a logical perspective as they confuse cause and effect. While it is true that being a person is one thing and being recognised as a person in one's dealings with other human beings is another thing, in order to be practically confirmable (by, e.g., acts of love, respect, esteem, solidarity), there must be something that can be confirmed. So, personhood cannot be the result of acts of recognition but is their precondition. One cannot confirm that something is the case if there is nothing that is the case.

Capability-based definitions ground personhood in the possession of the capability rationality. According to this class of definitions, to be a person, a being must be capable of rational behaviour. Capability-based definitions usually come in one of the three following argumentation patterns:

(AP1b) A being B is a person P only if B possesses the capability rationality and—although not currently doing so—can potentially make use of it.
(AP1c) A being B is a person P only if B is displaying rational behaviour in an empirically observable fashion.
(AP1d) A being B is a person P only if B belongs to a biological species whose typical members naturally actualise the capability rationality in an empirically observable fashion.

Ontological definitions of personhood hold that personhood is constituted by the possession of a rational nature. To be a person, a being must possess a rational nature. Ontological definitions usually come in either of two argumentation patterns:

(AP3) A being B is a person only if B possesses a rational nature.
(AP3d) A being B is a person if B belongs to a biological species whose typical members naturally display rational behaviour in an empirically observable fashion.

Having mapped the field, we could now content ourselves with this overview that helps organise the definitions of personhood found in the literature—and move on. But this would leave unanswered the utmost important question: "Are all human beings persons?" The multitude of contradicting opinions on the question of what a person is should not discourage us; rather, it should spur us on to find out which of the capability-based or ontological definitions we should actually follow.

Accordingly, the rest of this chapter tries to settle this question. It does so based on the commonalities of the two definitional approaches. For as much as they differ, the capability-based and ontological definitions of personhood also have several aspects in common:

- One of these common features is the central role of rationality in grounding personhood. For both, that which must be present in a certain form for a being to be a person is rationality. A being is a person if it possesses rationality in a form considered relevant by the respective definitional approach. Although they argue about whether rationality must be possessed as capability or as nature, both capability-based and ontological definitions of personhood are in full agreement that rationality plays a foundational role and is the element causing personhood.

- A second commonality is that both capability-based and ontological definitions take a person to be a special kind of individual, that is, a particular entity that exists for itself and separately from others. An individual is a being that does not exist in something else but is capable of existing on its own.
- Capability-based and ontological definitions also agree that persons are special bodily entities. As the concept of personhood is commonly used in ethical debate, it only applies to certain kinds of material substances.
- Last but not least, both definitional approaches share another commonality. Virtually all philosophers and authors who think about the concept of personhood agree that a person is a certain kind of living being. As far as I can tell, there is no definition of personhood in the ethical debate that applies personhood to inanimate matter such as printers, houses, cars or chairs. Nobody has ever claimed that a stone is a person. There is a general consensus that only an animated being, that is, a being that possesses life, can be a person. Only a life-form—or as it is also called in biology, an organism—can be a person.

Analysing the similarities between the definitional approaches, it becomes obvious that both ontological and capability-based definitions of personhood have a common core: In order to be a person, a being must be a living, bodily, individual being that is in some way rational.

By almost exclusively debating the relevant form of possession of rationality, we tend to forget that the most fundamental aspect of all definitions of personhood is the presence of life. At its core, the ethical debate on personhood revolves around the question whether all animated beings are persons or not, that is, whether only (all or some) humans or (all or some) animals and/or plants are persons. The fact that this question is so controversial shows that personhood is not tied to life as such but to a certain kind of life, namely life that is characterised by rationality. In other words, personhood is grounded in the possession of what can be called rational life. Under the assumption that the corporeality and individuality of persons are matters of course and need not be mentioned separately, the condition for personhood can be reformulated as follows: In order to be a person, a being must possess rational life.

This is the most fundamental and general condition of personhood that should be acceptable to both sides of the debate—regardless of whether they subscribe to a capability-based or an ontological definition of personhood. We can formulate this general condition of personhood as follows:

(AP0) A being B is a person P only if B possesses rational life.

Although they argue whether rational life consists in the possession of a rational nature or in the possession of rational capabilities, (AP0) is what both capability-based and ontological definitions of personhood should be able to agree to. As a common point of reference (*tertium comparationis*), (AP0) can therefore play a mediating role and help us decide which of the two definitional approaches to actually follow. However, in order to be able to do so, we must first understand what is meant by "rational life". Once we do, we can then decide whether rational life consists in the possession of rational faculties or in the possession of a rational nature. So, the key to understanding what a person is ultimately lies in understanding what rational life is—and the key to understanding rational life lies in understanding the concept of life.

2.1. Rational Life

We have already considered what life is in Chap. IV. This foundation allows us, in a next step, to deal with the question of what makes life *rational* as opposed to non-rational. Based on the insights into the phenomenon of life gained so far, this question should be relatively easy to answer.

As we have learned in Chap. IV, life is the capacity for endogenous activity. The empirically observable signs of life are actualisations of this capacity. A living being possesses life both in actuality and in potentiality. A living being is capable of endogenous activity if and insofar as it possesses a soul. It is the immaterial life-principle named "soul" that grants living beings their capacity for endogenous activity. As such, the soul is not only the principle of life but also the principle and efficient cause of all activity shown by a living being.

If life is the possession of the life-principle soul, rational life is the possession of a rational soul. We can therefore state:

(AP0a) A being B is a person P only if B possesses a rational soul.

As the soul is the capability-granting principle that enables its possessor to exhibit empirically observable signs of life, the rational soul is the principle that grants its possessor the specific capability for rational behaviour. Given the nature of soul, this capability must be possessed both in actuality and in potentiality. So, if a being possesses a rational soul, it possesses rationality as capability, that is, both in actuality and in potentiality. A being who possesses rational life possesses the capability to display the behaviour typically associated with rationality. This capability is possessed in actuality even if its potentiality is not actualised—either because the being chooses not to do so or cannot to do so (e.g. as its bodily infrastructure hinders an actualisation temporarily or permanently).

However, once again, we run into a verification problem. How can we ascertain that a being possesses a rational soul? The simplest case is if the being is currently actualising the capabilities that are associated with the possession of a rational soul. In this case, we know for certain that it possesses a rational soul. But what if the being—while exhibiting some empirically observable behaviour—does not currently display *rational* behaviour? How can we know that such a being possesses rational and not merely vegetative or sensitive life?

In order to be able to handle this latter case, we must bear in mind that every material being must and can only have one substantial form, that is, one organisational principle by virtue of which it is the kind of being it is. A being cannot simultaneously be something and something else; a stone cannot simultaneously be a stone and a light bulb.

If a being cannot possess less but also not more than one substantial form, then a being who possesses a rational soul as substantial form cannot have any other substantial form. If so, this allows us to conclude that the rational soul is not only the principle of this rational being's rational capabilities but must be the principle of all its capabilities. This has also been stated in Chap. IV, Sect. 4 where it was noted that a rational soul not only grants the capability for rational behaviour but for vegetative, sensitive as well as rational activity. The rational soul is therefore not only the basis of rational behaviour, but of all empirically observable signs of life (i.e. metabolism, growth and development, regeneration, regulation, reproduction, etc.). It should therefore be obvious that a rational being who has displayed but no longer displays rational activity must still possesses a rational soul as long as it exhibits any signs of life. This solves the case in which a being has already exhibited signs of rational activity but does not currently do so—either because it chooses not to do so or because its body does not allow to do so.

But what if a being has not yet displayed rational behaviour? This question can be resolved by reference to an explanatory device that we have already used in Chap. V, Sect. 1: species membership. As they can only possess one rational form, all members of a biological species whose typical members display rational behaviour in their natural development must possess a rational soul. If they were to possess a different kind of soul, they would not be members of this biological species. So, a living being can be regarded as possessing a rational soul if it belongs to a biological species whose members actualise the capability rationality in the course of their natural development. This insight is expressed in the expanded version of argumentation pattern (AP0a), namely (AP0b):

> **(AP0b)** A being B is a person P only if B possesses a rational soul. B possesses a rational soul if it shows any signs of life and B belongs to a biological species whose typical members naturally actualise the capability rationality in an empirically observable fashion.

With this last concretisation in place, we have now reached a point where we can draw a conclusion and answer the question of what a person is, and which beings can be persons.

2.2. What and Who Is a Person?

The above chapter has made it its mission to find a definition of personhood that is based on the commonalities between capability-based and ontological definitions of personhood. Working with what the proponents of these definitional approaches agree on should ideally result in a definition of personhood that is acceptable to both. As is commonly acknowledged, a person is a living rational kind of being. In other words, persons possess rational life. It was therefore suggested that all sides should be able to agree on the following premise:

> **(AP0)** A being B is a person P only if B possesses rational life.

Possessing life means possessing a soul. It is the soul that enlivens living beings and grants them the capability to exhibit empirically observable signs of life. Possessing rational life means possessing a rational soul, that is, the kind of soul that not only grants the capability to exhibit signs of life but also—amongst others—to display rational behaviour. (AP0) has therefore been modified to (AP0a):

(AP0a) A being B is a person P only if B possesses a rational soul.

We know that a being B possesses a rational soul if B exhibits rational behaviour, that is, actualises the capability rationality by rational acts such as acts of self-awareness, self-consciousness, self-distance, reasoning, reflection, valuation of existence, valuation of the future, communication, agency, memory, second-level preferences and autonomy. Given the nature of soul as the substantial form of a living being, we also know that a being B possesses a rational soul if B is alive, that is, does not show the certain signs of death, and is a member of a biological species whose typical members naturally actualise the capability rationality in an empirically observable fashion. (AP0a) has therefore been further developed into (AP0b):

> **(AP0b)** A being B is a person P if B shows any signs of life and belongs to a biological species whose typical members naturally actualise the capability rationality in an empirically observable fashion.

Having started from the generally acceptable premise (AP0), with premise (AP0b) we have now arrived at the definition of personhood proposed by this book—a definition that should be equally generally acceptable as the premise from which it has been derived.

As mentioned above, the question "What is a person?" logically precedes the questions "Who is a person?" or "Who can be a person?" Now that we know the answer to the first question, we can begin to look at the latter two questions. What we will be especially interested in is the extensional relation between the terms "person" and "human being".[7] While the intension of the former term is given by the definition of personhood (AP0b), a human being is best defined as a member of the biological species "homo sapiens sapiens". So, is the extension of these two concepts identical or not? Generally speaking, the terms "person" and "human being" can stand in five distinct extensional relations (ER):

- **(ER1)** Every human being is a person and every person is a human being.

[7] While the so-called intension of a term signifies the term's content as given by its formal definition, the so called extension of a term indicates the range of its applicability, that is, what is covered by it or to what it applies.

- **(ER2)** Every human being is a person, but not every person is a human being.
- **(ER3)** Not every human being is a person, but every person is a human being.
- **(ER4)** Not every human being is a person and not every person is a human being.
- **(ER5)** No human being is a person and no person is a human being.

The question we must answer next is which of these five options is actually the true statement about the extensional relation between persons and human beings. A first step to take is to find out whether all, some or no human beings meet the condition for personhood contained in (AP0b). This means that we need to find out whether all, some or no human beings belong to a biological species whose typical members naturally actualise the capability rationality in an empirically observable fashion. The answer should be quite straightforward: By definition, human beings belong to the biological species homo sapiens sapiens. In the course of their natural development, the typical members of this biological species do indeed display rational behaviour in an empirically observable fashion. Accordingly, all human beings belong to a biological species whose typical members naturally actualise the capability rationality in an empirically observable fashion.

This means that all living human beings are persons as soon as they show any signs of life and as long as they do not exhibit the certain signs of death. Or in short, all living human beings are persons. This also includes human beings in their earliest stages of development. From a biological perspective, a human being starts to exist as a member of the biological species homo sapiens sapiens at the moment of conception:

> A human ovum possesses human chromosomes as genetic carriers, not chicken or fish; This is now manifest; the evidence no longer allows a discussion as to if and when and in what month of ontogenesis a human being is formed. To be a human being is decided for an organism at the moment of fertilization of the ovum. For this reason we have to regard the intrinsic quality of the fertilized ovum as an essential prerequisite, decisive for all future ontogenesis. (Blechschmidt 1977: 17, cf. 2004: 7)
>
> In short, a fertilized human egg (conceptus) is already a human being. We appreciate that each egg contains species-specific chromosomes and further,

> we know that during ontogeny there are no mutations that would have the effect, say, of changing one species into another. [...] The following important statement is valid: that which changes during development is only the phenotype but not the essence. (Blechschmidt 2004: 8)[8]

The human zygote[9] does not develop into a living human being but is a living human being right from the beginning. This is not up to ethical debate, but a biological matter of fact determined by the possession of a distinctly human genetic code (DNA). Accordingly, as the human zygote is biologically human, every human zygote is a person.

We now know that all living human beings are persons from the moment of conception. This makes three of the five distinct extensional relations identified above false, namely (ER3), (ER4) and (ER5). So, we have considerably reduced the number of possible extensional relations between the terms "person" and "human being". But, which of the two remaining options of extensional relations is the correct one? (ER1) or (ER2)?

To be able to decide, we have to deal with the question whether only members of the biological species homo sapiens sapiens do and can fulfil the condition for personhood set out by (AP0b). The answer to this question is also rather straightforward. On our planet, the only biological species whose typical members can and do display not only aspects of rationality but rationality in the true and full sense of the word is the species homo sapiens sapiens. Of course, there are other animal species whose members possess astounding cognitive abilities and display behaviour that resembles rationality, but from the perspective of rationality, this behaviour is of a completely different quality. Animal "rationality" and human

[8] Although less concise in their wording, more recent standard textbooks on embryology and developmental biology clearly corroborate Blechschmidt's position: "Human development begins at fertilization when a sperm fuses with an oocyte to form a single cell, the zygote. This highly specialized, totipotent cell (capable of giving rise to any cell type) marks the beginning of each of us as a unique individual" (Moore et al. 2020: 11; cf. Sadler 2019: 1; Schoenwolf et al. 2015: 1).

[9] The human zygote is the first (diploid) cell of the human embryo; it is a single-cell or one-cell embryo. The zygote is formed during the process called "fertilisation" (also: conception). Generally speaking, fertilisation is the joining of two opposite-sex human haploid gametes (germ cells), that is, an egg (oocyte or ovum; female) and a sperm (spermatozoon; male). For a description of this complex process that takes about 24 hours cf. Condic (2009: 135ff), Condic (2013: 48ff), Schoenwolf et al. (2015: 33ff) or Moore et al. (2020: 25ff)

rationality are worlds apart: In contrast to human beings, non-human animals do not understand the laws of their being and existence, cannot form concepts or ideas and lack what is also expressive of true rationality, such as language, culture, political organisation, law and religion.[10] It is, in consequence, safe to say that the non-human animals on our planet do not possess personhood.[11]

It would be going too far, however, to conclude from the fact that—apart from the biological species homo sapiens sapiens—there is no known biological species whose typical representatives display rational behaviour that all persons must necessarily be human. In principle, the definition of a person given above also includes living beings that belong to a non-human species, whose members naturally actualise the capability rationality in an empirically observable fashion.[12]

[10] For a discussion of the "rationality" of non-human animals as well as an analysis of experiments attempting to teach language to primates cf. Jensen (2018: 159–175). For an additional overview of the difference in kind between human beings and other animals cf. Lee & George (2008a: 52ff).

When we come across authors who claim that some non-human animals are persons, we must critically scrutinise the understanding of rationality on which they base their claim. Doing so will usually do the trick of disenchanting their respective position. For example, Singer (2011: 94; cf. 84) takes persons to be "rational and self-conscious beings", that is, beings which are "aware of themselves as distinct entities with a past and a future". However, if we compare Singer's preference-utilitarian understanding of rationality with the actual intension of the concept as described above, we can easily see that his position is based on an absolutisation of one aspect of rationality that sacrifices the richness of the concept. Distorting the concept might do the trick for Singer but does not do justice to the world of difference between the rationality of human beings and the "rationality" of non-human animals.

[11] However, this in no way implies that non-human animals (or plants, for that matter) can and should be treated as mere things such as stones. Human persons do have moral duties in their dealings with the rest of animated creation.

[12] Given this general openness, the definition of personhood advanced here cannot be accused of "speciecism", that is, "an unjustified preference for the members of the human species, on the basis of the mere fact that they are human" (Reichlin 1997: 20; cf. Singer 1975: 21, 9). When it comes to the definition of personhood given by (AP0b), there is no such preference. Human beings are not persons because they are human beings but because they possess rational life. This condition does not privilege the human species but sets a standard that equally applies to all species. So, if rational life is discovered, for example, on other planets, then the living members of this newly discovered species are persons, too: "While membership on the species Homo sapiens is sufficient for full moral worth, it is not on any direct sense the criterion for moral worth. If we discovered extraterrestrial beings of a rational nature, or that some other terrestrial species did have a rational nature, then we would owe such beings full moral respect" (Lee & George 2008a: 82).

The preceding explanations necessitate the conclusion that the only correct option of the five possible extensional relations is (ER2). So, every human being is a person but not all persons are necessarily human beings. This insight allows us to formulate conclusion (C1) that answers the question "Who is a person?":

> **(C1)** Every human being is a person but not all persons are necessarily human beings.

(C1) can now be combined with (FP), that is, the foundational premise of this chapter as introduced at its beginning:

> **(FP)** A being B can possess moral rights and duties only if B is a person.

Combining (C1) and (FP) allows us to conclude that there is no stage of human life during which human beings can be said to not possess moral duties and rights. So, there is no justification to exclude any human being from the class of beings who can possess moral duties and rights. There might be phases in the life of a human being during which he might not be able to exercise all of his moral rights or meet all of his moral duties. But not being able to exercise a moral right or fulfil a moral duty is different from not possessing it. So, while it might be possible to hold a human being exempt from a specific moral duty or right, it is not possible to state that he cannot be subject to it due to a purported lack of personhood.

3. The Dignity of Human Beings

Besides making a being the subject of moral duties and rights, personhood has another feature. It also is a value-granting property that grounds a moral right of its own. Persons possess a special kind of value named "dignity" that sets limits on what is and is not permitted in dealing with a person. To be precise, dignity grounds the negative moral claim-right to have one's dignity as a person respected. In order to understand how this moral right comes about and what it entails, we must take a look at the concept of dignity and the different facets of dignity. So, what is human dignity?

3.1. Facets of Dignity[13]

Dignity is not something accessible to empirical measurements or something that can be found in our body such as a bone or an organ. Rather, it refers to a property or properties which "cause one to excel, and thus elicit or merit respect from others" (Lee & George 2008b: 410) and by which one is considered worthy, honourable or estimable. As a special kind of value, it thus implies and connotes terms such as worthiness for honour or esteem, elevation, excellence and distinction.

Value as opposed to disvalue generally indicates positive importance, whereby the term "importance" designates "that quality or characteristic of a thing that makes it not neutral and, lifting it out of neutrality, provides the ground of meaningful motivation" (Seifert 1997: 95). Dignity therefore conveys "a special standing for the beings that possess or display it [...] something elevated, something deserving of respect" (Kass 2008: 308). Human dignity refers to those properties of man which lift him out of neutrality and elicit or merit respect from others.

So far so good. But what is it that gives the bearer of human dignity special worth and standing? What is it that makes human beings worthy of our respect? Generally speaking, there are four distinct kinds of dignity that human beings can possess (cf. Fig. 2): The dignity of human beings can be divided into ontological and contingent dignity; the latter kind of dignity can be further distinguished into inflorescent as well as attributed dignity; inflorescent dignity can be further subdivided into dignity of actual rational consciousness and moral dignity.

These four facets of human dignity shall be explicated in the course of the following paragraphs.

- ***Ontological** (also: inherent, intrinsic) **dignity***
  The first form of human dignity—ontological dignity—is grounded in the personhood of human beings. A being B is a person P if B shows any signs of life and belongs to a biological species whose typical members naturally actualise the capability rationality in an empirically observable fashion. As all living human beings fit this description, all living human beings are persons from the moment of conception (although not all persons are necessarily human beings).

[13] The content of this chapter draws on, and in part reproduces verbatim, material published in Erk (2019: 211–222).

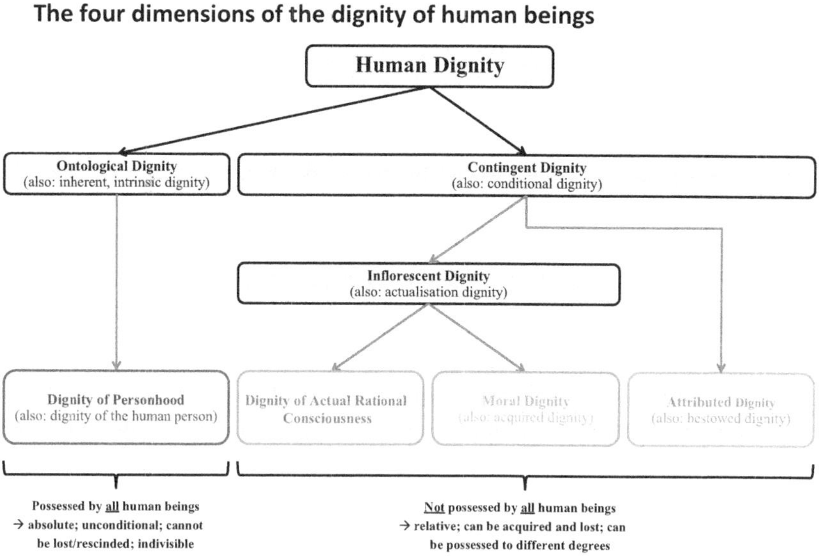

Fig. 2 The dignity of human beings

It is the intellectual aristocracy of possessing the capacity for rationality that lifts man out of the neutral and gives him an "inherent, intrinsic preciousness" (Seifert 1997: 96). This is why the term "person" is a ***nomen dignitatis***, that is, a dignity-conferring name: "To be a person, whether healthy or sick, whether male or female, whether old or young, whether conscious or in coma—is the first ground of human dignity" (Seifert 2003).

Ontological dignity signifies an inherent respect-worthiness which is absolutely independent from our subjective preferences, that is, our subjective likes or dislikes. While some things may be lifted out of the neutral only subjectively (i.e. insofar as they, e.g., please or displease us or are agreeable or disagreeable to us), persons are objectively lifted out of the neutral by the in-dwelling positive importance arising from their personhood—this is why the dignity of personhood is also called ontological, intrinsic or inherent dignity.

Ontological dignity is possessed by all (living) human beings of all times and at all times. Its possession is not contingent on age, consciousness, illness or our subjective inclinations; it is an inherent endowment of all human beings that is absolute, unconditional and indivisible and that cannot be lost or rescinded.[14]

As long as human beings live, they possess ontological dignity. Accordingly, no state of existence can be devoid of ontological dignity, not even the state of being sick, suffering or dying. No matter how a person dies, there can be no death without (ontological) dignity. There can only be death in conditions that do not do justice to the human person as an intrinsically valuable being, by, for example, leaving her alone with her pain and fear of death.

- ***Inflorescent dignity***

  Sometimes dignity is used to refer to a state of flourishing. This kind of dignity—inflorescent dignity—is possessed by human beings who are flourishing as human persons, that is, who are actually living lives that are consistent with and expressive of their intrinsic dignity. While ontological dignity is based upon endowment, inflorescent dignity is based on performance.

  - On the one hand, this flourishing consists in the actualisation of the person's capability for rational behaviour, a potential that grounds her ontological dignity. The actualisation of this potential mainly consists in the performance of rational acts, such as self-awareness, self-consciousness, self-distance, reasoning, reflection, valuation of existence, valuation of the future, communication, agency, memory, second-level preferences and autonomy.

    This facet of dignity is called the ***dignity of awakened personhood*** or the ***dignity of actual rational consciousness***. In contrast to the first source of human dignity, this second source of dignity is not possessed by all persons and human beings. There are persons with and persons without this kind of dignity. The human zygote, embryo or foetus do not yet possess dignity of awakened personhood, a comatose person has—either temporarily or permanently—lost this aspect of dignity. Therefore, the dignity of awakened personhood cannot be considered inalienable.

[14] Although Aquinas seems to have thought otherwise (cf. II^{a}-IIae q. 64 a. 2 ad 3), I agree with Grisez (1970: 69) that sinning does not make human beings lose their ontological dignity: "A person does in some sense degrade himself by his wrongdoing. Yet such self-degradation, even if it is conceived as a kind of existential suicide, cannot alter one's human nature or detract from one's inherent dignity as a human person".

- On the other hand, inflorescent dignity consists in what can be called ***moral dignity*** (also: acquired dignity). This aspect of inflorescent dignity is all about the way a person utilises her (awakened) personhood. As such, moral dignity is not automatically possessed by all human beings; it has to be acquired through morally good conduct. Moral dignity is not inalienable and can be lost.

- ***Attributed** (also: bestowed) **dignity***

  This last aspect of dignity refers to the value "that human beings confer upon others by acts of attribution" (Sulmasy 2008: 473). Attributed dignity results from holding and performing social roles and functions (e.g. the dignity of a judge). It is a created, conventional and subjective form of value. Bestowed dignity also comprises dignity that results from gifts not attributed by man (i.e. natural gifts, e.g. beauty, intelligence, genius or charm, strength of character) or gifts attributed by God (e.g. dignity of a religious office, grace).

  Attributed dignity is not inalienable and there are innumerable degrees of attributed dignity according to the respective talents, roles, functions and so on it is based on. Not only can it vary within a society, but also between societies, which might attribute a different "amount" of social dignity to social roles such as judges or teachers.

In summary, the dignity of human beings is a fourfold concept that comprises ontological dignity, dignity of actual rational consciousness, moral dignity and attributed dignity. The fact that all forms of dignity but ontological dignity are relative, that is, can be gained, lost and possessed to varying degrees, both over time and between persons, allows us to conclude two things:

- No person—and therefore no human being—can fall below a certain level of dignity, namely the level determined by ontological dignity.
- Apart from that, virtually no two persons—and therefore no two human beings—possess the same level of dignity.

These insights are also depicted in Fig. 3.

When it comes to the notion of levels of dignity, this must not be interpreted to mean that a human being's level of dignity can actually be calculated. But although we cannot put a number on the dignity of human beings, we have to accept the idea that not all human beings share the same level of dignity. When we say that all human beings are equal in

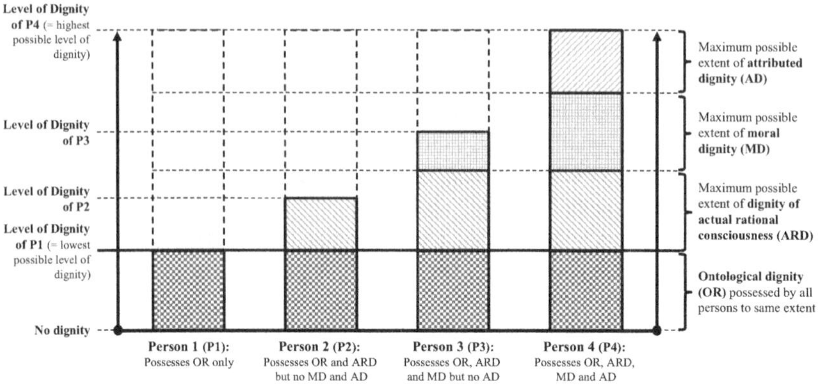

Fig. 3 The dimensions and levels of dignity

dignity this only refers to their ontological dignity; this kind of dignity is indeed possessed by all human beings to the same degree. But apart from their ontological dignity as persons, the dignity of human beings can vary from human being to human being. Nevertheless, the concept of differing degrees of dignity—although sound from a theoretical perspective—should be applied in our mutual dealings with due caution.

3.2. Respecting Dignity

So far, we have enquired into the nature and facets of the dignity that human beings can possess. We have learned that dignity is essentially a non-neutral, positive and exalted value that signifies positive importance. As such, it asks us to respond to it adequately. The positive value we call dignity calls for an affirmation on our part. It is not left to our arbitrary decision how to behave towards the dignity of a human being. Rather, human dignity exerts a sublime demand on our behaviour; it requires us to accord some appreciative form of recognition to its bearer. The recognition required to be accorded to the bearers of dignity is respect. In short, dignity is something deserving of respect.

> One of the central concerns of moral reasoning is that we recognise and respect the dignity and worth of persons. It can be argued that it is this concern which underlies the idea that persons possess certain inalienable rights, even in the absence of those rights' being acknowledged by society. (Larmer 1995: 241)

> Dignity is bound up with respect: to say that people have dignity is to say that they ought to be respected. (Dan-Cohen 2002: 290)

The dignity of human beings demands to be respected. However, as long as we do not know what is meant by respect, we cannot know what respect for the dignity of human beings actually requires. So, what is respect? The term respect is derived from the Latin verb *respicere*, which can be translated to "to look (back) at", but also "to regard" or "to consider". Respect is not only an attitude or feeling (such as esteem or deference), but rather a proper regard for or recognition of something that translates into action. Detailing the concept, Feinberg (1973) has identified three distinct aspects which have been associated with the term "respect":

- Firstly, there is the "respekt"-aspect of respect. Feinberg (1973: 1) defines this kind of respect as an "uneasy and watchful attitude that has 'the element of fear' in it". Its objects are "dangerous things or things with power over the subject" (Dillon 2018). It is contrasted with contemptuous disregard. It is "respekt" that a surfer has for a monster wave or that woodworkers are encouraged to have for power tools. However, "respekt" is not to be mistaken for fear (which might be a result of it); rather, it is the awareness of the graveness of an action or situation and is not meant to hamstring us but lift our attention.
- The second aspect of respect according to Feinberg is "observantia". Observantia-respecting an object means to regard it "as making a rightful claim on our conduct, as deserving moral consideration in its own right" (Dillon 2018). Observantia is respect in a practical sense, which guides our actions and requires us to behave in a certain kind of way.
- "Reverentia", the third aspect of respect, is "the special feeling of profound awe and respect we have in the presence of something extraordinary or sublime, a feeling that both humbles and uplifts us" (Dillon 2018). Reverentia is a distinctive positive feeling of deference that one has in the presence of something one considers as having exalted value.

If we think about these three aspects of respect, we will soon see that reverentia and "respekt" are two sides of the same coin: Both signify feelings, the former positive and uplifting, the latter rather cautious and alerting. Observantia, on the other hand, is more than a feeling: It motivates and calls for action.

When we say that the dignity of human beings necessitates or demands respect, we can only mean that it demands observantia-respect, that is, that it must be respected by a certain form of behaviour. There can be no claim to experience a feeling, but there can be a claim to behave in a certain way—even if such behaviour would be contrary to our actual feelings. So, when thinking about respect for the dignity of human beings, what we are thinking about is observantia-respect.

Possessing dignity means to be entitled to observantia-respect. Those who possess dignity possess a claim-right to have their dignity observantia-respected. They have the claim-right to be treated in a way that does not violate their dignity. This claim-right is, however, not the right to *equal* concern and respect; it does not entitle all human beings to be treated in an equal fashion. Rather, it is the claim-right to be treated according to one's actual level of dignity. While all human beings possess the same degree of ontological dignity, not all human beings possess the same level of dignity of awakened personhood, moral dignity and/or attributed dignity. So, while all human beings are entitled to have their ontological dignity observantia-respected, a human being can only claim to have the other dimensions of his dignity observantia-respected if and insofar as he has them. A human being's claim-right to have his dignity observantia-respected corresponds with a duty on the part of all human beings, namely the duty to take their hats off to and observantia-respect the dignity of all human beings.

At this stage it is worth noting that the moral rights grounded in human dignity are usually called "human rights". The class of human rights therefore comprises all those moral rights that are grounded in the dignity of human beings. So far so good. However, things get a bit more complicated by the fact that—by their very nature—moral rights have to be both universal and general, that is, rights that are held by all human beings of all times at all times and against all human beings.

Bearing this in mind, it should be obvious that not all dimensions of human dignity can actually ground a moral right to have one's human dignity observantia-respected. For not all dimensions of dignity are necessarily possessed by all human beings. In a strict sense, the only dignity-based human right there is is a human being's claim-right to have his ontological dignity observantia-respected. Contingent dignity can only be the foundation for human rights understood in a rather broad and non-strict sense.

Bibliography

Beckwith, Francis J. "The Human Being, a Person of Substance: A Response to Dean Stretton." In: Napier, Stephen (Ed.). *Persons, Moral Worth, and Embryos: A Critical Analysis of Pro-Choice Arguments.* Dordrecht, Heidelberg, London & New York: Springer, 2011. pp. 67–83. (Series "Philosophy and Medicine", Volume 111)

Blechschmidt, Erich. *The Beginnings of Human Life.* New York: Springer-Verlag, 1977.

Blechschmidt, Erich. *The Ontogenetic Basis of Human Anatomy: A Biodynamic Approach to Development from Conception to Birth.* Revised, edited, and translated by Brian Freeman. Berkeley, CA: Atlantic Books, 2004.

Boëthius, Anicius Manlius Severinus. "Liber contra Eutychen et Nestorium." In: Boëthius, Anicius Manlius Severinus. *The Theological Tractates and the Consolation of Philosophy.* Latin/ English Edition. Translated and edited by Hugh Fraser Stewart & Edward Kennard Rand. London: The Loeb Classical Library, 1918. Available at: http://www.gutenberg.org/etext/13316 Accessed: March 28, 2022.

Condic, Maureen L. "When Does Human Life Begin? The Scientific Evidence and Terminology Revisited." *The National Catholic Bioethics Quarterly* 9.1 (2009): 129–149. Originally published in 2008 by The Westchester Institute for Ethics & the Human Person as part of their White Paper Series (Volume 1, Number 1).

Condic, Maureen L.. "When Does Human Life Begin? The Scientific Evidence and Terminology Revisited." *University of St. Thomas Journal of Law and Public Policy* VIII.1 (2013): 44–81.

Dan-Cohen, Meir. "The Value of Ownership." In: Dan-Cohen, Meir. *Harmful Thoughts: Essays on Law, Self, and Morality.* Princeton, NJ: Princeton University Press, 2002. pp. 264–301. Originally published as: Dan-Cohen, Meir. "The Value of Ownership." *Journal of Political Philosophy* 9.4 (2001): 404–434.

de Locht, Pierre. "Discussion." In: *L'Avortement: Actes du Xeme Colloque International de Sexologie. 2ème partie.* Louvain: Centre International Cardinal Suenens, 1968.

Dillon, Robin S. "Respect." *The Stanford Encyclopedia of Philosophy.* February 18, 2018. Available at: http://plato.stanford.edu/entries/respect Accessed: March 28, 2022.

Eberl, Jason T. & Brandon P. Brown. "Brain Life and the Argument from Potential: Affirming the Ontological Status of Human Embryos and Fetuses." In: Napier,

Stephen (Ed.). *Persons, Moral Worth, and Embryos: A Critical Analysis of Pro-Choice Arguments.* Dordrecht, Heidelberg, London & New York: Springer, 2011. pp. 43–66. (Series "Philosophy and Medicine", Volume 111)

Erk, Christian. *Rationierung im Gesundheitswesen. Eine wirtschafts- und sozialethische Analyse der Rationierung nach Selbstverschulden.* Berlin & Boston, MA: de Gruyter, 2015.

Erk, Christian. *Moral Philosophy. Fundamental Concepts.* Second, revised and expanded Edition. Heusenstamm: editiones scholasticae, 2019.

Feinberg, Joel. "Some Conjectures on the Concept of Respect." *Journal of Social Philosophy* 4.2 (1973): 1–3.

Feser, Edward. *Scholastic Metaphysics. A Contemporary Introduction.* Heusenstamm: editiones scholasticae, 2014.

Finnis, John. "The Fragile Case for Euthanasia: A Reply to John Harris." In: Keown, John (Ed.). *Euthanasia Examined: Ethical, Clinical, and Legal Perspectives.* Cam-bridge: Cambridge University Press, 1995. pp. 46–55.

Giubilini, Alberto & Francesca Minerva. "After-birth abortion: why should the baby live?" *Journal of Medical Ethics* 39.5 (2013): 261–263.

Grisez, Germain Gabriel. "Toward a Consistent Natural-Law Ethics of Killing." *American Journal of Jurisprudence* 15.1 (1970): 64–96.

Grisez, Germain. "When do People Begin?" *Proceedings of the American Catholic Philosophical Association* 63 (1989): 27–47.

Harris, John. *The Value of Life: An Introduction to Medical Ethics.* London: Routledge, 1985.

Hüntelmann, Rafael. *Grundkurs Philosophie VI: Natürliche Ethik.* Neunkirchen-Seelscheid: editiones scholasticae, 2017. (Reihe Philosophie des gesunden Menschenverstandes, Band 6).

Iglesias, Teresa. "In vitro fertilisation: the major issues." *Journal of Medical Ethics* 10 (1984): 32–37.

Jensen, Steven J.. *The Human Person. A Beginner's Thomistic Psychology.* Washington, D.C.: Catholic University of America Press, 2018.

Joyce, Robert E. "Personhood and the Conception Event." In: Goodman, Michael F. (Ed.). *What is a Person?* Clifton, NJ: The Humana Press, 1988. pp. 199–212. Originally published in: *The New Scholasticism* 52 (1978): 92–109.

Kass, Leon R. "Defending Human Dignity." In: The President's Council on Bioethics (Ed.). *Human Dignity and Bioethics. Essays Commissioned by the President's Council on Bioethics.* Washington, D.C.: The President's Council on Bioethics, March 2008. pp. 297–331. Available at: https://bioethicsarchive.georgetown.edu/pcbe/reports/human_dignity Accessed: March 28, 2022.

Kitwood, Tom M.. *Dementia Reconsidered: The Person Comes First.* Buckingham: Open University Press, 1997.

Laitinen, Arto. "Interpersonal Recognition: A Response to Value or a Precondition of Personhood?" *Inquiry* 45 (2002): 463–478.

Laitinen, Arto. "Sorting Out Aspects of Personhood Capacities, Normativity and Recognition." *Journal of Consciousness Studies* 14.5–6 (2007): 248–270.

Larmer, Robert. "Abortion, Personhood and the Potential for Consciousness." *Journal of Applied Philosophy* 12.3 (1995): 241–251.

Lee, Patrick. "Substantial Identity and the Right to Life: A Rejoinder to Dean Stretton." *Bioethics* 21.2 (2007): 93–97.

Lee, Patrick & Robert P. George. *Body-Self Dualism in Contemporary Ethics and Politics.* Cambridge: Cambridge University Press, 2008a.

Lee, Patrick & Robert P. George. "The Nature and Basis of Human Dignity." In: The President's Council on Bioethics (Ed.). *Human Dignity and Bioethics. Essays Commissioned by the President's Council on Bioethics.* Washington, DC: The President's Council on Bioethics, 2008b. pp. 409–433. Available at: https://bioethicsarchive.georgetown.edu/pcbe/reports/human_dignity Accessed: March 28, 2022.

Locke, John. *An Essay Concerning Human Understanding.* London: 1690. Available at: http://oll.libertyfund.org/title/762 Accessed: March 28, 2022.

Lockwood, Michael. "When does a life begin?" In: Lockwood, Michael (Ed.). *Moral Dilemmas in Modern Medicine.* Oxford & New York: Oxford University Press, 1985. pp. 9–31.

Maguire, Marjorie Reiley. "Symbiosis, Biology, and Personalization." In: Doerr, Edd & James W. Prescott. *Abortion Rights and Fetal 'Personhood'.* Long Beach, CA: Centerline Press, 1989. pp. 11–19.

Marquis, Don. "Why Abortion is Immoral." *The Journal of Philosophy* 86.4 (1989): 183–202.

McMahan, Jeff. *The Ethics of Killing: Problems at the Margins of Life.* Oxford: Oxford University Press, 2002.

Moore, Keith L., T. V. N. Persaud & Mark G. Torchia. *The Developing Human: Clinically Oriented Embryology.* Eleventh Edition. Edinburgh et al.: Elsevier, 2020.

O'Donovan, Oliver. "Again: Who is a Person." In: Channer, John Hugh (Ed.). *Abortion and the Sanctity of Human Life.* Exeter: The Paternoster Press, 1985, pp. 125–137.

Oderberg, David S.. *Real Essentialism.* London: Routledge, 2007.

Oderberg, David S.. *The Metaphysics of Good and Evil.* New York & London: Routledge, 2020.

Parfit, Derek. *Reasons and Persons.* Oxford: Oxford University Press, 1984.

Penrod, Janice, Fang Yu, Ann Kolanowski, Donna M. Fick, Susan J. Loeb & Judith E. Hupcey. "Reframing Person-Centered Nursing Care for Persons with Dementia." *Research and Theory for Nursing Practice* 21.1 (2007): 57–72.

Reichlin, Massimo. "The Argument from Potential: A Reappraisal." *Bioethics* 11.1 (1997): 1–23.

Sadler, T.W.. *Langman's Medical Embryology*. Fourteenth Edition. Philadelphia: Wolters Kluwer, 2019.

Scanlon, Thomas. *What We Owe to Each Other.* Cambridge, MA: Harvard University Press, 1998.

Schoenwolf, Gary C., Steven B. Bleyl, Philip R. Brauer & Philippa H. Francis-West. *Larsen's Human Embryology.* Fifth Edition. Edinburgh et al.: Elsevier, 2015.

Seifert, Josef. *What is Life? The Originality, Irreducibility and Value of Life.* Amsterdam and Atlanta, GA: Rodopi, 1997. (Value Inquiry Book Series (VIBS), Volume 51)

Seifert, Josef. "The right to life and the fourfold root of human dignity." In: de Dios Vial Correa, Juan & Elio Sgreccia (Eds.). *The Nature and Dignity of the Human Person as the Foundation of the Right to Life. The Challenges of the Contemporary Cultural Context. Proceedings of the VIIIthAssembly of the Pontifical Academy for Life (Vatican City, 25–27 February 2002).* Vatican City: Libreria Editrice Vaticana, 2003.

Singer, Peter. *Animal Liberation: A new Ethics for our Treatment of Animals.* New York: New York Review/ Random House, 1975.

Singer, Peter. *Practical Ethics.* Third Edition. Cambridge: Cambridge University Press, 2011.

Singer, Peter & Helga Kuhse. "Ethics and the Limits of Tolerance." *The Journal of Medicine and Philosophy* 19 (1994): 129–145.

Solomon, Robert C. "Reflections on the Meaning of (Fetal) Life." In: Bondeson, Willi-am B., Tristram H. Engelhardt, Stuart F. Spicker & Daniel H. Winship (Eds.). *Abortion and the Status of the Fetus.* Dordrecht, Boston & Lancaster: Reidel, 1983. pp. 209–226. (Series "Philosophy and Medicine", Volume 13)

Spaemann, Robert. "Begotten, Not Made." *Communio: International Catholic Review* 33 (Summer 2006): 290–297.

Sulmasy, Daniel P. "Dignity and Bioethics: History, Theory, and Selected Applications." In: The President's Council on Bioethics (Ed.). *Human Dignity and Bioethics. Essays Commissioned by the President's Council on Bioethics.* Washington, DC: The President's Council on Bioethics, 2008. pp. 469–501. Available at: https://bioethicsarchive.georgetown.edu/pcbe/reports/human_dignity Accessed: March 28, 2022.

Tooley, Michael. "Abortion and Infanticide." *Philosophy & Public Affairs* 2.1 (1972): 37–65.

Warren, Mary Anne. *Moral Status: Obligations to Persons and Other Living Things.* Oxford: Clarendon Press, 1997.

Wiggins, David. *Sameness and Substance.* Oxford: Oxford University Press, 1980.

VI: The General Ethics of Killing Human Beings

It took a little bit of preparation, but we are now ready to assess the ethical quality and (im)permissibility of the killing of human beings. What have we learned so far? In the chapters preceding the two interludes, we have dealt with the elements, the voluntariness, the sources of morality as well as the standard of morality of actions. We have also familiarised ourselves with the two basic kinds of killing actions, namely essential and accidental killing actions.

Combining the insights gained so far allows us to take the next step and assess the general ethical (im)permissibility of such essential killing actions on the one hand and accidental killing actions on the other hand. So, this chapter will deal with the following questions:

- Under which conditions, if any, is it morally permissible to perform an essential killing action?
- Under which conditions, if any, is it morally permissible to perform an accidental killing action?

The answer to these questions provides us with a general ethical theory of killing human beings. This theory will then be used to evaluate the concrete killing actions that are at the centre of this book, namely suicide, lethal self-defence, abortion, euthanasia, organ transplantation and cooperation in suicide. But first things first.

C. Erk, *The Ethics of Killing*,
https://doi.org/10.1007/978-3-031-07183-6_6

1. The General Ethics of Essential Killings

Let us start with the examination of ***the general moral (im)permissibility of essential killings***. The question we are trying to answer reads as follows:

> In principle, is an essential killing action morally permissible or is it impermissible?

The comprehensive preliminary work that has been done over the course of the preceding chapters provides us with the building blocks necessary to answer this question.

In general, an action is ethically impermissible if there is a moral duty to refrain from performing this action; it is permissible if there is not such duty. So, in order to know whether an essential killing action is ethically permissible or impermissible we need to find out whether there is a negative moral duty to not perform this action. As we have learned, the answer to this question depends on the action's ethical quality. Accordingly, we need to take a closer look at the general ethical quality of essential killing actions. Once we have done so we can then turn to the problem of these actions' general (im)permissibility.

However, before dealing with the morality of essential killings we must first remind ourselves of the elements and sources of the morality of actions as well as the nature of essential killing actions. This reminder is given by means of the following four premises:

P0	*In an essential killing, the death of a human being is the natural end (and therefore also essential consequence) of the killing action's exterior action.*
P1	An action consists of four elements, namely the end of the agent (*finis operantis*), the exterior action and its natural end (*finis operis*), the circumstances of the exterior action and the consequences of the exterior action.
P2	Albeit to varying degrees, all elements of an action can be a source of morality insofar as they are deliberately willed, that is, directly voluntary.
P3	An action is evil in its entirety if the directly voluntary natural end of its exterior action (*finis operis*) is evil, regardless of the ethical quality of the action's other elements.

Based on these premises we can now start to investigate the general morality of essential killings. This will be done by translating the insights gained over the course of the preceding chapters into further premises and by combining the above as well as the additional premises into intermediate conclusions and, finally, into the final conclusion that answers the question of the (im)permissibility of essential killing actions.

P4	*The end of the exterior action of an action is evil if it violates the standard of morality.*
	+ P3
→ C1	An action whose natural end of the exterior action violates the standard of morality in a directly voluntary fashion is an evil action.
+ P5	Amongst others (the common good, the virtues and the natural ends of the faculties of human beings), a fundamental part of the standard of morality is the basic human goods (BHGs).
→ C2	An action whose natural end of the exterior action violates a basic human good in a directly voluntary fashion is an evil action.
+ P6	When performing an exterior action, an agent necessarily wills the exterior action's natural end in a directly voluntary fashion.
→ C3	An agent who performs an exterior action whose natural end is the violation of a basic human good commits an evil action.
+ P7	Life is a basic human good, even the most fundamental basic human good.
→ C4	An agent who performs an exterior action whose natural end is the violation of the basic human good life commits an evil action.
	+ P0
→ C5	An agent who performs an essential killing performs an evil action.

As the above argumentative steps have shown, essential killing actions are evil because the natural end of their exterior action is evil and because an agent cannot not will the natural end of its exterior action to occur when performing such action. Willing the death of an innocent human being as the end of the exterior action (*finis operis*) suffices to make the overall human action evil—regardless of the ethical quality of the end of the agent, the exterior action's circumstances or the exterior action's consequences. The latter three elements can make the human action more or less evil, but they cannot alter the species of the human action's ethical quality from evil to good. Having established the general ethical quality of essential killing actions, we can now turn to the question of their general permissibility or impermissibility.

P8	*All human beings have the moral prohibition-duty D_{Good} to not perform an evil action in a directly voluntary fashion.*	
	+ C5	
→ C6	All human beings have the moral prohibition-duty to not perform an essential killing action.	

Having gone through these steps of argumentation, we have now arrived at an answer to the question we are interested in. We set out to find an answer to the question whether, in principle, essential killing actions are morally permissible or impermissible. As the above multi-level argument shows, the answer is clear:

> In principle, essential killing actions are morally impermissible.

The observant reader may interject that there would have been a more direct way to arrive at this conclusion. And this is somewhat correct. In Chap. II, Sect. 2.3 we have learned that all human beings possess the moral duty D_{BHG}. According to this duty, all human beings possess—amongst others—the negative moral duty to not do what hinders the realisation of a basic human good or violates a basic human good. Undoubtedly, life is a basic human good. In conclusion, all human beings possess—amongst others—the negative moral duty to not violate the basic human good life. And this duty is tantamount to the negative moral duty to not kill a human being. However, straightforward as it may be, this line of reasoning has the disadvantage that it does not explicitly talk of essential killings but killings in general.

2. The General Ethics of Accidental Killings

Having established the general moral impermissibility of essential killings, we can now turn to the ***general moral (im)permissibility of accidental killings***. So, the question that interests us now reads as follows:

> In principle, are accidental killing actions morally permissible or are they impermissible?

In order to answer this question, we must again first establish the general ethical quality of accidental killing actions. Once we have done so we can turn to the problem of the general (im)permissibility of this kind of killing action.

2.1. The General (Im)Permissibility of Accidental Killings

Again, we start our investigation by introducing some general premises concerning the elements and sources of the morality of actions as well as the nature of accidental killing actions:

P0	*In an accidental killing, the death of a human being is an accidental consequence of the killing action's exterior action, that is, results from the exterior action although it is not its natural end.*
P1	An action consists of four elements, namely the end of the agent (*finis operantis*), the exterior action and its natural end (*finis operis*), the circumstances of the exterior action and the consequences of the exterior action.
P2	Albeit to varying degrees, all elements of an action can be a source of morality insofar as they are deliberately willed, that is, directly voluntary.
P4	An unknowingly accepted accidental consequence of an exterior action does not add to the goodness or badness of the overall action.
P5	An indirectly voluntary, that is, knowingly accepted, accidental consequence of an exterior action can increase or decrease the goodness or badness of the overall action but cannot make an otherwise evil action good or an otherwise good action evil.

Given these general premises, we can now start to investigate the general ethical quality of accidental killings.

P6	*A consequence of an action is evil if it violates the standard of morality.*
+ P7	Amongst others (the common good, the virtues and the natural ends of the faculties of human beings), a fundamental part of the standard of morality is the basic human goods (BHGs).
+ P8	Life is a basic human good, even the most fundamental basic human good.
→ C4	A consequence of an action is evil if it violates the basic human good life.
	+ P0
→ C5	The death that results from the exterior action of an accidental killing as an accidental consequence is an evil consequence.

	+ P4		**+ P5**
→ C6	If the death that results from an accidental killing is an unknowingly accepted accidental consequence, the evilness of this consequence does not—in principle—add to the goodness or badness of the overall action.	**→ C7**	If the death that results from an accidental killing is an indirectly voluntary, that is, knowingly accepted, accidental consequence, the evilness of this consequence decreases the goodness or increases the badness of the overall action but cannot make an otherwise evil action good or an otherwise good action evil.

As these considerations show, accidental killings are not made evil by the mere fact that a human being has died as an accidental consequence of their exterior action. One must distinguish whether the accidental consequence is an indirectly voluntary or an unknowingly accepted consequence. If indirectly voluntary, such a consequence can have a non-specifying influence on the ethical quality of the action, that is, can make it better or worse but not good or evil. If unknowingly accepted, such a consequence does not influence the ethical quality of the overall action.

Regardless of the degree of voluntariness of their accidental consequence, accidental killings are, however, made evil if the natural end of their exterior action or the end of the agent is evil. What do these insights mean for the problem of whether accidental killing actions are morally permissible or impermissible? This is the question we are going to address next.

P9	*All human beings have the moral prohibition-duty D_{Good} to not perform an evil action in a directly voluntary fashion.*		
	+ *C6*		+ *C7*
→ **C8**	The fact that the death of a human being is an unknowingly accepted accidental consequence of the exterior action of an accidental killing, has—in principle (i.e. unless it should have been, for example, foreseen by the agent)—no influence on the permissibility of the overall action. An otherwise permissible action remains permissible and an otherwise impermissible action remains impermissible.	→ **C9**	The fact that the death of a human being is an indirectly voluntary, that is, knowingly accepted, accidental consequence of the exterior action of an accidental killing does not make an otherwise permissible action impermissible or an otherwise impermissible action permissible. It does, however, make an otherwise impermissible action more impermissible and an otherwise permissible action less permissible.

The above steps of argumentation have produced a first answer to the question of the (im)permissibility of accidental killings: While essential killing actions are morally impermissible in general, the fact that the death of a human being results as an accidental consequence of the exterior action of an accidental killing does not yet make this action impermissible or permissible as such.

So, unfortunately, there is no one-size-fits-all answer to the question of the moral (im)permissibility of accidental killings. Whether an accidental killing action is permissible in its entirety is not determined by the accidental consequence—regardless of whether the death of a human being results as an unknowingly or a knowingly accepted consequence. Rather, the permissibility of accidental killings must be assessed case by case based on a set of conditions that will be detailed in the next chapter.

2.2. Accidental Killings as Double Effect Actions[1]

When it comes to assessing the general (im)permissibility of accidental killing actions we have to bear in mind that an accidental killing action has at least two consequences:

- The first consequence of an accidental killing is the death of a human being that results from the exterior action either as an indirectly voluntary or an unknowingly accepted accidental consequence. As the death of a human being violates the basic human good life, the ethical quality of this consequence is undoubtedly evil.
  While the accidental consequence's evilness has no influence when it is an unknowingly accepted accidental consequence, an indirectly voluntary accidental consequence has a non-specifying influence that makes an otherwise evil action more evil, an otherwise good action less good and, therefore, an otherwise permissible action less permissible and an otherwise impermissible action more impermissible.
- The second consequence that necessarily results from an accidental killing action is the natural end of its exterior action that materialises as the necessary result of performing the exterior action. If the natural end of its exterior action is evil, the accidental killing action is evil as a whole. In this case, the accidental action is morally impermissible.

Given their nature as possessing at least two consequences, accidental killings are prototypical instances of actions that have what is commonly called a "double effect". What is usually characteristic of this kind of action

[1] The content of this chapter draws on, and in part reproduces verbatim, material published in Erk (2019: 240–245).

is that its exterior action's accidental consequence is evil (or at least morally neutral) while the natural end of the exterior action is good. This is also true of the accidental killing actions whose moral (im)permissibility is typically discussed in the ethical literature, such as the following examples:

- A military pilot is bombing military targets in warfare. By doing so he is not only destroying a strategically important target (e.g. an ammunitions factory or a bridge) but also causes collateral damage.
- A surgeon is removing a woman's cancerous womb during a pregnancy or a woman's fallopian tube during an ectopic pregnancy. By doing so he saves the woman's life but kills the foetus.
- A doctor prescribes morphine or some other kind of analgesic for one of her patients. By doing so she helps relieve her patient's pain but risks hastening the patient's death.

However, knowing that the accidental consequence of an accidental killing is evil (or at least morally neutral) and the natural end of its exterior action is good does not suffice to make a definitive statement about the ethical quality of the overall action. In its entirety, such an accidental killing action can either be evil or good. In order to find out whether such an action is good or evil as a whole, one has to not only consider the ethical quality of its exterior action but also the ethical quality of the action's end of the agent.

As we have learned in a previous chapter, the evilness of an action's end of the agent suffices to make the entire action evil. However, the goodness of its end of the agent is merely a necessary but not a sufficient condition for the overall goodness of an action. A good end of the agent increases the goodness of an otherwise good action but does not suffice to make the action good in its entirety. So, an accidental killing can only be a morally permissible action if the end of its agent is good. It is evil and therefore impermissible in its entirety if the end of its agent is evil.

So far, we have established two conditions for the moral permissibility of accidental killings:

- ***The exterior-action condition***: The exterior action of the accidental killing (as defined by its natural end) must be either morally good or at least morally indifferent.

- ***The end-of-the-agent condition***: The end of the agent performing the accidental killing action must be good. The death of the human being that results from the exterior action must not be the end of agent.

Taken together, these two conditions require that the evil consequence of the exterior action may only be accidental in nature. The death of the human being that results from the exterior action may only be an accidental consequence of the exterior action.

If an accidental killing action does not meet either one of these conditions, the action is evil in its entirety. However, if it meets both conditions there is still no guarantee that the overall action is morally good. There are two further conditions that must be met.

Even if the end of the agent of an accidental killing action is good, it would be problematic if the action's evil accidental consequence were the means to realise the end of the agent. It must not be the case that the agent achieves his purpose because of the evil accidental consequence; this must be achieved by means of the natural end of the exterior action. The fact that a human being has died as an accidental consequence must not play a causal role in the realisation of the end of the agent who performs an accidental killing action. This basically means that the good consequence must follow from the exterior action at least as causally immediately as the evil accidental consequence.[2] So, although it may occur after the evil accidental consequence, the good consequence, that is, the realisation of the natural end of the exterior action, must not result from the evil accidental consequence. In short, the evil accidental consequence of an exterior action must not be a means to the realisation of the good consequence. The third condition for the permissibility of an accidental killing can therefore be stated as follows:

- ***The means-end condition***: The evil accidental consequence of the exterior action must not be a means to realise the action's end of the

[2] It is important to bear in mind that "the notion of immediacy here is causal, not temporal" (Oderberg, 2010: 326). This condition does not enshrine a temporal but a causal order between the natural end of the exterior action, that is, the good consequence, and the evil accidental consequence. From a temporal perspective, the former can be and often is realised after the latter. However, the former must not be realised because of the latter. If the evil consequence were the cause of the good consequence, the agent would have to will the evil consequence directly to attain the good—which would make the action evil as a whole.

agent. Therefore, the good consequence, that is, the realisation of the natural end of the exterior action, must follow from the exterior action at least as causally immediately as the evil accidental consequence. The natural end of the exterior action must not be produced by the evil accidental consequence.

The fourth condition concerns the relation between the accidental killing action's good consequence and its evil accidental consequence. Even if the means-end condition is met, it would be problematic if the evil consequence were to somehow outweigh the good consequence. The permissibility of an action with a double effect also requires that the evil accidental consequence must be proportionate to the good consequence. If the good consequence is not at least equivalent in importance to the evil accidental consequence, this suffices to make the overall action impermissible. This gives us the fourth condition for the (im)permissibility of accidental killings:

- ***The proportionality condition***: There must be a proportionately grave reason for permitting the exterior action's evil accidental consequence. The action's good consequence must be at least equivalent in importance to the evil accidental consequence. It is—as Cronin (1922: 58) puts it—necessary that "the good effect produced in some way counterbalances the ever grave evil of death".

While the application of the first three conditions is rather straightforward, the proportionality condition requires more guidance. For proportionality is a concept that is hard to measure. Luckily, McHugh & Callan (1958: §105; cf. Prümmer 1961: 48; Noldin 1936: 91f) have compiled a list of rules that can help us decide whether there is a sufficiently grave reason for permitting an evil consequence: "To judge whether a reason for permitting an evil effect is proportionately grave, the following rules should be kept in mind:

- (a) the greater the evil that results, the greater must be the good that is intended. Thus, it is not lawful to kill a robber in order to save a small amount of money; but it is lawful to kill an aggressor, if this is necessary in order to save one's life;
- (b) the greater the dependence of the evil effect on one's act, the greater must be the reason for performing the act. Example: Titus gives permission to his class to play a game against another class, foreseeing quarrels and disputes between the teams. Less reason is

required for granting the permission, if Titus knows that higher authority will grant it, should he refuse it;

- (c) the more nearly the evil effect follows upon the act, the greater must be the reason for the act. Thus, less reason is required to direct a person who looks like a heavy drinker to the city than to direct him to a bottle of strong drink;
- (d) the more certain it is that the evil effect will follow, the greater is the reason required for placing its cause. For example, one who speeds in an automobile on an unfrequented road, does not require the same excusing cause as one who speeds on a thoroughfare where many other cars are passing;
- (e) the more obligation one has to prevent the evil effect, the graver is the reason required for placing its cause. Thus, since parish-priests, lawgivers, superiors and policemen are bound by their office to prevent moral disorders, a far greater cause is required in them, than in persons who have no such charge, for doing what will have an evil consequence.

Let us draw an interim conclusion at this point: Over the course of the preceding chapters we have established four conditions that help assess the ethical (im)permissibility of accidental killings[3]:

- The exterior-action condition
- The end-of-the-agent condition
- The means-end condition
- The proportionality condition

An accidental killing action can only be morally permissible if it meets all four of these conditions; if it does not meet all four of them, it is impermissible. In this context it does not matter whether the evil accidental consequence is an unknowingly accepted or an indirectly voluntary consequence. The only difference between these kinds of consequences is that the latter case makes an otherwise impermissible action more impermissible and an otherwise permissible action less permissible, while the former has no influence on the action's overall (im)permissibility.

[3] These conditions are in principle congruent with the conditions set out by the so-called "doctrine of double effect" (also: principle of double effect; rule of double effect). For an overview of different formulations of the doctrine of double effect cf. Erk (2019: 356–364).

Before continuing, there is one last thing on which we should go into more detail, namely the difference between foresight and intention. Accidental killings are characterised by the fact that the death that results as its exterior action's accidental consequence is not intended; it must only be foreseen. This is part of the definition of accidental killing. At first sight, this sounds pretty clear. But what does it actually mean to foresee but not intend a consequence? How do we know which is when?

In this respect Kaczor (2005) offers a list of criteria that can be used to determine whether a bad consequence is merely foreseen or premeditated, that is, foreseeable, foreseen and intended. The criteria are based on the general insight that "intention, and not foresight, characteristically gives rise to deliberation about means to be taken in order to achieve the end" (Kaczor 2005: 77). So, in order to learn whether a bad consequence is foreseen or intended we have to check whether and to what extent it makes the agent deliberate about the means to achieve it. This general condition can be broken down into four criteria (cf. Kaczor 2005: 77ff):

- *Deliberation*: If the achievement of the bad consequence presents a problem for the agent and prompts deliberation about the means to achieve this consequence, then the bad consequence is intended.
- *Constraint*: If the achievement of the bad effect constrains other intentions of the agent, that is, limits the options for which the agent can give consent, then the bad consequence falls within the domain of intention. So, if the agent is open only for options for action that do not contradict or even make the achievement of the evil consequence impossible, then the bad consequence is not merely foreseen but intended.
- *Endeavour*: A bad consequence is also intended if the agent endeavours to achieve it in a way that makes him deliberate about how to achieve it and forces him to return to deliberation if circumstances change.
- *Plan*: Intention can furthermore be distinguished from foresight according to how success is measured. If a failure to realise the bad consequence is a failure in the agent's plan, that is, if the agent considers the failed realisation of the evil consequence a failure, then the bad consequence is not foreseen but intended.

So, it is by the criteria of deliberation, constraint, endeavour and plan that we can set apart intention from foresight.

3. Summary

This chapter was devoted to establishing the general (im)morality and (im)permissibility of both essential killings and accidental killings. These are the answers that have been developed over the course of the preceding pages:

- With respect to the first kind of killings we have learned that essential killings are morally impermissible in general. Accordingly, there is a negative moral duty, that is, a prohibition-duty, that forbids all human beings to commit essential killings.
- When it comes to accidental killings, we have learned that the fact that the death of a human being results as an evil accidental consequence of an exterior action does not yet make this accidental killing impermissible or permissible as such.
  Whether an accidental killing is ethically permissible or impermissible has to be established by reference to four conditions, namely (a) the exterior-action condition, (b) the end-of-the-agent condition, (c) the means-end condition and (d) the proportionality condition. Only accidental killings that meet all four conditions are ethically permissible. If one of these conditions is not met, the action is ethically impermissible; in this case, all human beings possess a moral prohibition-duty that forbids them to perform this action.
  An accidental killing is characterised by the fact that the death of a human being is an accidental consequence of the killing action's exterior action. This means that this evil consequence must have merely been foreseen but not intended. Whether this additional condition is met can be checked by reference to the criteria deliberation, constraint, endeavour and plan.

These insights set out the rules and conditions that govern the general moral (im)permissibility of essential and accidental killing actions. These rules and conditions can now be applied to the concrete killing actions that are at the centre of this book, namely suicide, lethal self-defence, abortion, euthanasia, organ transplantation and cooperation in suicide. This task will be tackled in the following chapter.

Bibliography

Cronin, Michael. *The Science of Ethics. Volume II: Special Ethics.* New York et al.: Benziger Brothers, 1922.

Erk, Christian. *Moral Philosophy. Fundamental Concepts.* Second, revised and expanded Edition. Heusenstamm: editiones scholasticae, 2019.

Kaczor, Christopher. *The Edge of Life. Human Dignity and Contemporary Bioethics.* Dordrecht: Springer, 2005. (Series "Medicine and Philosophy", Volume 85; Subseries "Catholic Studies in Bioethics")

McHugh, John A. & Charles J. Callan. *Moral Theology: A Complete Course Based on St. Thomas Aquinas and the Best Modern Authorities.* Revised and enlarged by Edward P. Farrell. New York City: Joseph F. Wagner; London: B. Herder, 1958. Available at: http://www.gutenberg.org/ebooks/35354 Accessed: March 28, 2022.

Noldin, Hieronymus. *Summa Theologiae Moralis: I. De Principiis.* Editio XXIV. Oeniponte: F. Rauch, 1936.

Oderberg, David S.. "The Doctrine of Double Effect." In: O'Connor, Timothy & Constantine Sandis (Eds.). *A Companion to the Philosophy of Action.* Oxford: Wiley-Blackwell, 2010. pp. 324–330.

Prümmer, Dominicus M.. *Manuale Theologiae Moralis Secundum Principia S. Thomae Aquinatis in Usum Scholarum.* Tomus I. Editio Quinta Decima. Barcinone, Friburgi Brisgoviae & Romae: Herder, 1961.

VII: The Special Ethics of Killing Human Beings

Assessing the general (im)permissibility of essential and accidental killings has allowed us to set out the general ethics of killing actions. This ethical theory of killing human beings can now be applied to the selected killing actions that are at the centre of this book, namely suicide, lethal self-defence, abortion, euthanasia, organ transplantation and cooperation in suicide. So, while the previous chapter has provided us with the general ethical yardstick for the assessment of killing actions, this chapter zooms in on the ethics of selected cases of the killing of human beings and asks under which conditions, if any, the respective killings are ethically permissible. So, let's get started.

1. The Ethics of Killing Oneself (Suicide)

This chapter discusses the ethical (im)permissibility of suicide. In order to be able to do so it is necessary to define this specific kind of killing action. In a general sense, suicide could be taken as any killing of oneself. However, this definition would be too broad. In a stricter and ethically more correct sense, suicide is the directly voluntary killing of oneself by one's own authority. To make it a case of suicide, a human being must intend his death to occur as his exterior action's essential consequence or as an intended accidental consequence. This means that, by definition, a suicide is an essential killing action.

C. Erk, *The Ethics of Killing*,
https://doi.org/10.1007/978-3-031-07183-6_7

- If a human being performs an exterior action from which his death follows as a foreseen and unavoidable but unintended accidental consequence, this is not a case of suicide. Such accidental self-killing is not what is meant by suicide. It is true that an agent puts his life in danger in an accidental self-killing, but putting one's life in danger without intending to kill oneself is not the same as suicide.
- A killing of oneself cannot be suicide unless it is committed on one's own authority. A samurai who has been sentenced to death and is allowed (e.g. in order to regain his honour) to carry out his punishment by "Seppukku", that is, self-inflicted disembowelment, does not commit suicide. And neither did Socrates commit suicide by refusing to escape but rather executing his death sentence and drinking the poison hemlock cup as condemned at trial.

Suicide can be committed by commission or omission; the exterior action of a suicide can either be a positive or a negative action. "Suicide can be committed positively or negatively: positively, by the performance of some death-dealing act against oneself; negatively, by omitting to use the ordinary means of preserving one's life" (Fagothey 1963: 236). Shooting oneself, taking poison, starving oneself to death[1] or refusing to step out of the way of an oncoming train—all of these are instances of suicide by

[1] In general, the action of starving oneself to death is referred to as "voluntary stopping of eating and drinking (VSED)" (also: voluntary refusal of food and fluids (VRFF)). In VSED, a competent person who is still physically able to take nourishment, voluntarily decides to stop eating and drinking in order to end her life (usually due to a persisting unacceptable suffering). So, the refusal of nutrition is not just a natural symptom of an underlying disease, the result of a mental illness, the corollary of an already ongoing dying process or the result of a condition that makes the intake of nourishment overly burdensome. VSED is also different from the refusal or removal of artificial nutrition.

Strictly speaking, a person engaging in VSED does not starve to death but dies of terminal dehydration or some intervening complication. Depending on the person's pre-existing condition, death will usually set in after one to three weeks (cf. Chabot & Walther 2017: 47f), mostly due to renal failure; if the person does not strictly abstain from fluids, the process can take longer. However, if death occurs within the first seven days, the cause of death is likely to be an underlying disease.

Although it has unique features (in contrast to other kinds of suicide, it is non-invasive, initially reversible, results from an omission and resembles the natural dying process), VSED is suicide. Therefore, it is morally impermissible for a person to engage in VSED. However, this does not automatically mean that—although an argument could be made for the palliative care of the symptoms of VSED—other persons are allowed to invade her body with artificial forms of nutrition or forced feeding.

commission or omission. With respect to their ethical character, "there is no difference either in kind or degree between negative and positive suicide. In both there is a positive aiming at death. The difference is only in the means chosen" (Cronin 1922: 52).

Now, what about the ethical (im)permissibility of suicide? Given the general ethics of killing actions detailed in the previous chapter, the answer is rather straightforward:

- By definition, suicide is an essential killing action.
- Essential killing actions are ethically impermissible.
- Therefore, suicide is ethically impermissible.

As human beings possess the moral duty to preserve and not violate the basic human good life, they are morally not allowed to commit suicide. Suicide violates our most fundamental natural inclination, namely the natural inclination to preserve ourselves in existence[2]:

> Each thing strives by nature to preserve its being. Each thing therefore has a natural inclination, which in man becomes mental love, to preserve himself in existence and to avoid the destructive to offer resistance to influences as far as possible. So, man too has this natural inclination or desire to preserve his life; and hence comes the natural fear of death and of every danger against bodily life, because the destruction of it is a natural evil. This inclination, which springs from the natural order, thus becomes a natural duty for man endowed with reason and free will, because natural law requires the preservation of the natural order. Therefore, man has the duty to preserve his life. (Commer 1886: 141f)[3]

[2] This is why authors such as Thomas Aquinas (II^{a}-IIae q. 64 a. 5 co.), Prümmer (1960: 106), Cathrein (1905: 224) or Commer (1886: 141f) hold that suicide is contrary to love of oneself ("caritas sui ipsius").

[3] In its original German version this quotation reads as follows: "Jedes Ding strebt von Natur nach der Erhaltung seines Seins. Jedes Ding hat daher ein naturgemässes Begehren, was beim Menschen zur geistigen Liebe wird, sich selbst im Dasein zu erhalten und den zerstörenden Einflüssen Widerstand zu bieten, so weit es möglich ist. Also hat auch der Mensch dieses Naturstreben oder Begehren, sein Leben zu erhalten; und daher kommt die natürliche Furcht vor dem Tode und vor jeder Gefahr gegen das leibliche Leben, weil die Zerstörung desselben ein natürliches Uebel ist. Diese Neigung, die aus der Naturordnung entspringt, wird also für den mit Verstand und freiem Willen begabten Menschen zur natürlichen Pflicht, weil das Naturgesetz die Erhaltung der Naturordnung gebietet. Der Mensch hat also die Pflicht, sein Leben zu erhalten" (Commer 1886: 141f).

Given the general ethics of killing actions introduced in the previous chapter, the negative assessment of the ethical permissibility of the essential killing action named suicide should be comprehensible and come as no surprise.

The problem in dealing with suicide from an ethical perspective is not so much the ethical assessment of suicide itself but being able to distinguish suicide from killings of oneself that are not suicide. Two such kinds of self-killing are of special interest:

- The moral duty to not kill a human being has two aspects: It comprises the negative moral duty to refrain from doing what kills a human being and the positive moral duty to do what is necessary to preserve a human being's life. While the first of these duties is rather straightforward, the positive nature of the second duty raises some practical questions. For how far do we have to actually go to preserve our own lives? Does this duty require us to do everything in our power to preserve our lives? These and similar questions have to be answered by recourse to the distinction between the ordinary and extraordinary means of the preservation of life.
- As has been mentioned above, we must distinguish suicide from self-killings that resemble suicide but actually are instances of a self-endangerment that results in death. What is characteristic of these cases of accidental self-killings is that the death of the agent does not result as an essential consequence from his exterior action but merely as an unintended, foreseen and unavoidable accidental consequence. While the ethical impermissibility of the essential killing action suicide has been established, we need to have a closer look at the problem of the ethical (im)permissibility of such accidental self-killings.

The following section will familiarise the reader with the distinction between the ordinary and extraordinary means of preserving one's life. Subsequently, the second section will provide us with the ethical tools needed to deal with accidental self-killings.

1.1. Ordinary vs. Extraordinary Means of Preserving Life[4]

This chapter sheds light on the question how far we are bound to go in order to fulfil our positive moral duty to keep ourselves alive. The answer to this question takes recourse to an important ethical principle at the starting point of which stands the distinction between positive and negative

[4] The content of this chapter draws on, and in part reproduces verbatim, material published in Erk (2019: 230ff).

moral duties. As we know, a positive duty is a duty to perform a positive action; such duty commands the commission and prohibits the omission of an action. A negative duty is a duty to perform a negative action; it commands the omission and prohibits the commission of an action.

1.1.1. Positive Duties, Negative Duties, Ordinary Means, Extraordinary Means

Although both a positive and a negative moral duty are a necessity binding the will of the human person, they oblige the duty-bearer in a different way and to a different extent. While negative duties oblige unconditionally, that is, under all circumstances and always, positive duties only oblige under certain circumstances and to a certain extent. Obviously, nobody can be under a duty that he cannot fulfil—to say otherwise would be nonsensical.[5] Although fulfilling it might result in harm and even death to the agent, a duty that requires its bearer to omit an action can always be met. As fulfilling it cannot involve an impossibility, a negative moral duty binds always and for all times.

Praecepta negativa obligant semper et ad semper.	Negative precepts bind always and for all times.

(IIª-IIae q. 33 a. 2 co.; cf. Iª-IIae q. 71 a. 5 ad 3, De Malo, q. 7 a. 1 ad 8)

Illud quod prohibetur praecepto negativo, a nullo, nec aliquo modo faciendum est.	That which is prohibited by a negative precept, must not be done by anybody not in any way.

(De Virtutibus, q. 3 a. 1 co.)

> A negative duty arising from the natural law admits of no excuse whatever. Such a duty concerns matters forbidden by the natural law as intrinsically wrong. Not even God can give anyone permission to do such things, and we are obliged under pain of losing our last end to choose death[6] rather than commit them. (Fagothey 1963: 222)

[5] This insight is expressed by the so-called "Ought Implies Can-Principle". This principle basically states that it cannot be true that you ought to do something, unless it is true that you could do it (if you chose to). The Romans expressed this principle as follows: "ultra posse nemo obligatur" (this translates as: no one is obligated beyond what he is able to do) or "ad impossibilia nemo tenetur" (this translates as: no one is obliged to do the impossible, or more literally: no one is held to impossible things).

[6] For the sake of caution, it should be noted that "to choose death" must not be misunderstood as "committing suicide". We may not commit suicide rather than commit an action that violates a negative moral duty. But presented with the option of being killed or commit, for example, adultery, we are obliged to choose the former option and by "choosing death" in this fashion allow ourselves to be killed or not avoid getting killed.

With respect to their fulfilment, negative moral duties do not admit for limits. They are binding always, everywhere and under all circumstances. But when it comes to positive moral duties, things are different. While we can constantly fulfil our negative moral duties, it would be impossible to continuously fulfil a single positive moral duty—let alone all of our positive duties:

> Unlike negative precepts (commandments that require us to refrain from doing certain things), positive precepts (commandments that require us to do certain things) are not to be acted upon all the time and in every possible way. For one thing, it would be impossible to do so. I cannot take every available opportunity for honouring my father and mother and for giving to the poor and for worshipping God and for the many other things I am obliged by positive precept to do. (Atkins & Williams 2005: XXVf)

Positive moral duties bind always but not for always, that is, not in all circumstances. There is a limit to what we are required to do in order to fulfil a positive duty. While it remains binding in principle there can be specific instances when the bearer of a positive moral duty might not have to satisfy it. Simply knowing that we possess a positive moral duty is not the end of the story.

Praecepta autem affirmativa […] non obligant ad semper, etsi semper obligent, obligant autem pro loco et tempore et secundum alias circumstantias debitas.	Affirmative precepts do not bind for always, although they are always binding; but they bind as to place and time and according to other due circumstances.

(IIª-IIae q. 3 a. 2 co.; cf. Iª-IIae q. 71 a. 5 ad 3; IIª-IIae q. 140 a. 2 ad 2; De Malo, q. 2 a. 1 ad 11; Super Sent., lib. 4 d. 15 q. 2 a. 1 qc. 4 co.)

Ad praecepta negativa tenemur semper et ad semper […]. Sed ad praecepta affirmativa tenetur homo semper, sed non ad semper, sed loco et tempore determinato.	We are held to negative precepts always and for always. […] Man is held to affirmative precepts always, but nor for always, but for a specified time and place.

(Super Sent., lib. 3 d. 25 q. 2 a. 1 qc. 2 ad 3; cf. IIª-IIae q. 79 a. 3 ad 3; De Virtutibus, q. 3 a. 1 co.; Quodlibet X, q. 5 a. 1 s.c.)

So while negative precepts (commandments that forbid action) bind always and for all times (semper et pro semper) even under danger of death, positive precepts bind always in general but not in every single instance (semper sed non pro semper). One is not always and everywhere and under every circumstance obligated to conduct a good action. Even though the basic obli-

> gation to do good always remains, the actuality of these actions binds only at a certain time and under certain conditions. (Sullivan 2007: 387)

While all negative moral duties are concretions of the fundamental moral duty to avoid doing evil, all positive moral duties are explications of the fundamental positive moral duty to do and realise the good. While this duty—and all duties derivative of it—requires us to do everything that is reasonable and proportionate in a specific situation, it does not require extraordinary efforts. We do not have to set the whole world in motion in order to fulfil a positive moral duty.

> One must, at all costs, avoid doing what is intrinsically evil; but there are reasonable and proportionate limits to one's duty of doing good. (Kelly 1958: 131; cf. Fagothey 1963: 222)

In general, a positive moral duty only obliges us to fulfil it by using the ordinary means (also: proportionate means) at our disposal. It does not normally oblige us to use the extraordinary means (also: disproportionate means) available to us. The bearer of a positive moral duty is only duty-bound to use the ordinary means available. When fulfilling a positive duty, the duty-bearer can therefore take into account the means available to him:

Es gibt aber in moralischer Hinsicht einen Unterschied zwischen Handlungsgeboten und Unterlassungsgeboten. Nur Unterlassungsgebote können unbedingt sein, Handlungsgebote nie. Handlungsgebote unterliegen immer einer Abwägung der Gesamtsituation, und dazu gehören auch die zur Verfügung stehenden Mittel. Sie sind nicht beliebig vermehrbar.	There is, however, a moral difference between positive duties and negative duties. Only negative duties can be unconditional, positive duties never. Positive duties are always subject to consideration of the overall situation, and this also includes the means available. They cannot be multiplied at will.

(Spaemann 1997: 30, 2006: 87)

While this principle is straightforward in theory, its practical application is anything but simple. For what is the difference between ordinary and extraordinary means? In general, extraordinary means are those means which cannot be applied without relatively grave inconvenience or hardship:

> At its most general, ordinary means are those that do not entail grave hardships. If the means do involve grave hardship, they are to be classed as extraordinary and hence usually not obligatory. (Oderberg 2004b: 50; cf. Fagothey 1963: 223)

Unfortunately, there is no universally valid concretion of this general definition of ordinariness and extraordinariness. As is often the case, universal validity comes at the expense of a certain degree of sketchiness. Whether something constitutes an ordinary or an extraordinary means must be assessed on an individual case-by-case basis and according to circumstances of persons, places, times and culture. Usually, the ordinary means will be those means that are commonly used under given circumstances and that the agent in his present physical, psychological and economic condition can reasonably employ with definite hope of proportionate success.[7] But in the end, the determination of a means' ordinariness or extraordinariness is a matter of prudent judgement and also common sense.

> In concrete cases it is not always easy to determine when a given procedure is an extraordinary means. It is not computed according to a mathematical formula, but according to the reasonable judgments of prudent and conscientious men. (Kelly 1958: 135; cf. Noldin 1935: 308; Prümmer 1961: 408)

Means that are extraordinary for one person can be ordinary for another, and means that are extraordinary for a person at a certain point in time might have become ordinary at a later point in time—and vice versa.

1.1.2. The Ordinary and Extraordinary Means of the Preservation of Life

The general notion of extraordinary means developed above can now be applied to our moral duty to preserve our life. As has been stated above, the duty to realise the basic human goods (D_{BHG}) contains the moral sub-duty to realise the basic human good "life" ($D_{BHG\text{-}Life}$). Just like D_{BHG}, the latter duty has both a positive and a negative aspect to it. $D_{BHG\text{-}Life}$ comprises (a) the positive duty to do what furthers the realisation of the basic human good "life" and (b) the negative duty to not do what hinders its realisation or violates it. While the negative duty requires us to not harm or take the life of a human person (including our own life), the positive duty basically requires us to utilise the ordinary means necessary to

[7] The acquisition of a competence that is needed to fulfil a positive duty can also be counted among the means. If a person has to display highly unreasonable efforts to appropriate the knowledge and competences needed to fulfil a duty, doing so would constitute an extraordinary means.

preserve and further our life and health. However, it does not normally extend to the application of the extraordinary means.[8]

Obligatio vitae sanitatisque conservandae se extendit ad ordinaria, non autem ad extraordinaria media.	The obligation to preserve life and health extends to the ordinary, but not the extraordinary means.

(Prümmer 1961: 408; cf. Cathrein 1905: 230; Noldin 1935: 307; Kälin 1954: 185)

Although we are not forbidden to use them, not applying extraordinary means to conserve or to regain our health is a signal of accepting the human condition in the face of death. Not applying the ordinary means to preserve one's life would, however, be tantamount to suicide or euthanasia.[9]

[8] The word "normally" is added here because there can be circumstances under which we are required to also apply the extraordinary means to fulfil a duty:

Aliquando tamen tenetur quis adhibere etiam media extraordinaria, si scil. eius vita est valde necessaria pro familia aut pro bono communi, e. g. mater debet media extraordinaria ad vitam conservandam adhibere, si secus eius filii remanerent orphani in magnis angustiis versantes.	Sometimes one is held to also apply extraordinary means, e.g. if one's life is very necessary for the family or the common good. For example, a mother must apply extraordinary means to preserve her life, if otherwise her children would remain as orphans living in great poverty.

(Prümmer 1961: 408; cf. Cathrein 1905: 230)

This position is also corroborated by Oderberg (2004b: 50): "There are exceptions. A person whose continued existence is vital for the common good is obliged to submit to more than ordinary attempts to keep him alive: for instance, a gravely ill president or military leader should submit to extraordinary treatment, especially at a time of national crisis. Also, if a person is not spiritually prepared for death, he should submit to extraordinary means to keep him alive until he can be so prepared".

[9] It should be noted that the distinction between ordinary and extraordinary means of preserving life must not be mixed up with the distinction between ordinary and extraordinary medical procedures. (Extra-)Ordinariness in a moral sense is not tantamount to medico-technological (extra-)ordinariness. While medical ordinariness refers to accepted routine, that is, what is scientifically established, statistically successful, commonly followed and readily available, the ordinariness that is in question here—and that is especially important when it comes to end-of-life decisions—is an ethical category and is characterised differently.

Furthermore, extraordinary means must not be mixed up with artificial treatments: "The mere fact that the treatment is artificial rather than natural makes no difference in itself—why should it? Every pill, every injection, every operation is artificial treatment, but many such treatments are ordinary means within the capacity of the patient to undergo and doctor to supply without excessive hardship" (Oderberg 2004b: 51).

So, what are the ordinary means of the preservation of life? As Henke (2005: 557) points out, the distinction between ordinary and extraordinary means of preserving life goes back to Thomas Aquinas with early contributions by Francisco de Vitoria, Domingo Banez, Dominic Soto and Juan de Lugo. As a result of this long tradition there has emerged a widely agreed-upon list of criteria that help in drawing this distinction.

> Positive obligations to preserve life and health do not go beyond the use of ordinary means to that end. If the means necessary in any particular case are extraordinary on account of danger or pain, immoderate expense, or for any other good reason, there is no strict obligation to use them. Thus, one may refuse to undergo an operation if it would be extremely painful, or if it would impoverish him too much. Nor is one bound to undergo an operation for which one has great horror, as a woman for certain surgical operations at the hands of a man. (Ross 1919: 183; cf. Prümmer 1961: 408; Noldin 1935: 307; Cathrein 1905: 230)

> Ordinary means of conserving life are those means commonly used in given circumstances, which this individual in his present physical, psychological and economic condition can reasonably employ with definite hope of proportionate benefit. Extraordinary means of conserving life are those means not commonly used in given circumstances, or those means in common use which this individual in his present physical, psychological and economic condition cannot reasonably employ, or if he can, will not give him definite hope of proportionate benefit. (Cronin 1989: 112)

> We can define ordinary means as including not only food, drink, shelter, and rest, but also all treatments and medicines that offer a reasonable hope of benefit for the patient and can be obtained and used without excessive pain, suffering or other inconvenience, including expense, on the part of the patient, his family and loved ones, and the doctors concerned. Means that are not ordinary are extraordinary. Ordinary means are absolutely binding. (Oderberg 2004b: 51, cf. Kelly 1958: 129)

Whether some means (e.g. a medicine, treatment or operation) that is generally suited to preserve or restore one's life is to be classified as ordinary or extraordinary is the result of a weighing between the benefits and burdens that result from the application of that means. As the references above show, this weighing has to take into account the following elements:

1. *On the one side of the scale there are the benefits*
   The benefits are the beneficial effects that result from the application of the means. These benefits are weighed according to the extent and likelihood of their occurrence.
   The beneficial effects of interest for the question at hand can be cure, improvement of symptoms, recovery of function and/or prolongation of life.
   The beneficial effects are expressive of the extent to which the application of the means fulfils the positive duty to preserve one's life.
2. *On the other side of the scale there are the burdens*
   The burdens are the burdensome effects of obtaining and/or applying the means. Just like the benefits, the burdens are also weighed according to the extent and likelihood of their occurrence.
   The relevant burdens can be intense and constant pain, excessive suffering, grave danger to life, great efforts or some other grave inconvenience to procure or use the means, including expenses that constitute an excessive hardship (*secundum proportionem* status, i.e. depending on social position), on the part of the patient, his family and loved ones, but also the physician. Furthermore, fear and psychological discomfort (e.g. repugnance or disgust),[10] moral objections or conflict with important moral principles also constitute burdens. If the application of a means results in an interference with important activities and experience-desires in the time remaining,[11] this can also be considered as a burden. Such burdens can arise during the application or as a result of the application of a certain measure.

If a treatment is extraordinary this means that the beneficial and the burdensome effects are gravely disproportionate, that is, that the latter outweigh the former to a large extent. The application of an extraordinary means is optional, that is, one is excused from applying it. Even if a positive moral duty can only be met by the application of extraordinary means, this does not allow for the conclusion that the duty-bearer does not actually bear this duty or is freed from this duty in the future. As it is in the nature of a moral duty to be held by all human persons of all times and at

[10] Examples for psychological discomfort are disgust with a deformity or an amputation but also repugnance at being cut open or having a tube inserted.

[11] Besides spending time with and saying goodbye to one's family and close friends or settling old quarrels, these activities also comprise the pursuit of spiritual goals, especially preparing oneself to meet God.

all times, a positive moral duty cannot be lost. Its bearer is merely exempted from fulfilling the duty under this set of circumstances. However, if the circumstances change and the duty can be fulfilled by ordinary means this exemption becomes invalid.

It has to be kept in mind that it is first and foremost the person whose life is at stake who has to do the weighing and make the decision. Only if she is incapable of doing so can other persons make the weighing for her. In assessing benefits and burdens to another person, surrogate decision-makers—often family members—have to be very careful to not import their own feelings about the patient's situation or privilege their own interests and desires. In this regard, it cannot be emphasised enough that what is being weighed are the benefits and burdens of a treatment *for the patient* person who is receiving the treatment. What must not be weighed are the benefits and burdens *of that patient* for others, such as the patient's family, the patient's insurance company, the economy or society as a whole.

The weighing up of benefits and burdens does not amount to a quality-of-life judgement. Human life is always of quality. Concluding that an available means to preserve one's life is extraordinary does not mean that one's life is no longer worth living. Although one is excused from applying extraordinary means to preserve one's life, not applying them with the intent to die makes such action evil due to its evil intention.

> If a treatment would be considered ordinary for a virile young man in his 20's, then it should not be deemed extraordinary for a woman with Down Syndrome or an elderly man with Alzheimer's disease. And just as it would be unethical to withhold even extraordinary treatment because of an intent to cause or hasten death, it would also be wrong to withhold even extraordinary treatment for the reason that the patient's quality of life is said to be poor—because that would amount to the same thing. If the reason life-preserving treatment of any type is denied is because the patient's quality of life is considered poor, that is tantamount to withholding it with the motivation of shortening that person's life because it is allegedly not worth living and the patient (or the patient's family or society) would be better off if she or he were dead. (Balch 2016: 5f)

1.2. Accidental Self-Killings

As has been mentioned above, there are cases of self-killings in which the death of the agent does not result as an essential consequence from his exterior action but merely as an unintended, foreseen and unavoidable accidental consequence. In contrast to the essential killing suicide, these self-killings are accidental killings. In such an action, the agent accepts the risk of dying and dies but does not intend to die.[12]

Let us look at prototypical examples of an accidental self-killing:

- Apollonia sits in her office which is located on the second floor of an office building. Suddenly, she realises that the building has caught fire and that the fire has already spread so far that she cannot use the stairs or the elevator to save herself. The fire comes closer and closer and the only way to save herself is to jump out of the window onto grass. She jumps, fatally injures herself on landing and dies before the ambulance arrives. Has Apollonia committed suicide?
- Justin is on a sailing trip and, unfortunately, the boat sinks in the middle of the high seas. Besides Justin, only a young girl, Monika, survives. All that is left from the boat is a plank that can support one person only. Justin has managed to grab hold of the plank. However, seeing that Monika is not a good swimmer he leaves the plank to her. As a result of letting go of the plank, Justin drifts away and eventually drowns. Has Justin committed suicide?

We could list many more such examples. However, these two suffice to establish the ethics of accidental self-killings.

When it comes to these kinds of killings, we have learned that the fact that the death of a human being results as an evil accidental consequence of an exterior action does not yet make this accidental killing impermissible or permissible as such. Whether an accidental killing is ethically permissible or impermissible has to be established by reference to the exterior-action condition, the end-of-the-agent condition, the means-end condition and the proportionality condition. Only accidental killings that meet all four conditions are ethically permissible. If one condition is not met, the action is ethically impermissible. In addition, the permissibility of an accidental killing also requires that the death that results as an evil accidental consequence of an exterior action has indeed merely been foreseen but not intended. This can be tested by means of the criteria deliberation, constraint, endeavour and plan.

[12] Therefore, Fagothey (1963: 236) suggests calling accidental killings "the exposure of one's life to serious danger".

So, let us apply these criteria to the first example, that is, the example of a person jumping out of the window of a burning building. This action is only ethically permissible if all the following questions are answered with "yes":

- **Is the action an accidental killing action?** *This question asks whether the death that results from the action's exterior action is an accidental or an essential consequence of this exterior action. If it is an essential killing, we are dealing with an ethically impermissible killing. If it is an accidental killing, the action can be ethically permissible.*
  So, what is the situation in the present example? This action's exterior action is the jump out of a window. In general, people do not usually die jumping out of windows; they only die if the window is too high above the ground. Assuming a floor height of three meters, jumping out of a second-floor window means jumping from a height of about seven meters. In general, people who jump from such heights necessarily hurt themselves in some way, but they do not necessarily die. If they do die, death has occurred as an evil accidental consequence. So, considering the exterior action in its circumstances, we can affirm that the action in question is an accidental killing action.
- **Has the evil accidental consequence merely been foreseen but not intended?** *An accidental killing can only be ethically permissible if the exterior action's evil accidental consequence is merely foreseen but not intended. If the consequence is intended, this makes the action ethically impermissible; if it is merely foreseen, the action can be ethically permissible.*
  How does our example fare in this respect? Unfortunately, we cannot look into the head of the Apollonia and see her train of thought. So, we have to content ourselves with assuming that she knows there is a certain—albeit low—probability of dying from the jump. Her jump out of the window can only be ethically permissible if she does not intend her death, that is, if in the given situation she (a) does not deliberate about how to kill herself, (b) is open to consider alternative ways to escape the fire that do not risk her death and (c) does not consider the action a failure if she does not die. Or put differently: Apollonia must not jump because but despite the fact that the window is on the second floor.
- **Does the action meet the exterior-action condition, that is, is the exterior action good or at least neutral?** *An accidental killing can*

only be ethically permissible if the exterior action of the accidental killing (as defined by its natural end) is either morally good or at least morally indifferent. So, in a third step we have to establish whether the exterior action of the accidental killing in question meets this condition.
In our example, the action's exterior action is the jump out of a window or—taking the circumstances into account—the jump out of a second-floor window of a burning building. Considered in itself, this exterior action is an ethically good action. Although the agent performing such an exterior action cannot not will to get hurt, its overall natural end consists in the saving of the jumping person's life.
In order to escape a fire, one may only jump out of a window provided there is hope of surviving the fall. If Apollonia were not to jump from a second-floor window but from a window on the fiftieth floor of a burning building, the natural end of this exterior action would not be the saving of her life but death. If she were to jump from a second-floor window of a building that is not burning, the natural end of this exterior action would not be the saving of her life but a more or less serious injury. In both these cases, the exterior action would violate the basic human good life and would therefore be an ethically evil exterior action.

- **Does the action meet the end-of-the-agent condition, that is, is the end of the agent good?** *An accidental killing can only be ethically permissible if the end of the agent performing the accidental killing action is good. So, a fourth step requires us to assess the ethical quality of the accidental killing's end of the agent, that is, the reason why the agent performs the accidental killing's exterior action.*
The end of the agent is the reason why the agent performs an action's exterior action. Once more, we must acknowledge that we cannot read other people's minds. Given this limitation, let us assume that Apollonia jumps out of the window in order to save her life. Certainly, this is a good motive. If this is indeed her purpose for jumping, we can hold this condition fulfilled.
If Apollonia were to jump in order to kill herself, that is, because jumping out of the window could or would kill her, this would be an ethically evil end of the agent; in that case the condition would not be met.
- **Does the action meet the means-end condition, that is, is the natural end of the exterior action realised at least as causally immediately as the evil accidental consequence?** *An accidental*

killing can only be ethically permissible if the evil accidental consequence is not a means to realise the action's end of the agent. The good consequence, that is, the realisation of the natural end of the exterior action, must follow from the exterior action at least as causally immediately as the evil accidental consequence. The natural end of the exterior action must not be produced by the evil accidental consequence.
This condition is fulfilled in the present example. If Apollonia dies as a result of jumping out of the window, her death is not what realises her end of the agent and it does not cause the natural end of the exterior action. It is not because she dies that she escapes death by fire. It is after having escaped death by fire that she dies.

- **Does the action meet the proportionality condition, that is, is there a proportionately grave reason for permitting the exterior action's evil accidental consequence?** *An accidental killing can only be ethically permissible if the exterior action's good consequence is at least equivalent in importance to and counterbalances the action's evil accidental consequence.*
In Apollonia's case what must be counterbalanced is her death that results from the jump. The action's good consequence, that is, the natural end of the exterior action, must be at least equivalent in importance to this evil accidental consequence. The general rule is that the more likely death results as the action's evil accidental consequence, the graver must be the action's good consequence; the graver the danger to one's life, the graver must be the reason to permit the action (cf. Noldin 1935: 312). Obviously, saving one's life is a sufficiently grave reason to endanger one's life.[13] This condition can, therefore, also be considered met.

According to the above considerations, all conditions for the ethical permissibility of Apollonia's jump out of the second-floor window of a burning building have been met. So, all in all, this accidental killing is an ethically permissible action. It does not constitute a case of suicide, but the ethically permissible endangerment of one's life that accidentally results in the agent's death.

[13] Reasons that are proportionate to the endangerment of one's life would also be saving the life of one or more other human beings or the common good. Sometimes, the list of goods that are proportionately grave to endangerment of one's life also include goods of great value (e.g. material value, but also chastity) or significant scientific research.

Given the high degree of detail of the above explanations, we can save ourselves the trouble of applying the conditions of the ethical permissibility of accidental killings to the second example. For those who are interested in the solution, scrutinising the second example would show that it is an ethically permissible action to leave somebody else the only plank in case of shipwreck.

The key message of this chapter is that the ethical (im)permissibility of an accidental killing is evaluated by reference to six conditions. An accidental killing is ethically permissibly only if all six questions listed above are answered with "yes".

1.3. Suicide and the Perverted Faculty Argument

Besides the argument given at the beginning of this chapter, there is a second argument than can be advanced against the ethical permissibility of suicide. This argument is based on the so-called Perverted Faculty Argument.[14] In its general version, this argument can be formulated as follows:

> Where some faculty F is natural to a rational agent A and by nature exists for the sake of some end E (and exists in A precisely so that A might pursue E), then it is metaphysically impossible for it to be good for A to use F in a manner contrary to E. (Feser 2015: 398)
>
> Any act in which a faculty is used for an end or object which is opposed to its natural end or object is unnatural, and being unnatural is morally bad. (Cronin 1922: 75)

Whenever an agent uses one of his faculties it cannot be good for the agent to use this faculty "for the sake of actively frustrating the realization of E or in a manner which of its nature tends actively to frustrate the realization of E" (Feser 2015: 399). All the faculties of human beings have for their ultimate end the maintenance and perfection of their possessor. When a human being uses his faculties to commit suicide, he is using them unnaturally and wrongfully.

[14] For an explanation of the Perverted Faculty Argument cf. Erk (2019: 154–157).

> Now in suicide a man makes voluntary use of his own powers, and by his act those powers are directed to attain an object the very contrary of that which, by their own nature, they are directed to attain—they are used, viz. not for the welfare but for the destruction of the agent. There could be no more direct or unequivocal violation of nature than this. To use a power and to use it for the accomplishment of what is most directly opposed to its own natural end is the most complete perversion that is possible of nature's purposes and aims. (Cronin 1922: 53; cf. Ross 1919: 65)

As it is metaphysically impossible for it to be good for a human being to use his faculties in a manner that is contrary to and frustrates his ultimate end of self-preservation and perfection, such action cannot be good—and is, therefore, ethically impermissible.

2. The Ethics of Killing Others

In the previous chapter we have dealt with the ethics of suicide, that is, cases in which a human being kills himself. In this chapter we turn to the ethical evaluation of actions in which a human being kills another human being. Specifically, we deal with the ethics of lethal self-defence, abortion and euthanasia. Applying a killing perspective, we also have a look at the ethical problems surrounding the transplantation of organic material from an organ donor.

The question that will guide our discussion of the issues is as follows: Under which conditions, if any, is it ethically permissible to perform the respective action?

2.1. Lethal Self-Defence

The first topic on which we focus is lethal self-defence. By self-defence we understand an action that a victim is taking in order to protect himself against an unjust aggressor in an emergency situation created by the latter's attack. Although running away can also be an act of self-defence, in general, an act of self-defence is characterised by the victim using some kind of force up to and including injury to life and limb of the unjust aggressor. Lethal self-defence is an act of self-defence which results in the death of the aggressor. As such, lethal self-defence has a double effect, namely the preservation of the victim's life (good consequence) and the death of the aggressor (evil consequence).

2.1.1. Accidental Killing in Self-Defence

So, under which conditions, if any, is lethal self-defence ethically permissible? The first step towards finding an answer is to distinguish between lethal self-defence as an essential killing action and lethal self-defence as an accidental killing action. As has been shown in Chap. VI, Sect. 1, essential killings are ethically impermissible. This general ethical impermissibility of essential killings also applies to lethal self-defence. The victim of an unjust attack is still under the negative moral duty to not commit an essential killing. The fact that somebody is transgressing the boundaries of morality does not justify others doing so as well. In order to avoid confusion, we should therefore not call an essential killing act by which a victim kills an aggressor lethal self-defence. While suicide is ethically impermissible because it is an essential killing by definition, there can be no lethal self-defence that is an essential killing; such an action would not be lethal self-defence but homicide.

Strictly speaking, an act of lethal self-defence can only be an accidental killing. As such, it can only be ethically permissible if it meets the conditions for the ethical permissibility of an accidental killing. This allows us to conclude that lethal self-defence is ethically permissible if it (a) is an accidental killing and (b) meets the conditions of the ethical permissibility of accidental killings. So, in order to establish whether lethal self-defence is ethically permissible we must answer all of the following questions in the affirmative (for the complete formulation of these questions cf. Chap. VII, Sect. 1.2):

- **Is the action an accidental killing action?**
  How does lethal self-defence fare with respect to this criterion? Lethal self-defence can come in many shapes and forms—too many to be discussed here. So, the only thing we can do here is to establish the general rule: Lethal self-defence is ethically permissible only if the natural end of the exterior action is not the death of the aggressor; rather, the exterior action by which the victim defends himself must be a kind of action from which the death of the aggressor does not naturally follow. This means that the exterior action may at most be an action the natural end of which is the incapacitation of the aggressor: "All that I as a private person may do in defending myself is to produce a state of quiet or nonactivity in the assailant so that he cannot continue his attack. [...] If my attempts to produce this nonactivity, a thing that is morally indifferent, result in death, that is incidental and regrettable" (Fagothey 1963: 247).

Accordingly, it would not be permissible to shoot the aggressor in the heart as this action naturally results in death. It would, however, be permissible to shoot him in the leg to immobilise him.

- **Has the evil accidental consequence merely been foreseen but not intended?**
  Whether the death that accidentally results from an act of self-defence is foreseen or intended can be tested by means of several questions: Does the victim deliberate about how to kill the aggressor? Is the victim open to consider alternative ways to successfully defend herself that do not involve the death of the aggressor? Does the victim consider her self-defence a failure if the aggressor does not die? If either the first or the last question is answered with yes or if the second question is answered with no, then the action is ethically impermissible. If the first and the third questions are answered with no and the second question with yes, then the lethal self-defence can be ethically permissible.
- **Does the action meet the exterior-action condition, that is, is the exterior action good or at least neutral?**
  The exterior action of an accidental self-defence action meets this condition if its natural end is not the death of the aggressor but the stopping of the aggressor's attack on the victim. The victim has the moral right to use force against the aggressor to defend herself. However, the actual means chosen to save her life may only have the natural end of "effectively stopping the movement directed against us or producing quiescence in the aggressor" (Cronin 1922: 99).
  While it is true that a victim has the moral right to use force to defend herself, she possesses that right only on the condition that there is no other way of repelling the attack. This also implies that the victim is only allowed to apply as much force as is necessary to fend off the aggression[15]: "The person attacked must not reply with force at all, if this is possible. He should escape, or call a policeman, or throw the weapon out of the window, etc., if these means will suffice. [...] The person attacked should use only such force as is necessary, if force must be employed. Thus, if the aggressor can be made helpless by the use of gas, permanent bodily harm should not be done him; if he can

[15] The ethical technical term for this condition is "moderamen inculpatae tutelae", that is, controlled amount of blameless force (cf. Noldin 1935: 320; Prümmer 1960: 113; Cathrein 1905: 241).

be subdued with the fists, knife or pistol wounds should not be resorted to; if wounds will suffice to hold him at bay (e.g., by black-jacking), killing should not be resorted to. In the heat of a fight, however, the person assailed sometimes unintentionally goes beyond what moderation requires" (McHugh & Callan 1958: §1833).
It should be a logical matter of course that the exterior action of a self-defence action can only have the natural end of incapacitating an aggressor if there is an aggressor. A human being is only an aggressor while actually committing an act of aggression, that is, while inflicting violence on a victim. Without an actual act of aggression there can be no aggressor and therefore no self-defence. Somebody does not yet count as an aggressor because he might potentially commit an act of violence. You can only defend yourself against somebody who is strangling you or is pointing a gun at you, but you cannot defend yourself against somebody whom you foresee will attack you several days from now or against somebody who has attacked you yesterday: "The person must be actually an aggressor, i.e. must be really attacking or performing an act from which he can be judged to already morally to have attacked—as loading or drawing a revolver. But it is not lawful to kill an aggressor after an attack, as this would be vengeance, not defense" (Ross 2012: 212 (Nr. 541)). Ethically permissible lethal self-defence requires an element of urgency, that is, is permissible only at the moment of the attack or if the attack is imminent.[16]

- **Does the action meet the end-of-the-agent condition, that is, is the end of the agent good?**
  When it comes to lethal self-defence, the end of the agent may only be self-defence, that is, it must not be the death of the aggressor but the saving of one's life and limb: "Thus, as his end he must intend only the preservation of his own life and look upon the death of his

[16]As Fagothey (1963: 245f; cf. McHugh & Callan 1958: §1830; Higgins 1949: 247) points out, this condition is the most difficult to apply: "In concrete cases it is often hard to determine just when preparation for attack turns into actual aggression. Mere purchase of a gun by my enemy with declaration that he intends to shoot me is not aggression. But I do not have to wait until he has actually shot at me, for then no further defense may be possible. Defense may begin as soon as the aggressor lifts his gun or even approaches to the oncoming fray, depending very much on the circumstances. Here the common estimate of men must be taken into consideration. In a matter as crucial as this, when the action to be efficient must be swift, it is impossible to draw fine lines and delicate discriminations, and it is better to favor the defender of life than the attacker".

neighbor as a misfortune" (McHugh & Callan 1958: §1834). If the victim were to lethally defend herself in order to kill the aggressor, this would be an ethically evil end of the agent; in that case this condition would not be met.

- **Does the action meet the means-end condition, that is, is the natural end of the exterior action realised at least as causally immediately as the evil accidental consequence?**
  An act of lethal self-defence meets this criterion if the death of the agent is not the means by which the victim incapacitates the aggressor and saves her life. This condition is met as the incapacitation of the aggressor follows from the exterior action at least as causally immediately as his death. He is not incapacitated because he is dead, but he is incapacitated and dead.
- **Does the action meet the proportionality condition, that is, is there a proportionately grave reason for permitting the exterior action's evil accidental consequence?**
  The good that is defended by an act of lethal self-defence must be proportionate to the death the results from it. A small material loss does not justify the killing of the aggressor. Only if one's life (or the life of somebody else) is at stake is lethal self-defence ethically permissible. However, limbs, sanity, liberty, material goods of great value and chastity can also be counted among the goods that can be defended by lethal force (cf. Fagothey 1963: 248f).[17]
  Although they are important goods as well, honour and reputation do not feature in the list of goods that are equivalent to life. When defending against insult, we must not go so far as to defend our honour or reputation by lethal force.[18]

Provided that a specific act of lethal self-defence meets all of the above criteria, it is ethically permissible. In this case, the victim is ethically permitted to perform this act of self-defence. However, while lethal

[17] Although it is not her life that is threatened, a woman who is about to be raped can—if she has no other option—defend herself with lethal force. In case of a foreseeable and unavoidable rape she is also ethically permitted to use (non-abortifacient) contraceptives in order to prevent pregnancy from arising from rape. However, she is ethically not allowed to kill herself in order to escape the rape.

[18] This does, however, not mean that we have to stand idly by or that we are not allowed to use any force in order to defend our or another person's honour or reputation: "[F]orce short of killing may be the only way of closing the mouth of a reviler or slanderer who refuses to stop. There must be some remedy against a perpetual tirade" (Fagothey 1963: 249; cf. Higgins 1949: 249).

self-defence can be permissible, we do not usually have a duty to defend ourselves by lethal means. Killing a human being—even if he is an aggressor—is usually taken to be an extraordinary means to preserve one's life. And as we have learned in the previous chapter, we are only ethically required to use ordinary means to preserve our life. Unless our survival is very necessary to the state or we possess very serious duties (e.g. as parent or child), our duty to preserve our life "ceases if such hard and abhorrent means are necessary" (Ross 2012: 213 (Nr. 546)).

> Therefore nothing prevents a man from choosing the heroic course of giving up his own life rather than taking another's. Only in unusual circumstances could self-defense become a duty, for example, in the case of a public personage indispensable to the community's welfare. (Fagothey 1963: 248)

When thinking about the ethical (im)permissibility of lethal self-defence there is another important point that needs to be emphasised: Self-defence is only allowed against an *unjust* aggressor. If the aggressor acts justly, it is not permissible to kill him; a burglar is ethically not permitted to kill an officer who is trying to arrest him. As far as the injustice of the aggression is concerned, "it makes no difference whether the aggression is formally unjust (e.g., when the aggressor attacks you because he wishes to wreak revenge, or because he fears you), or only materially unjust (e.g., when you are a stranger to the aggressor, but he is drunk, or a dope fiend, or a maniac)" (McHugh & Callan 1958: §1831; cf. Prümmer 1960: 114; Higgins 1949: 247). For self-defence to be licit, it suffices that the aggressor intends to inflict a material injustice. Provided that the relevant conditions are met, a victim is ethically permitted to lethally defend himself against a drunk, love-crazed or furious aggressor who intends to kill him—the mere fact that he is unjustly attacked suffices. The ethical permissibility of lethal self-defence is not influenced by the degree of voluntariness on the part of the aggressor.

Last but not least, it has to be pointed out that the conditions for the ethical permissibility of lethal self-defence also apply to the life of an innocent third party. One is not only ethically permitted to defend one's own life, limbs, sanity and so on; one is also allowed—and sometimes even duty-bound[19]—to defend these goods for another person who is victimised by an unjust aggression.

2.1.2. Deputised Essential Killing in Self-Defence

One might interject that the account of the ethical permissibility of lethal self-defence given above is overly rigid and unrealistic as it seems to prohibit most instances of killing in self-defence. In that respect, Jensen (2010: 65f; 219ff) outlines an argument that is based on Thomistic principles and both complements the above account of self-defence and mitigates its application. According to this "deputization"-argument, "in the state of emergency provoked by the attack we are granted public authority" (Jensen 2010: 222)—and as such we are permitted to intend to injure and even kill our assailant. Why is that?

For Aquinas, it is not ethically permissible for a man to intend to kill a man in self-defence, "except for such as have public authority" (II^a-IIae q. 64 a. 7 co.; cf. a. 3). This position has also been advanced by G.E.M. Anscombe:

> The plea of self-defence (or the defence of someone else) made by a private man who has killed someone else must in conscience—even if not in law—be a plea that the death of the other was not intended, but was a side effect of the measures taken to ward off the attack. To shoot to kill, to set lethal man-traps, or, say, to lay poison for someone from whom one's life is in danger, are forbidden. The deliberate choice of inflicting death in a struggle is the right only of ruling authorities and their subordinates. (Anscombe 1961: 50)

According to this Thomistic position, only individuals endowed with public authority are ethically permitted to intend to kill. Even in cases of self-defence, an individual without such authority is only ethically permitted to kill his attacker in self-defence by means of an accidental killing. This is the principle that governs cases of lethal self-defence in general.

However, Aquinas also insists that "necessitas non subditur legi" (I^a-IIae q. 96 a. 6 co.), that is, that necessity is not subject to the law. When a—sufficiently grave or extreme—danger is so sudden and imminent as to not allow the delay involved by referring the matter to the

[19] Ross (2012: 213 (Nr. 547)) gives the following examples of cases in which we are obliged to defend the life of another: "(a) as when the life of the attacked person is very useful to the State; (b) or stands in some special relation to him, as father, or child, or wife; (c) or when one is by his office (as a policeman) obliged to defend others. And these exceptions bind even under the risk of one's own life".

requisite authority, then this necessity brings with it a dispensation from the law.[20]

Self-defence is a prime example of such extreme necessity. In a case of self-defence, the defender finds himself in a situation where "necessity prevents one from having recourse to the proper authority" (Jensen 2010: 66). And just as property laws break down in case of extreme necessity, the "laws of designating public officials may also 'break down'" in case of self-defence and "ordinary citizens may act as public officials" (Jensen 2010: 66):

> Through the virtue of equity, an individual can recognize the situations in which necessity demands that she act as an official. If she is being attacked, and she cannot expect help from designated public officials, and if the situation is such that 'repelling the attack' is an inadequate means of defense, then it seems that she herself serves as the official; she acts in an official capacity, and as such she may intend to kill in order to defend her own life. (Jensen 2010: 66)

A defender who intentionally kills an attacker in self-defence may act outside the letter of the law but is allowed to do so by "deputized authority" (Jensen 2010: 219) by which those who are otherwise private citizens are given the power to act with public authority in cases of emergency. In cases of self-defence, the defender can therefore be said to act "as a kind of emergency public official" (Jensen 2010: 64) and in this capacity is justified to defend his life even by an act of essential killing. While essential killings are ethically impermissible in general, such an act of deputised essential killing in self-defence becomes ethically permissible due to the emergency character of situations of self-defence.

2.2. Abortion

The next killing action to which we turn our attention is abortion, including infanticide. The literature on this highly controversial topic fills entire

[20] We can see this principle at work in Aquinas' treatise on the permissibility of stealing. Though a mortal sin in general (cf. II^{a}-IIae q. 66 a. 6), the taking of what belongs to another can become ethically permissible in case of extreme necessity because in cases of need all things become common property (cf. II^{a}-IIae q. 66 a. 7; II^{a}-IIae q. 110 a. 3 ad 4).

library shelves. The aim of this chapter is not to give an extensive overview of the positions and arguments advanced for or against its ethical permissibility. Rather, it is meant to apply the general ethics of killing established over the course of the preceding chapters and by doing so establish the moral (im)permissibility of this killing action.

2.2.1. Mapping the Concept

Before we delve into the subject, we should first define the term. Every abortion presupposes a pregnant woman. A woman is pregnant if and as long she has a human being developing inside her body.[21] As the term covers all early stages of a human being's life cycle, we will call the human being who is developing inside the woman's body the unborn child or the unborn.[22] In a broad sense, an abortion is the termination of a pregnancy that results in the death of an unborn human being.[23]

Abortions can be spontaneous, that is, "a miscarriage that is no one's fault" (Fagothey 1963: 243) or induced, that is, brought about by the decision of a human being. As spontaneous abortions are outside the scope of ethics, we will focus on cases of induced abortions only.[24] An

[21] Given that "human pregnancy begins with the fusion of an egg and a sperm within the female reproductive tract" (Carlson 2019: 1), it is therefore biologically wrong to stipulate that pregnancy begins at implantation which begins six to seven and is completed 11 to 12 days after fertilisation (cf. Carlson 2019: 65ff).

[22] Human beings start their life as a zygote and continue their development via the stages embryo and then foetus; after birth they are called infants. A human zygote is the cell that is formed by the fertilisation event between two human gametes, that is, a haploid sperm cell and a haploid oocyte. The developing human being is referred to as an embryo until the ninth week after conception (or eleventh week gestational age). From this point of development, the unborn human being is called a foetus.

[23] For Fagothey (1963: 241) and Higgins (1949: 254) abortion is the expulsion of an unviable human being, that is, a zygote, embryo or foetus too young to survive outside the womb. In contrast, the termination of a pregnancy that results in the death of a viable foetus, is not a case of abortion strictly speaking but foeticide. Although theoretically quite reasonable, we do not adapt this definition and do not include viability in the definition of abortion.

Viability is the point in time from which a foetus is capable of life outside the womb, either naturally or supported by artificial means. Given the state of neonatal care, a foetus developing in a first world country has a more than 50% chance of survival after 25 weeks gestational age. Of course, a foetus develops the potential to survive outside the womb without medical support (natural viability) only considerably later.

A means to avoid the death of the unborn after the termination of a pregnancy would be partial ectogenesis, that is, "the development of a human being during the typical gestational period outside the maternal womb but not during the entire gestational period" (Kaczor 2007: 309). For a discussion of the ethics of ectogenesis cf. Kaczor (2007) and (2005: 105ff).

induced abortion is a killing action because a human being—albeit a human being in a very early development stage—dies as a result of another human being's action.

Usually, an abortion involves four parties. The most obvious parties are the pregnant woman and the unborn human being developing inside her.[25] As there can be no zygote without an ovum and a sperm, every unborn also has a biological father. The father is the third party to an abortion. In addition, an abortion requires somebody who actually performs the abortion; this is the fourth party to an abortion. This can be the mother; but in most contemporary cases this person is a representative of the medical profession.

Given these four parties, abortions can come in many constellations. In one constellation, the pregnant woman decides to have an abortion without consulting the father and performs the abortion herself. For the purposes of this chapter it is assumed that it is the pregnant woman and the father who jointly decide to have an abortion and that it is a medical practitioner who performs the abortion. With this constellation in mind let us turn to the question under which conditions, if any, abortion is ethically permissible.

2.2.2. Essential Abortions

The general ethics of killing developed in the course of the previous chapters commits us to the position that an abortion is ethically impermissible

[24] As a sidenote: It is true that some pregnancies result in spontaneous miscarriage and that some embryos fail to develop successfully before the pregnant woman knows she is pregnant. However, this fact is completely irrelevant in the debate about the ethical permissibility of induced abortion. The spontaneous death of some human beings during the first stages of their development has no logical implications for the permissibility of induced abortions. The probability of survival does not have any influence on the question of whether a human being has the right to live.

[25] Out treatment of the ethical (im)permissibility of abortion is based on the classical situation in which the pregnant woman is also the biological mother of the unborn.

This identity between biological mother and pregnant woman is not given in cases of gestational surrogacy. In these ethically highly problematic cases, a human being is created in vitro, that is, by means of in vitro fertilisation (IVF), using an ovum and sperm of the biological parents (also: donors) and is then transferred to the surrogate (also: gestational carrier). As the ovum is not an ovum of the surrogate mother, she is therefore not biologically related to the child.

Besides this full surrogacy there also is partial surrogacy. In partial surrogacy the surrogate is also the biological mother of the child. In a typical case of partial surrogacy, the surrogate's ovum is fertilised in vivo by the intended father's or a donor's sperm through natural or artificial insemination.

if it is an essential killing, that is, if the death of the unborn is the natural end of the exterior action. Given the general ethics outlined above, the impermissibility of essential abortions is so patently obvious that not many more words need to be said on the subject.

This ethical impermissibility remains in force even if the unborn has been conceived by an act of rape. The way human beings come into existence does not have an influence on whether they possess the right to life. This point is even made by Judith Jarvis Thomson, a proponent of abortion: "Surely the question of whether you have a right to life at all, or how much of it you have, shouldn't turn on the question of whether or not you are the product of a rape" (Thomson 1971: 49). Human beings possess the right to life from the moment they become a member of a biological species whose typical members naturally actualise the capability rationality in an empirically observable fashion, that is, from the moment of conception. So, from the perspective of the unborn person all abortions are equal in the sense that all abortions violate the right to life of a person. A person conceived in rape is still a person.

Contradicting this position, it is sometimes argued that abortion is permissible in case of rape because rape causes extreme psychological trauma in the rape victim and, therefore, changes the ethical landscape. This is an idea advanced by, for example, Susan Sherwin: "If the fetus is a result of rape or incest, then the psychological pain of carrying it may be unbearable, and the woman may recognize that her attitude to the child after birth will be tinged with bitterness" (Sherwin 1992: 101). It is then argued that, as a result, in case of rape the pregnant woman's right to bodily autonomy outweighs the unborn person's right to life.

However, this argument is deficient from a logical perspective. The premise that the extreme violation of a woman's bodily autonomy which is rape necessarily causes psychological trauma simply does not entail the conclusion that this trauma overrides the unborn person's right to life. Moreover, such a conclusion also contradicts the argument's tacitly underlying premise that the right to life outweighs, that is, is stronger than, a right to bodily autonomy: "If a right to life ordinarily outweighs a right to bodily autonomy, one cannot coherently say that this does not hold if there is a right to bodily autonomy that has been violated. To do so is flatly incoherent [...]. (I)f some violations of autonomy are grounds for abortion then it cannot be true that a right to life is morally weightier than a right to bodily autonomy" (Scarfone 2022: 136).

When it comes to essential abortions, the impermissibility of abortion extends to abortions during all stages of pregnancy, starting from the moment of conception and spanning to birth. Insofar as they are essential killings, this moral prohibition also includes

- early abortions, that is, abortions in the first trimester of a pregnancy
- late-term abortions, that is, abortions in the third trimester of pregnancy
- partial-birth abortions, that is, abortions of partially delivered children
- after-birth abortions,[26] that is, the killing of the newborn

Therefore, the impermissibility of abortion covers cases of essential embryocide, essential foeticide and essential infanticide.[27] This impermissibility is not influenced by the circumstances in which the unborn person came into being.[28]

While they are never ethically permissible as essential killing actions, abortions can, however, be ethically permissible as accidental killings. In the next chapter we look at the ethical (im)permissibility of this latter kind of abortion.

[26] This kind of "abortion" is, for example, advocated by Giubilini & Minerva (2013) but has already been argued for by, for example, Tooley (1972), Kuhse & Singer (1985) and McMahan (2007).

[27] In human pregnancy, the developing human being is called an embryo until the ninth week fertilisation age. From this time on and until birth, the unborn human being is referred to as a (human) foetus.

[28] Although all essential abortions are ethically impermissible, not all abortions are of equal ethical gravity by necessity. As Kaczor (2020: 82f) lists three aggravating reasons that make later abortion usually more serious than early abortion: "First, in early abortion, the prenatal human being can experience no pain. In abortion just prior to birth, the prenatal human being can experience pain. To kill someone in a painful way adds additional gravity to a killing. Second, in early abortion, the burden of pregnancy has not yet been endured through the second and third trimester. By contrast, in a late-term abortion, most of the pregnancy has already been completed, so it is (relatively) easier to finish the pregnancy. To fail to discharge a duty that is (relatively) easier to do is worse than to fail to discharge a duty that is (relatively) more difficult to do. Third, early in pregnancy, the humanity of the prenatal human being is more easily not understood. Just prior to birth, it is obvious to virtually everyone that the individual in question is alive and human. The culpability of the agent is in part determined by the knowledge of the agent. Early in pregnancy, it is more likely that an agent has inculpable ignorance about the humanity of the prenatal human being than in later pregnancy."

2.2.3. Accidental Abortions

Provided that the respective conditions are met, an abortion can only be ethically permissible as an accidental killing, that is, if the death of the unborn is the foreseen but unintended and unavoidable accidental consequence of an exterior action whose natural end is not the death of the unborn. Prototypical examples of such accidental (also: indirect) abortions are so-called therapeutic abortions. These arise if the mother has some serious disease and the only workable treatment has two effects, namely the cure of the disease and the death of the unborn.

An example of such a case is an ectopic pregnancy, that is, a pregnancy in which the embryo does not implant in the uterus but outside the uterus, for example (and in the vast majority of cases) in a fallopian tube. Unless the situation is remedied by a spontaneous abortion, such tubal pregnancy can be remedied by means of either the surgical removal of a fallopian tube with the embryo inside it (salpingectomy), the surgical removal of the embryo from a fallopian tube without removing the fallopian tube (salpingostomy or salpingotomy)[29] or the administration of methotrexate (MTX).[30] What about the ethical permissibility of application of these means?

While salpingectomies, salpingotomies and salpingostomies all manage to remedy the situation of an ectopic pregnancy and result in the death of the unborn, the former is an ethically permissible while the latter two are an ethically impermissible accidental killing. The use of methotrexate—although both non-invasive and non-surgical and therefore preserving the pregnant woman's fertility—shares the fate of salpingotomies and salpingostomies: it is ethically impermissible, too. Why this is the case becomes apparent once we apply the criteria for the ethical permissibility of accidental killings to the respective methods.

[29] In both salpingostomy and salpingotomy the fallopian tube is opened with a linear incision and the embryo removed from the tube (cf. Heinberg 2007). The difference between the procedures is that in a salpingotomy the incision is sutured closed while in a salpingostomy the wound is left to heal naturally (i.e. by so-called second intention).

[30] Methotrexate is an antimetabolite that was originally developed and continues to be used as a chemotherapy agent. Methotrexate interferes with DNA synthesis, repair and cellular replication of tissues with high rates of cellular proliferation, which does not only comprise cancer cells but also fetal and trophoblastic cells (cf. Verberne et al. 2019: 125). So, when used in the case of an ectopic pregnancy, it "inhibits cellular reproduction in the embryo as well as the trophoblast" (Kaczor 2005: 100). Accordingly, "as a known teratogen, methotrexate is highly toxic to embryonic tissue" (Anderson et al. 2011: 69).

In general, an accidental killing—and, therefore, an accidental abortion, too—is ethically permissible if all of the following questions are answered in the affirmative (for the complete formulation of these questions cf. Chap. VII, Sect. 1.2):

- Is the action an accidental killing action?
- Has the evil accidental consequence merely been foreseen but not intended?
- Does the action meet the exterior-action condition, that is, is the exterior action good or at least neutral?
- Does the action meet the end-of-the-agent condition, that is, is the end of the agent good?
- Does the action meet the means-end condition, that is, is the natural end of the exterior action realised at least as causally immediately as the evil accidental consequence?
- Does the action meet the proportionality condition, that is, is there a proportionately grave reason for permitting the exterior action's evil accidental consequence?

If an abortion meets all of these conditions, it is ethically permissible. If it fails one of them, it is ethically impermissible. If one applies these conditions to the above-mentioned procedures, salpingectomy, salpingotomy and salpingostomy, one will see that the former can be ethically permissible while the two latter are not. As the natural end of the surgical removal of an embryo from a fallopian tube is the embryo's death, both salpingotomy and salpingostomy fail the first condition already.

The same holds true with respect to the administration of methotrexate. Whatever else it may also be (e.g. mutilation), the natural end of administering a sufficiently large dose of a drug that is highly toxic to a living human being—as is methotrexate to the embryo—to this human being is the death of this human being. As such, this kind of poisoning—and, accordingly, the administration of methotrexate in cases of ectopic pregnancy, too—can only be considered a case of essential killing.

2.2.4. Abortion and the My-Body-My-Choice Argument

When it comes to the debate about the ethical (im)permissibility of abortions, there are two main areas of argument. The focus of discussion in the first area is the moral status of the unborn: Are all human beings persons from the moment of conception or not? If they are, then direct abortion is always ethically impermissible; if they are not, then abortion is ethically permissible in some or all cases. Proponents of abortion try to argue that the unborn does not start as but develops into a person at some point during the pregnancy. Depending on the suggested criterion (e.g. implantation, gastrulation, onset of neurulation, human appearance, conscious desires/organised electrical activity in cerebral cortex, viability, sentience, quickening, birth, valuation of future or self-awareness), human beings become persons and therefore acquire a right to life sometime during their embryological or foetal development or even after their birth. Opponents of abortion show that human beings are persons from the moment of conception and that there is no point during the embryological and foetal development that would not be highly arbitrary as the absolute criterion for the possession of the right to life. As shown in the chapter on personhood (cf. Chap. V), we do not have to further concern ourselves with the moral status and personhood of the unborn. Based on the notion of personhood as the possession of rational life, the debate in the first area of argument is decided in favour of the personhood of human beings from the moment of their conception.

The discussion in the second area of argument acknowledges this and assumes that the unborn is a person from the moment of conception and therefore possesses the right to life. For both sides of this debate, the personhood of the unborn is not a problem to be solved but the jointly shared starting premise of the debate. For example, Judith Jarvis Thomson, a prominent pro-abortion thinker, concedes that already the zygote is a person[31]: "I propose, then, that we grant that the fetus is a person from the moment of conception. [...] Every person has a right to life. So the fetus has a right to life" (Thomson 1971: 48). Nevertheless, she holds that abortion is ethically permissible.

The following paragraphs will deal with the standard argument advanced by the proponents of abortion who concede the personhood of the unborn from the moment of conception. What the argument basically does is frame the problem of the ethical permissibility of abortion as a

[31] It has to pointed out, however, that Thomson grants the zygote personhood for the sake of argument only. Although she is "inclined to think also that we shall probably have to agree that the fetus has already become a human person well before birth" (Thomson 1971: 47), she does not actually think that either a human zygote or foetus are a person: "A newly fertilized ovum, a newly implanted clump of cells, is no more a person than an acorn is an oak tree" (Thomson 1971: 48).

clash of rights, namely the right to life of the unborn on the one side and the pregnant woman's right to decide what shall happen in and to her body on the other side. For thinkers such as Judith Jarvis Thomson—whose "A Defense of Abortion" is perhaps the most (in)famous article on abortion—the right of the pregnant woman prevails. She explains why she thinks so by means of her well-known violinist analogy:

> You wake up in the morning and find yourself back to back in bed with an unconscious violinist. A famous unconscious violinist. He has been found to have a fatal kidney ailment, and the Society of Music Lovers has canvassed all the available medical records and found that you alone have the right blood type to help. They have therefore kidnapped you, and last night the violinist's circulatory system was plugged into yours, so that your kidneys can be used to extract poisons from his blood as well as your own. The director of the hospital now tells you, "Look, we're sorry the Society of Music Lovers did this to you—we would never have permitted it if we had known. But still, they did it, and the violinist now is plugged into you. To unplug you would be to kill him. But never mind, it's only for nine months. By then he will have recovered from his ailment, and can safely be unplugged from you." Is it morally incumbent on you to accede to this situation? (Thomson 1971: 48f)

Thomson comments that one surely has the right to unplug oneself from the violinist, even though the violinist will die. According to her, it is true that the violinist possesses a right to life. But the violinist's right to life does not entail the right to the use of another person's body.

> I am arguing only that having a right to life does not guarantee having either a right to be given the use of or a right to be allowed continued use of another person's body—even if one needs it for life itself. (Thomson 1971: 56)

Thomson then goes on to conclude that—by analogy—the right to life of the unborn does not entail the right to the use of their mother's body. So, a mother does not have the duty to not unplug herself from the unborn—for example, by means of an abortion. As long as the mother does not grant her unborn child the permission to grow and develop in her body, the child's right to life cannot outweigh and has to yield to the mother's right to decide what happens in and to her body. The mother has the right to refuse her child the use of her body. Thomson's argument features in the abortion debate in many shapes and forms, but the main tenet of the original argument as well as its variations can be simply put as: My body, my choice.

What should we think of this argument? As Kaczor (2018: 150) rightly points out, Thomson's argument "begins with the premise that persons have the right not to have their bodies used by others". However, this right not only applies to the mother but to the unborn child as well. For let us remember that the argument's basic premise was that all human beings are persons from the moment of conception. Accordingly, unborn children possess the same set of moral rights as their mothers, including the right to life and the right not to have their bodies used by others to keep them alive (let alone for lesser purposes). And this is where Thomson's argument becomes self-contradictory:

> If the 'right to control your own body' means anything, it means that no one can dismember a person without that person's consent. But this is precisely what takes place in a typical abortion. [...] To mangle or poison a person's body without the person's consent is to violate that person's rights. (Kaczor 2018: 151)

Thomson's argument rests on the idea that is it ethically wrong to force a person to force her body in order to help another person. However, thinking through this idea, it becomes obvious that it actually supports a pro-life position:

> If it is wrong to force one human being to give up some of her blood in order to save the life of another human being, then it is even more problematic to force one human being (the one in utero) to give up all of her blood, all her organs, and her life itself, not to save the life of another person, but to render another person free of pregnancy. (Kaczor 2020: 40)

As mother and child possess the same set of moral rights, the weighing of rights has to be done slightly differently than Thomson does it. It is not the child's right to life that has to be weighed against the mother's right to decide what happens in and to her body. It is the child's right to life plus her right to decide what happens in and to her body that stands against the mother's right to decide what happens in and to her body. As the child's and mother's rights to decide what happens in and to their bodies cancel each other out, the only right that remains is the child's right to life. So, if we take the premises of Thomson's argument seriously, we have to conclude that she fails to prove her point and that abortion remains ethically not permissible.

Furthermore, the violinist analogy rests on a very strange notion of pregnancy. Comparing the violinist's case with a pregnancy unrealistically and unfairly exaggerates the burdens of pregnancy. A pregnancy is not a nine-month kidnapping, and neither is it a disease—it is a natural condition:

> [In the violinist case] I cannot do my job, I cannot go visit my sick mother, I cannot go to my sister's wedding, I cannot go to films, I cannot go swimming, I cannot read (well, perhaps the violinist is a great talker), I cannot have a confidential conversation with anyone and I cannot make love. And all of this for a whole nine months. But the usual pregnancy does not make one bed-ridden, and even when it does, very rarely for nine months; nor is the foetus, even assuming it to be a person, someone whose presence rules out reading, private conversations, and sex. (Hursthouse 1987: 203)

While it requires a (depending on the circumstances, more or less great) sacrifice on the part of the mother, a pregnancy does not require a complete sacrifice on the part of the mother. A pregnant woman is not in a prison cell or in a coma; she is not—as the violinist analogy seems to insinuate—completely deprived of the use of her body. In reality, during the first weeks and sometimes even months a woman may not even be aware that she is carrying a baby: "A pregnancy does not impose (at least in the vast majority of cases) anything like the burden of the violinist. Indeed, some women report feeling better than normal both psychologically and physically during pregnancy" (Kaczor 2015: 179). And there is one more thing to consider: When using the language of sacrifice, one must not forget that the only person who is making a complete sacrifice is actually the aborted child; aborted children give up use of their body not partially and for some months but completely and forever.

In the end, justifying the ethical permissibility of abortions by means of a "my body, my choice"-type argument can produce a stalemate at the very best. For the mother's claim that it is her body and her choice can be countered by the unborn child's claim that it is her or his body and her or his choice.

2.3. *Euthanasia*

After having dealt with killing actions at the beginning of a human being's life, we now turn to killing actions that take place at the other edge of life. While abortions involve the killing of an unborn or newborn human being, these killings result in the death of an already dying or at least seriously ill, suffering human being. Such killings are summarised by the somewhat euphemistic term euthanasia. In general, euthanasia can be defined as "the giving of an easy painless death to one suffering from an incurable or agonizing ailment" (Fagothey 1963: 240), that is, a patient.[32] This chapter is devoted to answering the question under which conditions, if any, euthanasia is ethically permissible. This question will be answered by reference to the general ethics of killing as it has been established over the course of the preceding chapters.

2.3.1. Mapping the Concept

Before addressing the ethical (im)permissibility of euthanasia, it is helpful to distinguish several kinds of euthanasia. A first distinction can be made between what is generally called active euthanasia and passive euthanasia:

- In ***active euthanasia*** a human being causes the death of an already dying or at least suffering patient. For example, this can be done by injecting the patient with a lethal drug but also by withholding or withdrawing the *ordinary* means for the preservation of a patient's life.
- ***Passive euthanasia*** generally refers to the withholding or withdrawing of life-prolonging medical treatment. To be more precise, passive euthanasia is not just the withholding or withdrawing of any life-prolonging medical treatment; it is the withholding or withdrawing of the *extraordinary* means for the preservation of a patient's life. Without this small but important limitation, passive euthanasia would run the risk of becoming partially indistinguishable from active euthanasia.

  As the term "life-prolonging" implies, the patient must be in a situation where she would die without the respective extraordinary medical treatment. Therefore, passive euthanasia requires a patient who is suffering not simply from a chronic but not deadly affliction but who would—as a proximate consequence—succumb to an already existing underlying deadly condition unless some extraordinary life-prolonging medical treatment is started or continued.

[32] It has to pointed out that euthanasia does not necessarily have to do with physical suffering and pain or fear of such suffering. There is also what can be termed "existential suffering". A person suffering existentially usually suffers the loss of autonomy, that is, the loss of the ability to do the things one used to do and loved (e.g. because of paraplegia).

> Passive euthanasia means forgoing therapeutic overzealousness in cases (a) when a patient's death cannot be prevented but only delayed and (b) when this prolongation of life can only be achieved by extraordinary means. In this sense, a patient can be said to be allowed to die by passive euthanasia. In passive euthanasia it is not the withholding or withdrawing of treatment that causes the patient to die; what causes her death is her underlying affliction that is allowed to take its deadly course.

The choice of terms for these kinds of euthanasia is somewhat unfortunate and requires some explanation that hopefully prevents confusion. Otherwise, the terms become—as they are sometimes accused of being—ambiguous, misleading and unhelpful, indeed. The term "active euthanasia" seems to suggest that the person performing this kind of euthanasia must play an active role, that is, must perform a positive action. Similarly, the term "passive euthanasia" seems to imply that the person performing this kind of euthanasia must remain passive, that is, must perform a negative action. Following this semblance, the distinction between active and passive euthanasia is sometimes falsely equated with and reduced to the distinction between killing by commission ("doing") and letting die ("allowing").[33]

However, both active and passive euthanasia can be performed by an omission or a commission. Withholding the extraordinary means for the preservation of her life from a patient is an omission; withdrawing these means from her is a commission. In both cases, the patient is not killed but allowed to die; both actions are instances of passive euthanasia. Injecting a patient with a lethal dose of drugs is a commission; letting a patient who has slipped and hit her head in the bathtub drown is an omission. In the first case the patient is killed, in the second case she is let die, but both actions are instances of active euthanasia.

As has been outlined above, we are only duty-bound to apply the ordinary means to preserve our life and the lives of other human beings. So, as defined above, passive euthanasia is ethically permissible—provided, of

[33] Such equivalence can be found, for example, in the following statement by Rachels (1975: 78): "The distinction between active and passive euthanasia is thought to be crucial for medical ethics. The idea is that it is permissible, at least in some cases, to withhold treatment and allow a patient to die, but it is never permissible to take direct action to kill a patient."

course, that the end of the agent performing it is not the death of the patient. For the rest of this chapter we can therefore focus our attention on the problem of the ethical (im)permissibility of active euthanasia.

To this end, it is helpful to shed some more light on the action named active euthanasia. For within the class of active euthanasia, we can make two basic distinctions:

- First, we can distinguish between voluntary, non-voluntary and involuntary active euthanasia. This distinction is based on the presence or absence of the killed patient's express consent.
  - In ***voluntary active euthanasia*** "the patient agreed to be killed or requested it" (Oderberg 2000: 50). So, voluntary active euthanasia is performed at the request or with the express informed consent of the killed patient.
  - In ***involuntary active euthanasia*** the person killed either "refuses or otherwise actively withholds consent (again, whether it is at the time of the killing or in an advance directive)" or "could have been asked whether he consented, but was not" (Oderberg 2000: 52). Involuntary active euthanasia is performed against the will of the killed patient. It means imposing a death-causing action on a patient who does not want to die against her will or would be able to express her will but is not asked.
  - In ***non-voluntary active euthanasia*** "the killing of someone who is not capable of being consulted as to his wishes" (Oderberg 2000: 53). Examples of this kind of euthanasia are the euthanisation of handicapped babies, comatose patients or patients with severe dementia. Active euthanasia is non-voluntary if the explicit consent of the killed patient is not only not available (e.g. the patient has simply not been asked) but unavailable (i.e. is incompetent to express her will).

As it goes against the will of the patient, involuntary active euthanasia is impermissible from an ethical perspective. Non-voluntary active euthanasia is ethically impermissible when it comes to the withholding or withdrawing of the ordinary means for the preservation of life; it can only be ethically permissible as regards the withholding or withdrawing of the extraordinary means for the preservation of the life of the patient. As the former two cases are ethically impermissibly and as the latter case is governed by the already known ethics of ordinary and extraordinary means, we shall only concern ourselves with voluntary active euthanasia hereinafter.

- Second, we can distinguish between direct and indirect active euthanasia.
  - ***Direct active euthanasia*** is the essential killing of a patient. In this kind of euthanasia, the death of the patient is brought about by an essential killing action.
  - ***Indirect active euthanasia*** is the accidental killing of a patient. In this kind of euthanasia, the death of the patient is brought about by an accidental killing action. An example of indirect active euthanasia is the administration of a palliative that not only eases the pain of the patient but can also result in the patient's death.

As has been stated at its outset, this chapter is devoted to answering the question under which conditions, if any, euthanasia is ethically permissible. The above distinctions now enable us to sharpen this question. As we have already excluded some forms of euthanasia from our discussion as either ethically permissible (passive euthanasia) or impermissible (involuntary active euthanasia, non-voluntary active euthanasia), the only forms whose ethical (im)permissibility remains to be assessed are direct voluntary active euthanasia and indirect voluntary active euthanasia. Let us start with the former kind of euthanasia.

2.3.2. Direct Voluntary Active Euthanasia

The general ethics of killing developed in the course of the previous chapters commits us to the position that euthanasia is ethically impermissible if it is an essential killing, that is, if the death of the patient is the natural end of the exterior action. This condition is met in the case of direct voluntary active euthanasia. So, from an ethical perspective, little need be said on this topic.

While one can sympathise with the sentiments of pity and mercy underlying the arguments for this kind of euthanasia, there is no ethical justification for direct voluntary active euthanasia. Or as Fagothey (1963: 240f) puts it: "Even if euthanasia should be made permissible by civil law, for which there is strong propaganda in some quarters, nothing can make it moral". Direct voluntary active euthanasia is an ethically impermissible action—regardless of whether the person euthanised is an elderly person or an infant.[34]

The ethical impermissibility of direct voluntary active euthanasia not only affects the person who is killing the patient by, for example, giving her a lethal injection, it also affects the patient who agrees to be killed or requests it. In direct voluntary active euthanasia, the mind of the patient is set on killing herself. Although she does not—technically speaking—commit suicide, her intention in partaking in the action is to kill herself. No matter what we call the sub-action that the patient performs as part of a direct voluntary active euthanasia, her end of the agent is her own death. So, both the person who performs direct voluntary active euthanasia as well as the patient who lets herself be euthanised thus commit an evil and ethically impermissible action. Direct voluntary active euthanasia is suicide and murder combined.

Advocates of this kind of euthanasia often argue that direct voluntary active euthanasia is not an ethically impermissible type of killing as the victim agrees or even requests to be killed. According to this autonomy argument, a patient's consent suffices to make direct voluntary active euthanasia ethically permissible. Unfortunately, the argument misses or ignores the fact that life is a basic human good that is not at the disposal of human beings—unfortunately, not even of a human being who is sick and suffers. Man is not master of but only steward of his corporeal life. In addition, a patient's request for euthanasia—just like suicide—is not the ultimate exercise of autonomy but rather its ultimate self-contradiction: a supposedly autonomous act that destroys the subject of autonomy and autonomy itself.

In the end, the arguments advanced for the ethical permissibility of direct voluntary active euthanasia are the same as those for justifying suicide. Both euthanasia and suicide force us to deal with the role of suffering in human life:

> There is nothing in this life, no matter how unwelcome to us, that is not better than annihilation. For annihilation is nothing, and in nothing there is no perfection and no "good." And if this reply, though it really strikes at the root of the present difficulty, is regarded as too abstract to afford comfort in bearing the trials of life, we answer that it is not meant to give comfort, but

[34] Some, such as Kuhse & Singer (1985: 143), have argued for the permissibility of the direct, and necessarily non-voluntary direct active euthanisation of newborns.

> only to represent the true facts of the case. But there are other considerations also that can supply all the comfort and sustaining power that are required. In every life, no matter how unhappy, there is much good. The evil of each one's life is but one of its many elements, and it is outweighed many times by the good. It is the very essence of sorrow, however, that it turns our attention away from the good and fixes it upon the evil, and thus we find it hard to realise that in an unhappy life there is any real good or happiness. (Cronin 1922: 55; cf. Commer 1886: 142)

Unlike what is suggested by those in favour of euthanasia, it is questionable whether euthanasia can be properly described as actually relieving suffering: "To be relieved of suffering, a person must be in a position to experience the relief of suffering. But if a person is dead, then the person no longer has any bodily experiences. The dead feel neither pain nor relief of the pain" (Kaczor 2020: 140).

Last but not least, we should also ask whether a patient's motivation for asking for direct voluntary active euthanasia is best served by actually killing her.

> Scientific studies reveal that most patients who seek euthanasia are motivated by psychological factors—often depression—not by unbearable physical suffering. [...] (P)hysical pain plays a very small role in motivating patients' interest in or requests for euthanasia. [...] (P)sychological factors are what commonly drive patients' interest in euthanasia and physician-assisted suicide. Depression, hopelessness, anxiety and the like are why patients request aid in dying. (Emanuel 1997; cf. Lübbe 2015: 156)

As in suicide, a patient's request for euthanasia is often the result of desperation, an unmet need for closeness and understanding, or the desire to not be a burden to or relieve one's loved ones—and sometimes the unfounded fear of unbearable suffering, depression and loneliness. Death is then seen as the alternative to a life no longer deemed worth living. But what the patient really wishes for is not death, but a different kind of life. Her wish to be euthanised does not translate into "I want to die" but rather "I do not want to live like this". If the patient's situation could be alleviated (e.g. by palliative medicine or social attention), death would not be an option for her. A patient susceptible to the idea of euthanasia or suicide usually does not wish for non-existence but rather not being miserable. We should, therefore, be careful to not throw the baby out with the bath water. In the vast majority of cases, direct voluntary active euthanasia does not solve the patient's actual problem—rather, it gets rid of the subject who has the problem.

2.3.3. Indirect Voluntary Active Euthanasia

Direct voluntary active euthanasia is ethically impermissible. But what about indirect voluntary active euthanasia? The standard textbook case for this kind of euthanasia is so-called palliative sedation. In palliative sedation a patient who suffers from persistent, refractory, that is, therapy-resistant, and intolerable pain is continuously administered an infusion with a sedative drug. The drug is administered in a dose and combination that reduces the patient's consciousness to a level that adequately relieves her suffering—even to the point of unconsciousness. Palliative sedation can be applied for a limited amount of time and then discontinued; and in some cases, it can continue until the death of the patient. However, while relieving the patient from otherwise unbearable pain, the practice can—depending on the dose of the medication—also hasten her death.[35]

Therefore, palliative sedation is a case of an action with a double effect, one good (the alleviation of pain) and one evil (death). As such, palliative sedation in particular as well as indirect voluntary active euthanasia in general are instances of accidental killing. This means that the question whether palliative sedation and indirect voluntary active euthanasia are ethically permissible can be answered by reference to the general conditions for the ethical permissibility of accidental killings.

In general, an accidental killing—and, therefore, accidental euthanasia, too—is ethically permissible if all of the following questions are answered in the affirmative (for the complete formulation of these questions cf. Chap. VII, Sect. 1.2):

[35] While sedation can slow breathing and lower blood pressure as well as heart rates to potentially dangerous levels, the question whether palliative sedation actually hastens death remains unsettled. A study by Sykes & Thorns (2003: 312) concludes that "there is no evidence that initiation of treatment, or increases in dose of opioids or sedatives, is associated with precipitation of death". This finding has also been confirmed by Maltoni et al. (2009: 1168) whose study attest "a lack of impact of PST [palliative sedation therapy] on survival in sedated and nonsedated hospice patients".

Pointing to "clearly a significant lack of research into this specific area" (Maltoni et al. 2009: 1168), the studies do not, however, categorically deny a possible life-shortening effect of palliative sedation in selected cases: "While studies have shown that palliative sedation does not seem to hasten death in the majority of patients, the retrospective nature of these data leaves some clinicians uncertain about the effects of palliative sedation on hastening death" (Patel et al. 2019: 841).

- Is the action an accidental killing action?
- Has the evil accidental consequence merely been foreseen but not intended?
- Does the action meet the exterior-action condition, that is, is the exterior action good or at least neutral?
- Does the action meet the end-of-the-agent condition, that is, is the end of the agent good?
- Does the action meet the means-end condition, that is, is the natural end of the exterior action realised at least as causally immediately as the evil accidental consequence?
- Does the action meet the proportionality condition, that is, is there a proportionately grave reason for permitting the exterior action's evil accidental consequence?

If a case of indirect voluntary active euthanasia meets all of these conditions, it is ethically permissible. If it fails one of them, it is ethically impermissible. So, what about the ethical permissibility of palliative sedation? If we apply the above conditions, we can see that palliative sedation is an ethically permissible practice. In palliative sedation, the death of the patient is a foreseen and unintended accidental but not an essential consequence of its exterior action. Considered for itself, the exterior action of alleviating pain by means of a sedative drug is a morally good action. Furthermore, the alleviation of pain is not achieved by means of the patient's death; rather, the sedation's good consequence (alleviation of pain) follows from the exterior action before its evil consequence (the patient's death). And last but not least, the alleviation of unbearable pain can be judged to be a reason that is proportionately grave to counterbalance the patient's death. However, should the person administering the sedative drug intend the death of the patient, this turns otherwise ethically permissible accidental euthanasia into an ethically impermissible action.

2.4. Organ Transplantation

Organ transplantation refers to the practice of taking, that is, explanting organic material (cells, tissue, organs and/or organ systems) from one human being (the explantee) and implanting this transplant into another living human being (the implantee). So, transplantation consists of the explantation and subsequent implantation of some organic material. The specific kind of transplantation that is of interest here is the transplantation of organic material from one human being to another human being.[36] Both explantation and implantation are usually performed by a surgeon. The object of transplantation is called transplant but can also be referred to as explant or implant, depending on whether one considers the transplant from the perspective of the explantee or implantee: From the perspective of the explantee, the transplant is called explant; from the perspective of the implantee it is called implant.

The practice of organ transplantation is governed by the so-called dead donor rule (DDR). In general terms, this rule prohibits the killing of organ donors in order to obtain their organs. As such, the DDR is an application of the general ethical prohibition-duty to not kill an innocent human being to the specific practice of the transplantation of organic material. As should be obvious, the DDR is in line and congruent with the general ethics of killing underlying this book.

> The "Dead Donor Rule" (DDR) lies at the heart of current organ procurement policy. It is not a legal statute; rather, it reflects the widely held belief that it is wrong to kill one person to save the life of another. On those grounds, an organ donor must already be dead before vital organs are removed. The DDR is therefore an ethical norm: vital organs may be removed only after the organ donor is dead. (Sade 2014: 1131)

According to the DDR, any explantation of organic material from an explantee destined for the subsequent implantation of this organic material into an implantee that results in and causes the death of the explantee is ethically impermissible. From the perspective of the dead donor rule, an organ transplantation can only be ethically problematic in case of a transplantation of vital organic material.

Vital organic material is organic material that is indispensable to the explantee's life. Examples of vital organic material are nonpaired organs such as the heart, liver or pancreas. Examples of organic material that is

[36] If the transplantation occurs between two genetically non-identical human beings, this kind of transplantation is called "allotransplantation". If it occurs between genetically identical human beings, it is called syngeneic transplantation or isotransplantation. Cases of autotransplantation, that is, transplantations of organic material from one part of the body of a human being to another part of the body of the same human being, are not within the scope of our analysis. What is also outside the scope of our analysis is cross-species transplantation (xenotransplantation), that is, the transplantation of organic material of a non-human animal into a human being.

		Kind of organic material explanted in transplantation	
		Vital organic material	**Non-vital organic material**
Status of the explantee	**Living**	**Premortal transplantation of vital organic material** → ***Causes death of explantee*** → ***Violates dead donor rule***	**Premortal transplantation of non-vital organic material** → *Does not violate dead donor rule*
	Dead	**Postmortal transplantation of vital organic material** → *Does not violate dead donor rule*	**Postmortal transplantation of non-vital organic material** → *Does not violate dead donor rule*

Fig. 1 Transplantation and the dead donor rule

not indispensable to the explantee's life are paired organs such as the kidneys and lungs, but also blood, bone marrow, tissue (e.g. cornea), parts of the intestines or parts of the liver. While an explantee can donate the latter (or at least parts of the latter) and continue life (albeit with more or fewer major restrictions), donating vital organic material necessarily results in the death of the explantee. By definition, explanting vital organic material from a living human being kills this human being.

So, given these facts, we can conclude the following: Whether an organ transplantation violates the dead donor rule depends on the answer to the following two questions (cf. Fig. 1):

- What kind of organic material is to be explanted?
- Is the explantee dead or alive?

Neither the explantation from a dead donor nor the explantation of non-vital organic material can constitute a violation of the DDR.

In fact, the DDR only prohibits the explantation of vital organic material from a living human being—or put differently: When it comes to the transplantation of vital organic material, the DDR only allows the explantation of vital organic material from a corpse, that is, a dead human being. The explantation of vital organic material from a living human being is an essential killing action and therefore ethically impermissible. Regardless of the end of the agent performing this explantation (e.g. saving another human being), the natural end of this exterior action is the death of the explantee. It is by necessity that a living explantee dies by the explantation of vital organic material.

Therefore, whether the explantation of some vital organic material constitutes a violation of the DDR depends on whether the explantee is dead before the explantation process begins: If the explantee is dead, the DDR is not violated; if the explantee is not dead, the DDR is violated. So, when it comes to the ethical permissibility of the transplantation of vital organic material the crucial question is: When is a human being truly dead?

We have dealt with this question in detail in Chap. IV. There it has been established that a previously living human being is dead, that is, has completely lost his life and can therefore no longer show any signs of life or endogenous activity, if he shows no empirically observable signs of life and if he shows the early certain signs of death. This definition and criteria of death have been derived from a definition of life that must logically precede every definition and criterion of death. As the presence of the certain signs of death is the strictest criterion of death and logically presupposes the presence of uncertain signs of death as well as the absence of any previously present empirically observable signs of life, the death of a human being can be considered validly established if the human being shows the early certain signs of death. Thus, the short version of the above definition and criteria of death reads as follows: A human being has completely lost his life if he shows the early certain signs of death.

So, what is the result if we apply the above definition and criteria of death to the problem of the explantation of vital organic material from a living human being? Given this definition and criteria of death, the explantation of vital organic material from a living explantee is ethically permissible only if the explantee shows the early certain signs of death. The two typical certain signs of death are postmortem lividity (*livores mortis*) and postmortem rigidity (*rigor mortis*). The latter signifies the stiffening of the limbs of the deceased; although the time of onset and the full development are highly variable, postmortem rigidity of larger muscle groups usually occurs two to six hours after circulatory death. The former refers to the bluish-purple discoloration of or under the skin of the lower or dependent body parts of the deceased that is caused by the gravitational settling of his blood; the first patches develop within 30 minutes to three hours after circulatory death.

However, the time periods until the appearance of the certain signs of death have serious consequences for the feasibility of an organ transplantation. While the organs of a corpse that shows the certain signs of death can still be explanted, they are not implantable to another human being. They can, of course, be put into the implantee but as they are dead they will not work again; due to severe ischaemic injury they are already no longer viable at the time of procurement.[37] So, organs harvested from a corpse that shows the certain signs of death cannot be of any use to the implantee. The transplantation of such organs defies the purpose of organ transplantation.

We are now facing a dilemma: If an organ is explanted from a dead human being, that is, from a human being who shows the certain signs of death, the organ is useless for implantation; this is true of both vital and non-vital organs. However, if a vital organ is explanted before a human being shows the certain signs of death, the human being is not yet dead at the time of the explantation of his vital organ and, therefore, killed by the explantation. This would be an essential killing that is, as such, ethically impermissible and in violation of the DDR. So, given the general ethics of killing and the definition as well as criteria of death outlined above, the explantation of vital organic material from a human being who does not

[37] Due to increasing irreversible ischaemic injury, the organic material of a circulatory-dead human being remains viable only for a certain period of time; for unpaired vital organs this time window of viability between circulatory death and explantation is usually less than 30 minutes. As they remain perfused with blood and therefore do not suffer ischaemic damage, organs that are harvested from a brain-dead human being are usually viable for implantation.

This difference is also highlighted by Dunne & Doherty (2011: 82): "The main difference between DCD [i.e. donation after circulatory death] and DBD [i.e. donation after brain death] organs is the duration of warm ischaemia. Warm ischaemia commences when there is inadequate oxygenation or perfusion of the organ as defined by an SAP <50 mm Hg, oxygen saturation <70%, or both such as during withdrawal of treatment or cardiac standstill. The period of warm ischaemia continues until the initiation of cold perfusion. Cold ischaemic time extends from initiation of cold preservation until restoration of warm circulation after transplantation. Cold perfusion will slow down tissue metabolism, resulting in a reduction in the rate of ischaemic injury. During DBD, organs undergo cold perfusion before organ retrieval and thus warm ischaemia is minimal. However, DCD by definition will incur some degree of warm ischaemia as there will be an interval after asystole where organs are not being perfused and have not yet been cooled. As a result, organs from DCD donors have the potential to develop irreversible damage due to the accumulation of ischaemic metabolites."

show the certain signs of death is ethically impermissible—it means committing an essential killing action.

Of course, one could start tampering with the definition of death in order to overcome this dilemma and create a time slot that allows the subsequent use of the explant. One such attempt is to define death as circulatory death and focus on circulatory death as the sole criterion of death; this is, for example, done by the proponents of the ethical permissibility of "donation after cardiac death" (DCD; also: non-heart-beating donation), that is, "the surgical removal of organs after the pronouncement of death based on the cessation of cardiopulmonary function" (Edwards et al. 1999: 293f). Another attempt is to redefine death as the loss, that is, absence, of previously present brain activity; this is done by proponents of the ethical permissibility of "donation after brain death" (DBB). But both such redefinitions would not be in line but rather contrary to reality, that is, result in a very strange and distorted notion of what it means to be a living being. This has already been shown in Chap. IV, Sect. 5.

One can and perhaps even should sympathise with the intention that might drive the efforts to redefine death, namely the saving of the lives of otherwise dying persons. But a good intention does not yet make a good action. Calling living human beings dead does not change the fact that they are killed by the explantation of vital organic material. While this linguistic trick might calm consciences, one would still be killing a living, though admittedly dying, human being to save the life of another—an action that cannot be approved in light of the ethical theory advanced in this book.

3. The Ethics of Helping Others Kill Themselves

So far, we have been dealing with the ethics of killing oneself and the ethics of selected cases of killing another human being. This chapter discusses the morality of a human action by which a human being helps another human being to kill herself. So, the action that is at the heart of this chapter is not a killing action in the strict sense. What we are actually discussing is the ethical impermissibility of assisted suicide, that is, the cooperation in the suicide of another human being.

Most proponents of this practice stipulate that such cooperation may only take place in a medical context and be rendered by a physician (or some other representative of the medical profession[38]). Therefore, this

[38] Such a representative could not only be a nurse, a pharmacist or even an orderly (in the case where the orderly, e.g., collects the lethal drug from the pharmacy) but also a hospital manager.

cooperation is usually called "physician-assisted suicide" (PAS). In PAS, a physician provides a patient who is able and willing to commit suicide with the knowledge and means (usually a lethal drug) to eventually commit suicide.

> A paradigm case of physician-assisted suicide is a patient's ending his or her life with a lethal dose of a medication requested of and provided by a physician for that purpose. A paradigm case of voluntary active euthanasia is a physician's administering the lethal dose, often because the patient is unable to do so. The only difference that need exist between the two is the person who actually administers the lethal dose-the physician or the patient. In each, the physician plays an active and necessary causal role. (Brock 1992: 10)

In PAS, it is not the physician but the patient who is committing the actual suicide. So, while the physician can prescribe and prepare the lethal drug (usually a barbiturate) on behalf of and even place it within range of the patient, he does not administer it. It is the patient who actually performs the self-killing action, that is, drinks the lethal drink or presses the plunger of a syringe or pushes a button that lets the poison trickle into her veins (if an intravenous drip is used) or into her stomach (in case a stomach tube is used). This is what sets PAS apart from euthanasia where it is the physician who is actually performing the killing action.

Physician-assisted suicide is a highly controversial topic. The debate about its ethical (im)permissibility has produced a vast amount of literature. This chapter is not, however, meant to review the arguments of the respective proponents and opponents of this practice. What it will rather do is to evaluate PAS by reference to an ethical argumentative device that is a logical extension of the general ethics of killing as it has been developed over the course of the first part of this book. This device is called "cooperation in wrongdoing" and its knowledge is vitally important for the assessment of the moral (im)permissibility of assisted suicide.

In cooperation in wrongdoing (also: cooperation in evil actions), a person ("cooperator") assists in the evil action ("primary action") of another person ("principal agent"). When thinking about such cases, the initial reaction tends to be rather straightforward: If an act is evil, assisting someone in its performance—be it by commission or omission—makes the cooperative action evil, too. However, while this is true in some or even many cases, it is far from true in all cases. What is therefore required is a concise understanding of the ethics of cooperation in wrongdoing. In order to develop such understanding, it is helpful first to distinguish between different kinds of cooperation and then enquire into the ethical evaluation of cooperation in wrongdoing. Only then and on this basis can we turn to the question of whether assisted suicide in general and physician-assisted suicide in particular is an ethically permissible or an impermissible action.

3.1. Mapping Cooperation in Wrongdoing[39]

When we think of evil actions, we tend to think of particular evil actions performed by an individual agent or sometimes by a group of agents, such as stealing, murdering, cheating or committing adultery. But as reality shows, there are also situations when we are not performing an evil action but are nevertheless somehow implicated in another person's evil action, namely by being an assistant or cooperator in that act.

> Being a cooperator is subsidiary to being a primary agent—the performer of the primary act—but it is a kind of partnership in that act that receives, as it were, some of its moral taint. (Oderberg 2018: 41)

Generally speaking, cooperating in another person's wrongdoing means helping a person, whom one has not seduced, to carry out or to carry out more easily an evil action.

> Co-operation is physical activity (or its omission) by which a person assists in the evil act of another who is the principal agent. (Higgins 1949: 353; cf. McHugh & Callan 1958: §1506)

What the above definition of cooperation actually captures is cooperation strictly understood. By cooperating, the cooperator is affording help to a primary agent to carry out an evil action. However, besides such "effective coöperation" (McHugh & Callan 1958: §1509) or cooperation in a strict sense there are other ways "in which a person can share the guilt of another's wrongdoing" (Oderberg 2004a: 203). These types of cooperation in a wide sense are called "occasional coöperation" (McHugh & Callan 1958: §1509) and comprise scandal, seduction and solicitation[40]:

[39] The content of this chapter draws on, and in part reproduces verbatim, material published in Erk (2019: 245–250).

[40] Occasional cooperation is called occasional not because it happens at irregular or infrequent intervals but because it presents somebody who did not intend to do so with an occasion to do evil. Avoiding this ambiguity, instead of effective and occasional cooperation, Ross (1919: 232) speaks of physical and moral cooperation "according as it influences the act or will of the agent".

- Scandal "refers to a fall into sin which one occasions for others by misconduct" (McHugh & Callan 1958: §1445). In scandal, a person leads another person into committing an evil action by bad example, be it in the form of word, deed or omission.
- Seduction is "a strong form of inducement to wrong, typified by command, counsel [...] and enticement through praise or provocation" (Oderberg 2004a: 203; cf. McHugh & Callan 1958: §1496). Besides these forms of seduction that primarily aim at intellect and will, there are also more subtle forms of seduction that act on body, senses and imagination, for example seduction by surrounding others with enticing pictures, companions, music and so on (cf. McHugh & Callan 1958: §1505).
- Solicitation refers to "a softer form of inducement typically involving requests, appeals, and invitations" (Oderberg 2004a: 203).

An occasional cooperator induces an agent who did not intend to do so before to perform an evil action; however, such a cooperator does not help the agent with the commission of the evil action. In contrast, an effective cooperator does not induce but helps in the commission of an evil action; effective cooperation supposes that the perpetrator of the evil action has already made up his mind before the cooperation takes place. The following explanations only focus on cooperation strictly speaking, that is, effective cooperation. Therefore, when the term "cooperation" is used in the following text, it means and is shorthand for "effective cooperation".

So, let us take a closer look at the concept of effective cooperation. As already indicated in the above definition of (effective) cooperation, a person can cooperate in the wrongdoing of another person by a negative or positive action. Whether somebody cooperates by an omission or a commission depends on whether the cooperator "does something to assist the principal or does nothing to hinder him" (Oderberg 2004a: 206). Positive cooperation is cooperation by means of a positive action (commission) while negative cooperation is cooperation by means of a negative action (omission).[41]

[41] Examples of negative cooperation are cooperation by silence or cooperation by not opposing. In the former case, a person cooperates by keeping silent in view of an evil action, rather than, for example, scaring off a thief by shouting or calling the police. In the latter case, a person cooperates if she does not impede the wrongdoing by deed, as, for example, when she does not frustrate a robbery by defending the property.

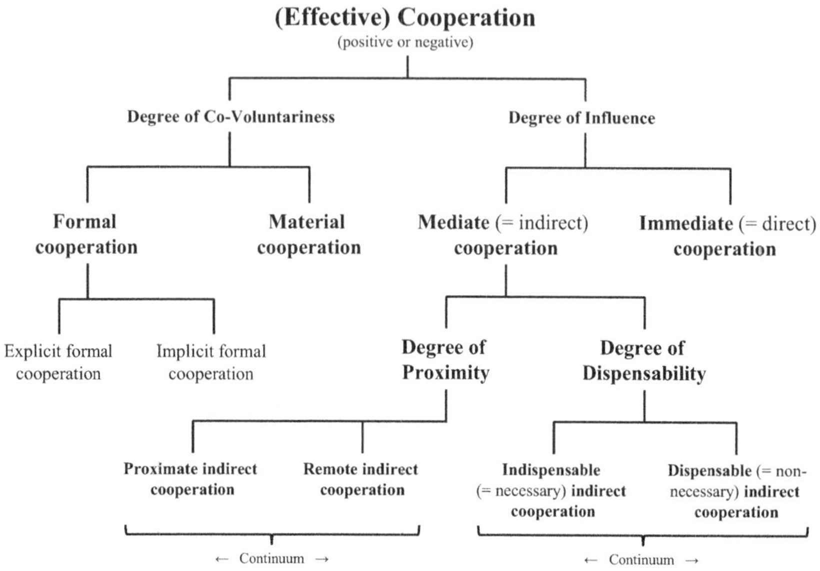

Fig. 2 Kinds of (effective) cooperation

Next, let us introduce two distinguishing criteria that will be important and the basis for the ethical evaluation of (effective) cooperative acts, namely their "degree of co-voluntariness" and their "degree of influence" (cf. Fig. 2):

A cooperative action's ***degree of co-voluntariness*** refers to the extent to which the cooperator approves of the evil action taken by and the purpose of the principal agent. A cooperative action's ***degree of influence*** concerns the question of how implicated the cooperator is in the evil act, that is, how tightly connected or near the cooperating action is to the evil action.

Depending on its degree of co-voluntariness, a cooperation can either be formal or material[42]:

[42] The locus classicus for the distinction between formal and material cooperation is usually taken to be Alphonsus Ligouri (1852: 180 (Liber 3, Tractatus 3, Caput II, Articulus III, §63)).

- ***Formal cooperation*** "occurs when a person explicitly approves of another's action or shares the principal agent intentions, either for its own sake or as a means to some other goal" (Melé 2009: 120). Formal cooperation is the willing participation by the cooperator. In this kind of cooperation the cooperator cooperates while approving of the evil action, that is, willing the evil action taken by the principal agent to occur. In other words, in formal cooperation both the principal agent and the cooperator want the same *finis operis*: "If you assist someone to rob a bank with the intention that the bank be robbed, you are a formal cooperator" (Oderberg 2018: 44).
- Formal cooperative actions can either be implicit or explicit. This distinction refers to the congruence of the cooperator's and the primary agent's end of the agent (*finis operantis*):
  - A formal cooperation is explicit "when the co-operator's end is precisely the wrong of the principal, such as joining a terrorist organization while agreeing with its aims" (Oderberg 2004a: 206; cf. McHugh & Callan 1958: §1511; Noldin 1935: 117). In ***explicit formal cooperation***, the *finis operantis* of the cooperator is the realisation of the evil act; the cooperator is assisting so that the evil action can be realised. An explicit formal cooperator not only wills the evil action's *finis operis* but does so for the same purpose as the primary agent. For example, a person who joins a terrorist organisation and cooperates in their evil actions because she is in agreement with its aims (e.g. to overthrow government) would be an explicit formal cooperator.
  - In ***implicit formal cooperation***, the end of the cooperator does not consist in the promotion of the evil action as such; he does not help the principal agent in order that the evil action be realised but as a means to realise some other purpose of his own. A formal cooperation is implicit "when the co-operator does not intend to associate himself with the wrong of the principal but nevertheless intends for some other reason that the principal should do what he is set upon doing (or consents to its being done)" (Oderberg 2004a: 206; cf. McHugh & Callan 1958: §1511; Noldin 1935: 117). For example, a person who joins a terrorist organisation and cooperates in their evil actions not because she subscribes to its aims (e.g. to overthrow government) but for the sake of some social or financial advantage is an implicit formal cooperator.

In order to cooperate formally, a cooperator does not have to be enthusiastic about the intention of the primary agent; it is enough for the cooperator to reluctantly will what the primary agent wills.

- In a case of ***material cooperation***, the cooperator does not intend that the evil action with which he cooperates be performed but nevertheless assists in its performance. This means that his cooperation cannot be perfectly voluntary but has an element of pressure to it; for why else would somebody who does not want an evil action to happen cooperate in it? It is characteristic of material cooperation that the cooperator is in some way pressured into helping the principal agent: "The material cooperator is a reluctant or unwilling assistant who might know what the principal agent intends but who cooperates out of fear, pressure, inducement and more generally a desire to avoid some loss or gain some benefit to herself or a third party" (Oderberg 2017a: 193, cf. 2018: 51; 44).
  Material cooperation therefore has two effects: "The evil effect of the wrong committed by the principal and the good effect of avoiding a loss and/ or retaining or procuring a good on the part of the co-operator" (Oderberg 2004a: 212). As we will see, this insight will be important for the ethical evaluation of material cooperation.

Depending on its degree of influence, a cooperative action can be either direct or indirect:

- ***Direct*** (also: immediate) ***cooperation*** either means that the cooperator performs some part of an evil action himself or that the cooperator and the principal agent jointly perform an evil action. So, in direct cooperation the cooperator directly participates in the performance of the evil action in such a fashion that he could be said to perform the evil action himself (at least partially)—and by doing so becomes an accomplice (or as McHugh & Callan (1958: §1507) put it: "an equiprincipal or coordinate agent") to the evil action. If a cooperating action, "described from an external perspective, completely overlaps with the action of the principal agent" (Kaveny 2000: 285), the cooperation can be assessed as direct. Given the direct involvement of the cooperator in the evil action, this kind of cooperation should be better called complicity or collaboration.
- In ***indirect*** (also: mediate) ***cooperation***, the cooperator supplies the general means or conditions for the evil action to take place. An indi-

rect cooperator makes a contribution that somehow enables the primary agent to perform the evil action.

The action of an indirect cooperator is not an essential part of the evil action itself but paves the way for the evil action, that is, is a condition that enables the principal agent to take the evil action: "Mediate cooperation should be thought of as the supply of means or conditions for the primary act to take place" (Oderberg 2004a: 207, 2017a: 194, cf. 2018: 54). In indirect cooperation, the cooperator supplies some peripheral assistance or preparation so that the principal agent can perform the primary act. Helping a thief carry another person's property out of the latter's house would be direct cooperation; giving the thief the keys to the house (assuming that there was no joint plan) would be indirect cooperation.

- Cases of indirect cooperation can be further distinguished according to the ***proximity*** and ***dispensability*** of the cooperative action to the evil action:
  - *Proximity of cooperation* concerns "where in the causal chain leading to the principal act the cooperator stands" (Oderberg 2017a: 194). The position of a mediate cooperation in the chain of actions preparing the way for the evil action is not only determined by the mediate cooperation's nearness, that is, (a) by the spatio-temporal distance from the mediate cooperative to the evil action and (b) the number of causal steps between the mediate cooperative and the evil action. It is also determined (c) by the mediate cooperation's definiteness, "which depends on how clearly the co-operation points to the primary act" (Oderberg 2004a: 207; McHugh & Callan 1958: §1512). The closer the cooperation to the evil action, the more the cooperator shares in the action of the primary agent.

    Proximity is a matter of degree; proximity judgements are not absolute but relative. Whether a mediate cooperation is proximate or remote depends on what else is being done before the evil action takes place: "When we say, for example, that handing over an instrument to a surgeon is proximate cooperation in the surgery, we mean relative to such an act as anaesthetising the patient" (Oderberg 2017a: 195). The same action can be proximate or not depending on the circumstances.
- In contrast to cases of indirect cooperation which can be proximate or remote, direct material cooperation is always proximate.

- *Dispensability* (also: necessity) *of cooperation* is related to how important or "practically necessary" (Oderberg 2017a: 194) a mediate cooperation is in the causal chain of actions preparing the way for the evil action. Dispensability determines whether the evil action would occur but for the cooperative action and to what extent the primary agent needs the cooperation of the cooperator. Mediate cooperation can either be dispensable (also: non-necessary) or indispensable (also: necessary) "according as the principal agent cannot act without it, or can" (McHugh & Callan 1958: §1509). An indispensable cooperation is essential, while a dispensable cooperation is not essential to the commission of the primary action. If the primary act causally depends on the cooperation, that is, if the "primary act, practically speaking, cannot be performed, or would be performed with great difficulty, or would have its probability significantly lowered without the cooperation" (Oderberg 2018: 47, cf. 2017b: 17), the mediate cooperation is indispensable. If an indirect cooperator is one of several or many available, that is, if there were others who could have performed the indirect cooperative action, "then the co-operation of any of them would be dispensable, or nor causally necessary for the principal act" (Oderberg 2004a: 207).

Having outlined the different kinds of cooperation in wrongdoing we are now in a position to assess the ethical quality of the respective cooperative acts.

3.2. The General Ethics of Cooperation in Wrongdoing[43]

Based on the distinctions outlined above we can now turn to the question of how the resulting kinds of cooperation in wrongdoing fare with respect to their moral quality. So, when does the cooperation in the evil action ("primary action") of another person ("principal agent") become so tainted by the evil action that it becomes an evil action itself?

[43] The content of this chapter draws on, and in part reproduces verbatim, material published in Erk (2019: 250–260).

Given the ethics of human actions developed over the first part of this book, it should be obvious that both formal and direct cooperative actions constitute evil actions and are never morally permissible.

In formal cooperation the cooperator cooperates out of an intent that an evil action be performed. Formal cooperation in an evil act—regardless of whether such cooperation is direct or indirect—constitutes a violation of the sources of morality, according to which one's intentions must be good or at least indifferent. Willing the performance of an action with an evil *finis operis* means willing an evil. And this intention is enough to make the entire cooperative act evil.

> Formal co-operation is always wrong. If performing an act is wrong, then so is conspiring to do it, attempting to do it, tolerating it, and a fortiori assisting someone to do it. (Oderberg 2004a: 209)

> Unlike material cooperation which can sometimes be justified, formal cooperation in the wrongdoing of another is always wrong. (Kaczor 2005: 88)

Cooperatio formalis nunquam licita est: etenim cum ipsa includat approbationem peccati alterius, semper in se mala est.	Formal cooperation is never licit: as it in fact includes the approval of another person's sin, it is always evil in itself.

(Noldin 1935: 118; cf. Prümmer 1961: 448)

> Formal coöperation is always sinful, for it includes the approval of the sin of another and the willing participation in the guilt of that sin. (McHugh & Callan 1958: §1513; cf. Ross 1919: 232)

A formal cooperator is equally as guilty as the principal agent. As formal cooperation is always "morally off limits" (Oderberg 2018: 53), this automatically holds true for both direct formal and indirect formal cooperation as well.

What about direct cooperation? Immediate cooperation in an evil action is also morally impermissible. The direct cooperator shares in the wrongdoing of the principal agent in such a way that he becomes the complicit perpetrator of the evil action and must necessarily will the evil end of the external action. As direct cooperation is "a kind of part-performance of the primary act itself" (Oderberg 2018: 60), that is, as a direct cooperator is co-performing the evil external action of the principal agent, direct cooperation in an evil action must necessarily be an evil action, too.[44]

> If the behaviour looks like whole or partial execution of the primary act, and the primary act is wrong, the immediate cooperator acts wrongly. (Oderberg 2017a: 193f)

A direct cooperator is so closely associated with the evil action of the primary agent that his cooperation is equivalent to a formal cooperation.

Having analysed the ethical permissibility of formal direct, formal indirect and material direct cooperation, our findings so far can be summarised as follows:

- A formal direct cooperation constitutes an evil action (due to its being both formal and direct in nature) and is therefore ethically impermissible.
- A formal indirect cooperation constitutes an evil action (due to its being formal in nature) and is therefore ethically impermissible.
- A material direct cooperation constitutes an evil action (due to its being direct in nature) and is therefore ethically impermissible.

If all direct and formal cooperation in an evil action—and therefore all direct-formal, indirect-formal as well as direct-material cooperation—is evil and therefore impermissible from an ethical perspective, this leaves only one candidate to be further scrutinised, namely indirect material cooperation. So, when is indirect material cooperation in an evil action an

[44] Although direct cooperation is ethically impermissible in general, one can think of rare constellations in which it might be permissible: "If the cooperation is immediate [...] then it might be justified if something about the circumstances meant that although the act was wrong for the primary agent to do, it was permissible if done (in part) by the assistant. Imagine a thief who is assisted by a starving man to take food from a supermarket—the starving man having been promised some of the food as an inducement. It is at least arguable that if someone is genuinely starving, it is not theft to take food from another, even though it is wrong for the thief to do so" (Oderberg 2018: 60). So, under threat of death one could directly cooperate in a theft "because the one so constrained is in extreme need and this use of his neighbor's goods becomes a unique means of saving his life" (Higgins 1949: 355; cf. Davis 1959: 342f).

evil action itself? When is indirect material cooperation ethically impermissible?

In general, "material coöperation is not lawful" (McHugh & Callan 1958: §978; cf. Prümmer 1961: 448). However, as moral philosophers agree (cf. Oderberg 2004a, 2017a; Prümmer 1961: 448; McHugh & Callan 1958: §978, §1515; Noldin 1935: 118; Merkelbach 1935: §489; Ross 1919: 232), an indirect material cooperation can become morally permissible provided that the following set of conditions is met:

- The material cooperation must itself be a good or at least an indifferent action, that is, the *finis operis* of the cooperative action must be good or at least indifferent.
- The intention of the material cooperator must be good, that is, the occurrence of the evil primary act (which is the bad consequence of the material cooperation) must only be foreseen, but not intended.
- The bad consequence must not be used by the material cooperator as a means to avoid some threatened harm or loss, retain some good and/or gain some promised benefit either to himself or a third party (which is the good consequence of the material cooperation), that is, the good consequence must not directly result from the bad consequence but must rather directly result from the cooperative action.
- There must be a reason for the cooperation proportionate to the gravity of the bad consequence, that is, a proportionate reason sufficient to justify allowing the bad consequence to occur.

The observant reader might have realised that these conditions are exactly the same conditions as the ones that govern the moral permissibility of an action with a double effect.

> The conditions necessary in order that material coöperation be lawful are the same as for any other act that has a double result; for from the coöperation follow two results, one that is bad (viz., the sin of the other person) and one that is good (viz., the avoidance of loss or the retention of good). (McHugh & Callan 1958: §1515)

In other words, cases of material cooperation are just "a special case or application of PDE" (Oderberg 2004a: 211), that is, the principle of double effect. The applicability of this principle—which has been introduced in Chap. VI, Sect. 2.2—is due to the fact that material cooperative actions have both a good and an evil consequence: The good consequence is that they bring about some good (to the cooperator or some other person) and/or avoid some harm (to the cooperator or some other person); the evil consequence is the occurrence of the evil primary act.

Such action can only be good if it complies with the nature-of-the-act condition, the end-of-the-agent condition, the means-end condition and the proportionality condition. So, how does a material cooperation fare with respect to these conditions? Answering this question is surprisingly uncomplicated when it comes to the first three conditions:

- The *nature-of-the-act condition* can be verified rather easily by looking at the natural end of the external action of the cooperative action. If the *finis operis* of the cooperative action is good or at least indifferent, this first condition is met. As the external action of the cooperative action, that is, what the cooperator does in cooperating, can usually be observed, this assessment should not pose too many problems.
- The *end-of-the-agent condition* is necessarily met in cases of material cooperation because a cooperator who intends the evil primary action is not a material but a formal cooperator. The very definition of material cooperation—which contains an element of duress—rules out the possibility that the cooperator intends the occurrence of the evil consequence of his cooperation.
- The *means-end condition*, that is, the condition that the good consequence, that is, the retention of a good and/or the avoidance of a loss, must not be secured through the evil consequence, that is, the occurrence of the evil primary action, is also automatically verified in cases of material cooperation. For a cooperator who intends the evil consequence as a means to bring about the good consequence of his cooperative action is not a material but an implicit formal cooperator by definition.

While the application of the first three conditions is somewhat unproblematic, the application of the fourth condition is often difficult and requires prudence in judgement. When it comes to the *proportionality condition*, there are some guidelines that help determine whether it is actually met by an indirect material cooperation. The general and overarching rule by which to make this assessment could be formulated as follows:

> The more serious the wrong with which a person cooperates and the more implicated they are, the greater the reason they need for their cooperation to be justified. (Oderberg 2018: 60)

Depending on the influencing factors mentioned in it, this general rule can be further concretised and translated into the following seven guidelines (cf. Oderberg 2004a: 215f, 2018: 60f; McHugh & Callan 1958: §1519; Prümmer 1961: 449; Noldin 1935: 118f; Merkelbach 1935: §489):

- *Evilness of the primary action*: The graver the wrong of the principal, that is, the more evil the primary action, the graver the reason required for the indirect material cooperation in the primary action.
- *Likelihood of the primary action*: The more likely the evil consequence of the cooperation, that is, the more certain it is that the principal agent will perform the evil primary action, the greater the reason required for cooperation: "You do not need as serious a reason for assisting someone who *might* do something to which you conscientiously object as for assisting someone who definitely will" (Oderberg 2018: 61).
- *Proximity of the cooperation (to primary act)*: The more proximate the indirect material cooperation to the evil primary action, the greater the reason required for cooperation. A graver reason is required for proximate than for remote cooperation.
- *Dispensability of the cooperation (for primary act)*: The greater the dependence of the primary action on one's indirect material cooperation, that is, the less dispensable the cooperation is for the performance of the primary action, the greater the reason must be for cooperation. A graver reason is required for indispensable than for dispensable cooperation.
- *Duty to impede the primary action*: The greater the duty to avoid the indirect material cooperation or to prevent the primary action, the greater the reason must be for cooperation. This means that those persons who are under a duty—for example, *ex officio* as a police officer—to avoid cooperation in an evil primary action or prevent the evil primary action from being performed require a graver reason for cooperation than those persons who are not under such a duty.

- *Potential for scandal*: Furthermore, a material cooperator should also consider the potential for scandal.[45] In a technical sense, scandal refers to the setting of a bad example and by doing so providing occasion for the wrongdoing of others (by, e.g., encouraging them in rationalisation and self-deception regarding the evil action).[46] Scandal is given by performing an evil action in the presence of others who are apt to be scandalised by the perpetrator of the evil action.[47] Scandal causes those others to come to a mistaken belief about the ethical quality of the evil action and thus induces persons who did not intend to do so to perform the evil action. Scandal is concerned with the problem that a material cooperation might suggest to persons who cannot appreciate, for example, the relevant ethical distinctions (e.g. the distinction between formal and material cooperation) that because of his cooperation the cooperator is indifferent to the ethical quality of the evil action or takes it to be permissible.
  As has been established above, a material cooperation can only be ethically permissible if it is a good or at least neutral action. If the cooperative action is evil, the cooperation is ethically impermissible anyway—regardless of its potential for scandal. So, a good or neutral material cooperation can only cause scandal if it is misinterpreted as an evil action, "either out of malice (Pharisaic scandal), or out of ignorance or frailty (scandal of little ones)" (McHugh & Callan 1958: §1464; cf. II^{a}-IIae q. 43 a. 7 co.). In both of these cases,

[45] The English word "scandal" derives from the Latin *scandalum* or the Ancient Greek σκάνδαλον which means trap, stumbling block but also temptation.

[46] For a helpful treatise on the concept of scandal cf. McHugh & Callan (1958: §1445ff) as well as Prümmer (1961: 437ff). Scandal occasions another person to an evil action by example; however, the actual cause of the evil action is the scandalised person who chooses to act on the suggestion offered by scandal.

[47] As McHugh & Callan (1958: §1461) point out, persons "who are habitually so bad or so good that anything disedifying done before them is not calculated to influence their attitude towards evil" are not readily subject to bad influence and are, therefore, not apt to be scandalised. Besides those who cannot be occasioned to an evil action by bad example, persons who are already inclined to do evil cannot be scandalised either.

In contrast, those who are peculiarly disposed to be led astray by bad example are "young persons whose character is yet unformed, the ignorant and well-meaning persons who are weak" (McHugh & Callan 1958: §1461).

Scandal can only be given by or taken from those who are considered an example by the respective witnesses of the cooperation. So, when assessing the potential for scandal we have to also examine the social standing of the cooperator.

scandal is taken although none has actually been given.[48] Since we need not be disturbed by Pharisaic scandal,[49] it suffices to focus on the potential for unintentional "scandal of little ones" (also: scandal of the (morally) weak).

So, a material cooperator has to be concerned with the message his involvement sends to onlookers of his cooperation. The question he has to answer in this respect is the following: To what extent is it possible that ignorant or frail people who happen to witness a material cooperation take—due to the exemplary role of the cooperator—scandal at this cooperation, because they mistake the material cooperation for a formal cooperation and are thereby, as the cooperator seems to approve of and sympathise with the evil action, led to think that (a) the formal cooperation with the evil action and/or (b) the commission of the evil action itself are ethically permissible—and in consequence are led to commit one or both of these actions?

- *Potential for corruption of moral sense (of cooperator)*: Regular material cooperation—for example, in the context of an employment relationship—might result in the cooperator's desensitisation to the evilness of the evil action. Even for an initially conscientious cooperator, what one routinely does is apt to seem morally routine. By regularly cooperating, one could start to approve of and condone the evil action or resign to the presence of this evil in society. Cooperation can be ethically deleterious to the cooperator and make him indifferent or even sympathetic to the evil action.

[48] Aquinas (cf. II^{a}-IIae q. 43 a. 2 co.) divides scandal into active scandal and passive scandal. The former refers to the scandal given and is tantamount to the evil action of the scandaliser; the latter refers to the scandal caused in the scandalised.

There are two ways in which a passive scandal can occur without an active scandal. On the one hand, a person is scandalised at another person's good action; this kind of scandal is called "Pharisaic scandal" or "scandal of the Pharisees" (cf. II^{a}-IIae q. 43 a. 7 co.). The second type of such scandal is called "scandal of the little ones" or "scandal of the weak". This scandal proceeds from weakness or ignorance, for example, when a person takes scandal at an action that she mistakes for an evil action.

[49] As Davis (1959: 334; cf. Prümmer 1961: 442) points out, we are allowed to act in spite of Pharisaic scandal "if we have even a small reasonable cause for doing so"; if we do not have such reason, we should, however, refrain for the sake of charity. In general, the right answer to a person who takes pharisaic scandal is: "If you are scandalized, that's your problem" (Oderberg 2004a: 213).

> Repeated cooperation carries the risk that the cooperator becomes indifferent to the evilness of the primary agent's action; it can also lead to a personal involvement with the primary agent that could dull the cooperator's moral sense (cf. Kaveny 2000: 305f). In the end, the fact that one's material cooperation is not good in itself but only excused might be pushed to the back of the cooperator's mind and forgotten.
> Therefore, one should also address the question whether a material cooperation carries the risk of corruption of the cooperator's moral sense. The greater this danger, the more a cooperator has a reason not to cooperate.

These seven guidelines must be combined in order to assess the gravity of the reason that is required to excuse the indirect material cooperation in an evil primary act. Any reduction or increase in these factors lowers or increases the overall threshold for the counterbalancing reason. As should be obvious there are many permutations that can obtain:

> At one extreme we have cooperation that is highly proximate and absolutely indispensable to a seriously immoral act, where the benefit gained/loss avoided by the cooperator is slight. [...] At the other extreme, we have cooperation that is very remote and quite dispensable to a minor wrong, where the benefit/loss is great. [...] Between these two extremes lie a spectrum of cases: the closer and more indispensable the cooperation and the graver the primary wrong, the greater the benefit/loss to the cooperator must be before cooperation is permissible. (Oderberg 2017a: 195, cf. 2018: 51)

Although the above quotation does not mention the possible existence of a duty to impede the primary action, the point it makes should be obvious: There are no easy answers when assessing how grave the reason for an indirect material cooperation must be. Sometimes the situation is clear; at other times it is more subtle. It is not always easy to ascertain, for example, the likelihood of the primary action, the proximity of the cooperation or the dispensability of the cooperation. And it is even harder to compare different cooperative actions. This is not simply an arithmetic exercise. In such situations, it is the task of the virtue of prudence to make the correct judgement.

There is one more thing to be addressed: Virtually all of the above guidelines mention an "excusing reason" which has to be more or less grave. But what exactly constitutes an excusing reason? And what makes an excusing reason more or less grave? In general, it is the good consequence of the indirect material cooperation that counts as excusing reason. In cases of material cooperation, the cooperator is cooperating because he or a third party would otherwise suffer some evil, lose some good or not gain some good; the material cooperator is cooperating in order to avoid these undesirable consequences of non-cooperation. And it is the avoidance of these undesirable consequences of non-cooperation that constitutes the good consequence of the cooperation: "Overall, the seriousness of the reason corresponds to the seriousness of the loss to be avoided" (Oderberg 2018: 61). Therefore, the good consequence of the indirect material cooperation can consist in the avoidance of some evil, that is, harm or loss, and/or the retention or gain of some good—either by the cooperator or a third party. In other words, the gravity of an excusing reason is a direct function of the gravity of the good consequence that obtains from cooperating.

The proportionality condition therefore requires us to weigh the evilness of the primary act, the likelihood of the primary action, the proximity of the cooperation, the dispensability of the cooperation, a possible duty to impede the primary action, the potential for scandal and the potential for the corruption of the moral sense of the cooperator against the gravity of the cooperating action's good consequence.

Summing up the insights gained so far, we can state the conditions for the ethical permissibility of a material indirect cooperation as follows: ***A material indirect cooperation can only be ethically permissible if the following conditions are met*:**

- ***Nature-of-the-act condition***: The natural end of the cooperative action's external action must be ethically good or at least indifferent.
- ***End-of-the-agent condition***: In a material indirect cooperation, this condition is necessarily met due to the cooperation's nature as material.
- ***Means-end condition***: In a material indirect cooperation, this condition is necessarily met due to the cooperation's nature as material.
- ***Proportionality condition***: This condition requires the comparative weighing of the following two sets of factors:
  - Gravity of cooperating action's good consequence vs.
  - Evilness of primary act, likelihood of primary action, proximity of cooperation, dispensability of cooperation, duty to impede primary action, potential for scandal and potential for corruption of moral sense

While the evaluation of the first condition should not pose too much of a problem and while conditions two and three are automatically fulfilled in case of a material indirect cooperation, assessing whether a material indirect cooperation meets the proportionality condition is a somewhat complicated task. This assessment necessitates us to weigh the gravity of the cooperating action's good consequence against the above-mentioned specific factors on a case-by-case basis. Unfortunately, there is no mathematical formula by means of which this weighing can be done. Nevertheless, there are some guidelines that facilitate this prudential weighing and help us establish whether a case of material indirect cooperation is ethically permissible or not:

1. A remote and dispensable indirect cooperation can be justified by a slight reason, for example fear of job loss.
2. A proximate indispensable indirect cooperation usually requires a serious reason, that is, is excused only in case of a grave good consequence. However, if the proximate indispensable indirect cooperation is a cooperation in a seriously evil primary action, a very serious reason is required; such cooperation is excused only in case of an extremely grave good consequence.

   > Highly proximate, indispensable cooperation in a very serious wrong—perhaps the taking of life or serious bodily harm—could only ever be justified by a very grave risk to the cooperator of a similar magnitude: a gun to the head, as it were. (Oderberg 2018: 62; cf. Noldin 1935: 122)

3. Both a proximate dispensable and a remote indispensable indirect cooperation are excusable only in case of a grave good consequence, that is, both require a serious reason. A proximate dispensable or a remote indispensable indirect cooperation in a seriously evil primary action requires a very serious reason; such cooperation can only be excused in case of a very grave good consequence. A remote and dispensable indirect material cooperation can be excused by a reasonably good consequence, for example a light loss or harm: "The more proximate and indispensable, the greater the reason needed for assistance. If the cooperation involves relatively minor acts such as cleaning the ward, serving the food, and the like, a minor ticking-off by one's employer is enough to justify involvement" (Oderberg 2018: 62f). However, if the primary action is seriously evil, the good consequence has to increase in gravity as well: "If the cooperation is neither indispensable nor proximate but still involves a serious primary wrong [...]

we should expect that a threat of loss that was perhaps not as serious as the primary wrong but still substantial, such as unemployment or destruction of reputation, might suffice" (Oderberg 2018: 62).

4. In general, scandal of the weak should be avoided unless this causes a "notabile incommodum" (Prümmer 1961: 442), that is, a notable inconvenience, to the agent. So, if a material cooperator otherwise faces a sufficiently grave inconvenience, he is permitted to cooperate despite the potential for scandal of the weak. Whether the inconvenience is sufficiently grave must be assessed in relation to (a) the number of persons scandalised, (b) the graveness of the evil action to which these persons are occasioned by one's cooperation and (c) the likelihood that these persons are not only occasioned but actually perform or formally cooperate with this evil action. To reduce the probability of a scandal of the weak occurring, the material cooperator should educate the relevant observers by, for example, explaining what is being done or not done and declaring his want of sympathy for the evil action with which he is materially cooperating. Furthermore, the social standing of the cooperator whose cooperation causes scandal has to also be taken into consideration. When given by somebody who by nature or office is obliged to teach and educate others, scandal is graver.
5. According to McHugh & Callan (1958: §1520; cf. Noldin 1935: 122), a grave reason for cooperation exists "when, if one refuses it, a great good will be lost or a great evil incurred". Examples of such grave reason are (fear of) great pain, great indignation, lost profit (Latin: *lucrum cessans*), loss sustained (Latin: *damnum emergens*)[50] or great shame. Similarly, a very grave reason for cooperation is the gain or retention of a very great good or the avoidance of a very great evil. Examples of a very grave reason are (fear of) long imprisonment, loss of employment (without being able to find a new job except with notable detriment or damage in reputation or goods of fortune) or notable drop in one's social position. And finally, a most grave reason for cooperation is the gain or retention of a supremely great good or the avoidance of a supremely great evil. Examples of a most grave (*gravissimi*) reason are death, loss of one's station in life, loss of a primary limb or a capacity

[50] When it comes to lost profit and loss sustained it has to be noted that the latter has more weight than the former. Not being able to realise a profit in the amount of $ 1000 is less bad—and therefore a less grave reason—than being actually deprived of already possessed goods in the same amount.

for sensing (e.g. loss of eyesight), incurable disease, very great defamation, severe or perpetual captivity/imprisonment, extreme disgrace, loss of all or almost all of one's property and the public safety of Church or State.[51]

6. When establishing the gravity of the good consequence, we need to be aware of the fact that there is not an absolute standard of gravity. Rather, the gravity of the good consequence of an indirect material cooperation is relative to the circumstances of the cooperator:
   We have to look at the goods and bads and where they reasonably figure in a person's welfare in their particular circumstances. The loss of a day's wages, say, is moderately serious for the average worker but trivial for a rich person. It might be catastrophic for a poor, third-world farmer. Similarly, the threat of physical pain is usually easier to bear for a soldier than a civilian, and the loss of reputation harder for someone whose livelihood depends on it than for someone whose livelihood does not. A specialist surgeon might find it harder to find alternative employment than a nurse or therapist (though in some situations they might find it easier, due to market demand). We need to take all the facts into consideration and ask whether, in the particular circumstances, the potential loss to the cooperator justified that particular level of cooperation in that particular wrongful act. (Oderberg 2004a: 216, 2018: 61f)
7. Within the classes of gravity, the following order applies: "Losses to life and limb are more important than financial losses; serious public harm is worse than serious private harm; certain harms are worse than merely probable or barely possible harms" (Oderberg 2018: 56). Private harm always weighs less than harm to the public good or public safety. Oderberg (2004a: 221, cf. 2018: 62) goes so far as to state "that when it comes to a real threat to public safety, such as war, terrorism, mass outbreak of illness, mass environmental damage, and so on, one ought not to cooperate in any way but resist, because the cost is too great".

[51] For McHugh & Callan (1958: §1520) the loss of one's station in life, incurable disease, loss of an eye or other principal member, severe or perpetual imprisonment constitute graver reasons for cooperation, that is, reasons "that surpass the very grave without being supreme". This would leave only the public safety of church or state, loss of all one's property, death, extreme disgrace, and the like as the most grave reasons.

3.3. The Ethics of (Physician-)Assisted Suicide

Having introduced the general ethics of cooperation in wrongdoing in the preceding chapter, this chapter applies this theory to the problem of (physician-)assisted suicide.[52] So, given the general ethics of cooperation in wrongdoing, is it ethically permissible for a human being (cooperator) to cooperate in the suicide (primary action) of another human being (principal agent)?

The kind of suicide we are dealing with in (physician-)assisted suicide is an essential self-killing. As has been the clear result of Chap. VII, Sect. 1, such direct suicide is an ethically impermissible action as the natural end of this exterior action is evil. Pointing this out is important as we are only dealing with a cooperation in wrongdoing if the primary action is an evil action. If the cooperator does not cooperate in an evil action, then there is no problem. In (physician-)assisted suicide, there is a problem because the primary action of the principal agent in which the cooperator cooperates is evil.

Furthermore, we have to outline the physician's cooperative action. The kind of assistance we are trying to evaluate is not only the prescription of a deadly dosage of a barbiturate but also the preparation of the drug for the ready use of the patient who then kills herself by means of this drug. So, the physician either prepares and puts within the patient's reach a glass with a lethal liquid that the patient then grabs and drinks or he sets up a suicide device where the patient has to simply push a button that causes the poison to trickle into her veins. It is this standard case of physician-assisted suicide whose ethical (im)permissibility shall concern us here.[53]

[52] Assisted suicide is only one application example for the concept of cooperation in wrongdoing. Further bioethical examples are the assistance in ethically impermissible procedures (such as direct abortions, direct sterilisations), the distribution of contraceptives (e.g. condoms, birth control) or the intake of a medication that contains material from aborted foetuses (e.g. smallpox vaccination). Supporting a person who is engaging in VSED (by, e.g., alleviating the symptoms brought about by VSED) also constitutes a case of cooperation in wrongdoing.

[53] Of course, this standard case of physician-assisted suicide is not the only way a physician can cooperate in a patient's suicide. All of these different ways of cooperation have to be assessed separately. Knowing whether the standard case of physician-assisted suicide is ethically permissible or impermissible does not automatically mean that other forms of cooperation are permissible or impermissible.

In order to find out whether this standard case of physician-assisted suicide is ethically impermissible we must ask whether such cooperative action is formal direct, formal indirect, material direct or material indirect. Only if the physician-assisted suicide qualifies as material indirect can it be ethically permissible. So, what about the physician who cooperates in the suicide of a patient? Is he cooperating materially or formally, that is, what is his cooperation's degree of co-voluntariness? And is he cooperating directly or indirectly, that is, what is his cooperation's degree of influence? Let us start with the first of these questions.

As has been stated above, a cooperator's cooperation with a principal agent's evil primary action is formal if the former approves of the latter's action, that is, if the former wills the evil primary action to occur. In this case, the cooperator cooperates willingly. If a physician partakes in the suicide of another human being while approving of the latter's action and willing its realisation, this makes the cooperation formal and, therefore, ethically impermissible. The standard case of physician-assisted suicide can only be ethically permissible if the cooperator does not approve of the principal agent's action and does not will its realisation, that is, if the physician is cooperating unwillingly. Physician-assisted suicide is material if, for example, a physician's employer requires him to assist his patients should they wish to commit suicide; the physician thinks that suicide is an evil action, does not approve of his patient's suicide and does not want him to kill himself; he only cooperates because otherwise he would lose his job.

What about the degree of influence of assisted suicide? As has been stated above, a cooperator's cooperation with a principal agent's evil primary action is direct if the cooperator performs some part of the evil primary action himself or if he and the principal agent jointly perform the evil primary action. A cooperation is indirect if the cooperator supplies the means or conditions for the evil action to take place. When it comes to the standard case of physician-assisted suicide, one could argue that the cooperation is only indirect. The physician's cooperative action usually consists in the counselling of the patient and the provision of the lethal drug. So, the physician might supply the means or conditions for the primary act to take place; the actual killing action is, however, performed by the patient alone.

This interpretation is, however, a bit too naïve. Injecting a lethal drug into a suicidal patient's hand is not simply providing her with the means to kill herself. Whether a cooperation is direct or indirect must be decided by comparison to clearly indirect forms of cooperation. In an assisted suicide, indirect cooperators are the cleaning staff who clean the room in which the suicide takes place as well as the corridors that lead to it, the janitor who is to make sure that the lights and the blinds in the room work, the producer of the drug which—in excessive dosage—can be used for suicide, the physician who is covering her colleague's duties while he attends to the patient's suicide request or the taxi driver who is driving the patient to her deadly appointment. Compared to these instances of clearly indirect cooperation, the cooperation of a physician who not only prescribes a deadly dosage of a barbiturate but also prepares it for the ready use of the patient who then kills herself by means of this drug is on another level. The physician does not simply make a contribution that somehow promotes the primary action; rather, he contributes to the emergence of circumstances that are essential to the principal agent's primary act. What a physician does in a physician-assisted suicide is so essential to the suicide that it is hard to qualify it as anything but a direct cooperation. So, in the standard case of physician-assisted suicide on which these considerations are based, the cooperation of the physician is a direct cooperation.

The above remarks suggest that the standard case of physician-assisted suicide is a formal direct cooperation at worst and a material direct cooperation at best. However, such cooperation is ethically impermissible in both cases because it is a form of direct cooperation in the evil action of another human being. Accordingly, the standard case of physician-assisted suicide is always ethically impermissible. This is the result we get when we analyse the standard case of physician-assisted suicide from the perspective of the ethical device named cooperation of wrongdoing.

It has to be noted that the above result is valid only for the standard case of physician-assisted suicide. It is this specific cooperation in the suicide of another human being that is ethically impermissible. There are, however, other ways in which a physician can cooperate in the suicide of another human being. Further examples of a physician cooperating in suicide are given below (cf. Miech 1999):

- A physician could allow her name to appear in an advertisement for a hospital that advertises physician-assisted suicide as one of its "services". Assuming that the physician refuses to participate in physician-assisted suicide herself, this would be an example of a material direct cooperation.

- A physician could take over work of one of his colleagues while the colleague assists in a patient's suicide. Assuming that the physician refuses to participate in physician-assisted suicide himself, this would be a general example of a material indirect cooperation.
- A physician could advise a friend to buy shares in a pharmaceutical firm that is about to come out with a new lethal drug to be used in physician-assisted suicide. Assuming that the physician refuses to participate in physician-assisted suicide himself, this would be an example of a proximate material indirect cooperation.
- A physician could write a prescription for a medication whose manufacturer also produces a lethal drug that is used in physician-assisted suicide. Assuming that the physician refuses to participate in physician-assisted suicide himself, this would be an example of a remote material indirect cooperation.

As it is a direct form of cooperation, the first of these examples describes an ethical impermissible kind of cooperation. The latter three examples contain different kinds of material indirect cooperations. Using the set of criteria for the evaluation of the ethical permissibility of material indirect cooperations as it has been outlined in the previous chapter, the ethical permissibility of each of these latter three forms of cooperation has to be assessed separately and independently from the standard case of physician-assisted suicide.

Bibliography

Anderson, Marie A., Robert L. Fastiggi, David E. Hargroder, Joseph C. Howard Jr. & C. Ward Kischer. "Ectopic Pregnancy and Catholic Morality. A Response to Recent Arguments in Favor of Salpingostomy and Methotrexate." *The National Catholic Bioethics Quarterly* 11.1 (2011): 65–82. 667–684.

Anscombe, Gertrude Elizabeth Margaret. "War and Murder." In: Stein, Walter (Ed.). *Nuclear Weapons: A Catholic Response.* London: Sheed & Ward, 1961. pp. 44–52.

Atkins, E. Margarete & Thomas Williams. "Introduction." In: Atkins, E. Margarete & Thomas Williams (Eds.). *Thomas Aquinas: Disputed Questions on the Virtues.* Translated by E. Margarete Atkins. Cambridge: Cambridge University Press, 2005. IX–XXX.

Balch, Thomas J. "Catholic Treatment Ethics and Secular Law: How Can They Cohere?" *Solidarity: The Journal of Catholic Social Thought and Secular Ethics* 6.1 (2016): 1–15 (Article 4). Available at: https://researchonline.nd.edu.au/solidarity/vol6/iss1 Accessed: March 28, 2022.

Brock, Dan Willets. "Voluntary Active Euthanasia." *The Hastings Center Report* 22.2 (1992): 10–22.

Carlson, Bruce M.. *Human Embryology and Developmental Biology.* Sixth edition. St. Louis, MO: Elsevier, 2019.

Cathrein, Victor. *Philosophia Moralis in Usum Scholarum.* Editio Quinta ab Auctore Recognita. Friburgi Brisgoviae: B. Herder, 1905.

Chabot, Boudewijn & Christian Walther. *Ausweg am Lebensende Sterbefasten—Selbstbestimmtes Sterben durch Verzicht auf Essen und Trinken.* 5., aktualisierte und erweiterte Auflage. München: Ernst Reinhardt Verlag, 2017.

Commer, Ernst. *System der Philosophie. Vierte (Schluss-) Abtheilung: Ethik.* Münster & Paderborn: Ferdinand Schöningh, 1886.

Cronin, Michael. *The Science of Ethics. Volume II: Special Ethics.* New York et al.: Benziger Brothers, 1922.

Cronin, Daniel Anthony. "The Moral Law in Regard to the Ordinary and Extraordinary Means of Conserving Life." In: Smith, Russell Edward (Ed.). *Conserving Human Life.* Braintree, MA: Pope John XXIII Medical-Moral Research and Education Center, 1989. pp. 3–116. Originally published in 1958 as Doctoral Dissertation at the Pontifical Gregorian University (Rome).

Davis, Henry. *Moral and Pastoral Theology in Four Volumes. Volume One: Human Acts, Law, Sin, Virtue.* Eighth Edition, revised and enlarged. Edited by L. W. Geddes. London & New York: Sheed and Ward, 1959.

Dunne, Kathryn & Pamela Doherty. "Donation after circulatory death." *Continuing Education in Anaesthesia Critical Care & Pain* 11.3 (2011): 82–86.

Edwards, John M., Richard D. Hass, Jr. & Virginia M. Robertson. "Non–Heart-Beating Organ Donation: Process and Review." *AACN Advanced Critical Care* 10.2 (1999): 293–300.

Emanuel, Ezekiel J.. "The Painful Truth About Euthanasia." *Wall Street Journal.* January 7, 1997. p. A16.

Erk, Christian. *Moral Philosophy. Fundamental Concepts.* Second, revised and expanded Edition. Heusenstamm: editiones scholasticae, 2019.

Fagothey, Austin. *Right and Reason. Ethics in Theory and Practice.* Third edition. Saint Louis: The C.V. Mosby Company, 1963.

Feser, Edward. "In Defense of the Perverted Faculty Argument." In: Feser, Edward. *Neo-Scholastic Essays.* South Bend, Indiana: St. Augustine's Press, 2015. pp. 378–415.

Giubilini, Alberto & Francesca Minerva. "After-birth abortion: why should the baby live?" *Journal of Medical Ethics* 39.5 (2013): 261–263.

Heinberg, Eric M.. "Ectopic Pregnancy." In: Sokol, Andrew I. & Eric R. Sokol (Eds.). *General Gynecology. The Requisites in Obstetrics and Gynecology.* Philadelphia, PA: Mosby, 2007. pp. 257–285.

Henke, Donald E.. "A History of Ordinary and Extraordinary Means." *The National Catholic Bioethics Quarterly* 5.3 (2005): 555–575.

Higgins, Thomas J.. *Man as Man: The Science and Art of Ethics.* Milwaukee: The Bruce Publishing Company, 1949.
Hursthouse, Rosalind. *Beginning Lives.* Oxford: Oxford University Press, 1987.
Jensen, Steven J.. *Good & Evil Actions. A Journey through Saint Thomas Aquinas.* Washington, D.C.: The Catholic University of America Press, 2010.
Kaczor, Christopher. *The Edge of Life. Human Dignity and Contemporary Bioethics.* Dordrecht: Springer, 2005. (Series "Medicine and Philosophy", Volume 85; Subseries "Catholic Studies in Bioethics")
Kaczor, Christopher. "Artificial Wombs and Embryo Adoption." In: Brakman, Sarah-Vaughan & Darlene Fozard Weaver (Eds.). *The Ethics of Embryo Adoption and the Catholic Tradition: Moral Arguments, Economic Reality and Social Analysis.* Dordrecht: Springer, 2007. pp. 307–322. (Series "Medicine and Philosophy", Volume 95; Subseries "Catholic Studies in Bioethics")
Kaczor, Christopher. *The Ethics of Abortion: Women's Rights, Human Life, and the Question of Justice.* Second Edition. New York & London: Routledge, 2015.
Kaczor, Christopher. "Abortion as Human Rights Violation." In: Greasley, Kate & Christopher Kaczor. *Abortion Rights. For and Against.* Cambridge: Cambridge University Press, 2018. pp. 86–164.
Kaczor, Christopher. *Disputes in Bioethics. Abortion, Euthanasia, and other Controversies.* Notre Dame, IN: University of Notre Dame Press, 2020.
Kälin, Bernard. *Lehrbuch der Philosophie. Band II: Einführung in die Ethik.* Zweite Auflage; umgearbeitet von Raphael Fäh. Sarnen: Selbstverlag Benediktinerkollegium Sarnen, 1954.
Kaveny, M. Cathleen. "Appropriation of Evil: Cooperation's Mirror Image." *Theological Studies* 61.2 (2000): 280–313.
Kelly, Gerald A.. *Medico-Moral Problems.* St. Louis, MO: The Catholic Hospital Association of the United States and Canada, 1958.
Kuhse, Helga & Peter Singer. *Should the Baby live? The Problem of Handicapped Infants.* Oxford: Oxford University Press, 1985.
Ligouri, Alphonsus. *Theologia Moralis S. Alphonsi de Ligorio. Tomus Secundus, Continens Tractatus de Actibus Humanis, de Peccatis, de Praeceptis Virtutum Theol. et 1. et 2. Praecepto Decalogi.* Edited by Mich. Heilig. Mechliniae: P. J. Hanicq, 1852.
Lübbe, Andreas S. "Palliativmedizin als Angebot gegen eine Normalisierung des Tötens." In: Hoffmann, Thomas Sören & Marcus Knaup (Hrsg.). *Was heißt: In Würde sterben? Wider die Normalisierung des Tötens.* Wiesbaden: Springer Fachmedien, 2015. pp. 151–170.
Maltoni, M., C. Pittureri, E. Scarpi, L. Piccinini, F. Martini, P. Turci, L. Montanari, O. Nanni & D. Amadori. "Palliative sedation therapy does not hasten death: Results from a prospective multicenter study." *Annals of Oncology* 20.7 (2009): 1163–1169.
McHugh, John A. & Charles J. Callan. *Moral Theology: A Complete Course Based on St. Thomas Aquinas and the Best Modern Authorities.* Revised and enlarged

by Edward P. Farrell. New York City: Joseph F. Wagner; London: B. Herder, 1958. Available at: http://www.gutenberg.org/ebooks/35354 Accessed: March 28, 2022.

McMahan, Jeff. "Infanticide." *Utilitas* 19.2 (2007): 131–159.

Melé, Domènec. *Business Ethics in Action: Seeking Human Excellence in Organizations*. Basingstoke, UK: Palgrave Macmillan, 2009.

Merkelbach, Benedictus Henricus. *Summa Theologiae Moralis ad Mentem D. Thomae et ad Normam Iuris Novi: I. De Principiis*. Parisiis: Typis Desclée de Brouwer et Soc., 1935.

Miech, Ralph P.. "Physician Cooperation in Patient Suicide." *Ethics & Medics* 24.7 (1999): 1–4.

Noldin, Hieronymus. *Summa Theologiae Moralis: II. De Praeceptis*. Editio XXIII. Oeniponte: F. Rauch, 1935.

Oderberg, David S.. *Applied Ethics. A Non-Consequentialist Approach*. Oxford: Blackwell, 2000.

Oderberg, David S. "The Ethics of Co-operation in Wrongdoing." In: O'Hear, Anthony (Ed.). *Modern Moral Philosophy*. Cambridge: Cambridge University Press, 2004a. pp. 203–227. (Royal Institute of Philosophy Supplement 54; Royal Institute of Philosophy Annual Lecture Series 2002–2003).

Oderberg, David. "Ordinary Duties, Extraordinary Means." *Human Life Review* 30.1 (2004b): 50–56.

Oderberg, David S. "Further Clarity on Cooperation and Morality." *Journal of Medical Ethics* 43.4 (2017a): 192–200.

Oderberg, David S. "Cooperation in the Age of *Hobby Lobby*: When Sincerity is not Enough." *Expositions* 11.1 (2017b): 15–30.

Oderberg, David S. *Opting Out. Conscience and Cooperation in a Pluralistic Society*. London: The Institute of Economic Affairs, 2018.

Patel, Chirag, Paul Kleinig, Michael Bakker & Paul Tait. "Palliative sedation: A safety net for the relief of refractory and intolerable symptoms at the end of life." Australian Journal of General Practice 48.12 (2019): 838–845.

Prümmer, Dominicus M.. *Manuale Theologiae Moralis Secundum Principia S. Thomae Aquinatis in Usum Scholarum*. Tomus II. Editio Quinta Decima. Barcinone, Friburgi Brisgoviae & Romae: Herder, 1960.

Prümmer, Dominicus M.. *Manuale Theologiae Moralis Secundum Principia S. Thomae Aquinatis in Usum Scholarum*. Tomus I. Editio Quinta Decima. Barcinone, Friburgi Brisgoviae & Romae: Herder, 1961.

Rachels, James. "Active and Passive Euthanasia." *The New England Journal of Medicine* 292.2 (1975): 78–80.

Ross, John Elliot. *Christian Ethics. A Textbook of Right Living*. New York: The Devin-Adair Company, 1919.

Ross, John Elliot. *Ethics. From the Standpoint of Scholastic Philosophy.* Heusenstamm: editiones scholasticae, 2012. (Reprint of the third edition, New York: The Devin-Adair Company, 1938.)

Sade, Robert M.. "Consequences of the Dead Donor Rule." *The Annals of Thoracic Surgery* 97.4 (2014): 1131–1132.

Scarfone, Matthew. "Incoherent Abortion Exceptions." *Journal of Social Philosophy* 53.1 (2022): 127–140.

Sherwin, Susan. *No Longer Patient: Feminist Ethics and Health Care.* Philadelphia, PA: Temple University Press, 1992.

Spaemann, Robert. "Es gibt kein gutes Töten." In: Spaemann, Robert & Thomas Fuchs (Hrsg.). Töten oder sterben lassen? Worum es in der Euthanasiedebatte geht. Freiburg, Basel & Wien: Verlag Herder, 1997. pp. 12–30.

Spaemann, Robert. "Töten oder sterbenlassen?" *Aufklärung und Kritik* 13 (2006): 80–88. (Sonderheft 11)

Sullivan, Scott M.. "The development and nature of the ordinary/extraordinary means distinction in the Roman Catholic tradition." *Bioethics* 21.7 (2007): 386–397.

Sykes, Nigel & Andrew Thorns. "The use of opioids and sedatives at the end of life." *The Lancet Oncology* 4.5 (2003): 312–318.

Thomson, Judith Jarvis. "A Defense of Abortion." *Philosophy and Public Affairs* 1.1 (1971): 47–66.

Tooley, Michael. "Abortion and Infanticide." *Philosophy & Public Affairs* 2.1 (1972): 37–65.

Verberne, Eline A., Emma de Haan, J. Peter van Tintelen, Dick Lindhout & Mieke M. van Haelst. "Fetal methotrexate syndrome: A systematic review of case reports." *Reproductive Toxicology Review* 87 (2019): 125–139.

VIII: Postlude: Specific Questions at the Margins of Human Life

Before this book concludes, the following pages address—some in more and some in less great detail—several rather specific and technical questions and concerns that are sometimes levelled against the definition and criteria of life and death developed in the book. They are meant for those readers who are interested in a more in-depth discussion of theoretical aspects concerning ensoulment in general and ensoulment with a rational soul in particular.

1. Delayed or Immediate Ensoulment?

In the chapter on human life (cf. Chap. IV, Sect. 4.1) it has been established that the principle of human life is the special kind of soul called rational soul. Living human beings possess a single rational soul that grants them the capacity for all vegetative, sensitive and rational activity. The rational soul is the principle of human life and the principle and efficient cause of all activity shown by living human beings.

So far, it has been implicitly taken for granted that human beings possess such a rational soul from the moment of conception, that is, the fusion of a human ovum (female) and a human sperm (male). We could do so because from a biological perspective—as has been touched upon in Chap. V, Sect. 2.2—human beings begin to exist and live at conception. The

C. Erk, *The Ethics of Killing*,
https://doi.org/10.1007/978-3-031-07183-6_8

zygote produced during this process is a living human being.[1] If this is the case and if being a living human being means possessing a rational soul, this means that the zygote must be ensouled by a rational soul from the first moment of its existence and is not ensouled with this kind of soul during a later stage of development. The position that the zygote already possesses a rational soul is called *immediate ensoulment* (also: animation; hominisation) *theory*.[2]

[1] While what biologically happens during the process of fertilisation is uncontested, the question of when exactly during fertilisation the zygote and with it a new living human being comes into existence is somewhat contested. Some embryologists take the zygote to come into existence at the end of the fertilisation process when "the pronuclei fuse into a single aggregation of chromosomes" (Moore et al. 2020: 27); it is only at syngamy that what they call the "ootid" turns into the zygote.

Condic (2009), however, begs to differ. For her, syngamy is little more than "the breakdown of the nuclear membranes that separate the two pronuclei" (Condic 2009: 139) and therefore "a relatively mundane event along an already progressing developmental trajectory" (Condic 2009: 141; cf. Condic & Condic 2018: 110–114) and "merely another step in an already ongoing developmental process" (Condic 2013: 56) that causes neither a change in the material composition of the cell nor its behaviour. It is therefore "clear that the zygote, or one-cell human embryo, forms immediately upon sperm-egg fusion" (Condic 2013: 67).

Condic (2009: 142) is correct in pointing out that it is problematic and scientifically inconsistent to stipulate that one type of cell (a "pre-zygote" that exists following sperm-egg fusion but prior to syngamy) can turn into another type of cell (the zygote that exists after syngamy) "without any actual change in the material state or behavioral trajectory of the cell. To assert that life begins at syngamy is to propose some form of mysticism: although a zygote cannot be distinguished in any significant manner from the 'pre-zygote' that precedes it, the cell is now a zygote simply because one asserts that it is".

In light of these considerations, we should therefore be careful with our choice of words: Once an egg has been penetrated by a sperm, the resulting cell is not merely a modified gamete—as the term "fertilised egg" indicates—but an ontologically new kind of being, namely a one-cell embryo or zygote. This thought is also corroborated by Joyce (1988: 203): "The so-called fertilisation process is not as passive as the terminology would suggest. The nuclei of the sperm and the ovum dynamically interact. In doing so they both cease to be. One might say they die together. They really should not be said to unite. That suggests that they remain and form a larger whole. But the new single-celled individual is not an in tandem combo of the two parent sex cells. In their interaction and mutual causation of the new being, the sperm and ovum are self-sacrificial. [...] There is neither sperm nor ovum once the process of interaction is completed, even though cytoplasmic matter from the ovum remains. It is really a misleading figure of speech to say that the ovum is 'fertilised' by the sperm, passively as a farmer's field is fertilised. It is proper rather to speak of the sperm-ovum interaction process. There is no such thing as a 'fertilised ovum'".

[2] Immediate ensoulment is defended by, for example, Gerber (1966), Ashley (1976), Heaney (1992), Johnson (1995), Ashley & Moraczewski (2001), Haldane & Lee (2003a), Eberl (2005a), Austriaco (2011a: 59f) and Condic & Condic (2018: 46ff).

However, there are some thinkers who object to this position and instead advocate a so-called delayed (also: mediate or belated) ensoulment theory. According to one of its most prominent proponents, this theory claims "that the embryo is alive, that it is animated from the very start, but the soul which animates it is not yet a human soul, is a vegetative or animal soul; the human soul comes later. Animation is immediate, hominization is delayed" (Donceel 1970: 76).[3] In order to support their position, modern proponents of delayed ensoulment point to the fact that this theory was not only adopted by many ancient and medieval (Catholic) philosophers (e.g. Aristotle, Augustine, Thomas Aquinas, Alphonsus Liguori) but is also at least not discarded as obviously false by more recent manuals of scholastic philosophy (cf. Mercier 1921: 318 (n.163); Noldin 1935: 328 (n. 342); Prümmer 1960: 128 (n. 138)).

The quantitative assessment underlying this appeal to authority is undisputed and also acknowledged by proponents of immediate ensoulment.

> Up to well past the Council of Trent, meager mention was made of the immediate theory of animation, except for Gregory of Nyssa's suggestion in the fourth century of its possibility. The principles and conclusions of Aristotle and Thomas found general acceptance then and well until the seventeenth century, when medical observations suddenly began to undermine the mediate theory [...]. (Gerber 1966: 238)

As the presumably most prominent and cited proponent of delayed ensoulment, Aquinas did indeed hold that the human embryo is first ensouled by a vegetative soul; this soul then perishes and is succeeded by a sensitive soul; as the embryo continues to develop, the sensitive soul also perishes and is succeeded by a rational soul (cf. Contra Gentiles, lib. 2 cap. 89 n. 11; I^{a} q. 76 a. 3 ad 3; I^{a} q. 118 a. 2 ad 2; Super De causis, l. 1). Aquinas also tells us when the rational soul is infused to the human body: Stating that this event marks the completion of conception, Aquinas holds that male human embryos receive a rational soul after 40 or 46 days and female human embryos only after 90 days (cf. Super Sent., lib. 3 d. 3 q. 5 a. 2 co.).

[3] Besides Donceel's now classic exposition of the theory of delayed ensoulment, this theory has also been defended more recently by, for example, Pasnau (2002), Dombrowski & Deltete (2000) and Wallace (1994). For a brief history of delayed ensoulment thinking cf. Jones (2013).

How should we deal with this situation? Should we take Aquinas' and the other authorities' word for it and reject immediate ensoulment? In order to be able to decide which of the two ensoulment theories to follow and which to reject, it is helpful to turn to Aquinas as the preeminent representative of delayed ensoulment and try to elaborate the key principles of his general theory of ensoulment.

So, what does Aquinas say about ensoulment? The usual way to approach this question is by pointing out that the soul—as Aquinas states on several occasions (cf. Contra Gentiles, lib. 2 cap. 89 n. 3; I[a] q. 76 a. 8 ad 2; Q. d. de anima, a. 10 co.)—is the act of a "corpus organicus", that is, an organic body. What does he mean by this expression? For Aquinas, what is characteristic of an organic body is that such a body is constituted of, that is, possesses, various organs (cf. Q. d. de anima, a. 10 co.; Sentencia De anima, lib. 2 l. 1 n. 20; Super I Cor., cap. 12 l. 3). In short, an organic body is an organ-bearing body. From his treatises on the soul (cf. Q. d. de anima, a. 10 ad 1; Sentencia De anima, lib. 2 l. 1 n. 22) we can learn that Aquinas takes an organ to be a body part that is suitable to a specific operation of the soul. An organ can therefore be said to be a bodily instrument of the soul, that is, something by means of which the soul operates.

Having settled these terminological issues, we can now turn our attention to the problem of the ensoulability of the human body. What must a body be like in order to become ensouled? According to Aquinas, the general condition for ensoulment is that the respective body must be "organised":

Cum anima sit actus corporis organici, ante qualemcumque organizationem corpus susceptivum animae esse non potest.	Because the soul is the act of an organic body, the body cannot be receptive to the soul before it is in any way whatever organised.

(De potentia, q. 3 a. 12 co.)

Animae non est naturale infundere vitam nisi tali corpori, scilicet quod est sic complexionatum et organizatum; et manente tali dispositione corporis, vita continuatur ab anima.	It is not natural for the soul to infuse life unless to a certain kind of body, namely, one that is put together and organised just so; and while this sort of disposition of the body remains, the soul continues to supply its life.

(Super Sent., lib. 2 d. 19 q. 1 a. 4 ad 1)

What Aquinas is saying here is that in order to be ensoulable a body has to be furnished with organs.[4] As the matter to which a form is united must be such that it suits the operations of that form, this means that a body can only be receptive to a soul if it is equipped with the organs that are suited to the different operations of this soul (cf. Q. d. de anima, a. 10 ad 1). In other words, the body must be put together in the way necessary to serve as a suitable instrument of the operations of the soul.

As a general criterion, this sounds good. But what does it actually mean with respect to the concrete problem at hand, namely rational ensoulment? Which organs must a body be equipped with in order to be ensoulable with a rational soul? The usual argument advanced by modern proponents of delayed ensoulment goes something like this: The operations distinctive of the rational soul are universal cognition and concept formation, judging as well as reasoning. The specific organic infrastructure necessary to perform those operations are the brain and the nervous system.[5] Therefore, the organs that need to be present for rational ensoulment are the brain and the nervous system. This reasoning is exemplified by the following statement:

> The human soul 'is infused at that point when the fetus is sufficiently developed, in its brain and sensory systems, to support the soul's intellectual operations. [...] But modern embryology has established that the parts of the brain responsible for higher-level thought processes are not active until after the midpoint of gestation. [...] There remains some uncertainty about the precise timetable for higher-level brain development. But it is quite clear that these events occur after the twentieth week of gestation [...]'. (Pasnau 2002: 111)

For Pasnau, the rational organs are sufficiently developed for ensoulment when they show empirically observable signs of activity. In his opinion

[4] Given the formulations used by Aquinas in the above quotations, it is a bit misleading when Donceel writes that "the human soul can exist only in a highly organized body" (1970: 79) or in "a fully organized human body" (1970: 80). Aquinas does not speak of a high or even full degree of organisation; he simply demands organisation, that is, sufficient organisation.

[5] As the functioning of these rational organs is dependent on inputs by the senses, especially the five external senses (touch, taste, vision, hearing, smell), rational operation also requires the presence of the senses. As the capability for sensation is not granted by the rational but the sensitive soul and is shared by all animals, the senses are not listed as organs directly required for rational operations.

ensoulment therefore happens some 126 days into the existence of the foetus.[6] If Pasnau—as he would surely claim of himself—is true to Aquinas, how can his account be squared with the fact that for Aquinas ensoulment takes place 40 days into the existence of a (male) child?

To be able to answer this question, it is helpful to revisit Aquinas and try to understand not only that but also why he advanced delayed rather than immediate ensoulment. The best way to start is by turning our attention to Aquinas' definition of soul. It is in this definition that we can find the key to his theory of ensoulment. According to Aquinas, the soul is the *first* act (*actus primus*) of a physical organic body (cf. Sentencia De anima, lib. 2 l. 1 n. 23). Why is this relevant? Let us explain the difference between first act and second act (*actus secundus*) with an example: If Socrates possesses the capability to talk in the sense of *actus primus*, then this means that Socrates can talk, but does not (currently) do so—either because he chooses not to or because he is temporarily or permanently hindered (e.g. because his tongue has been cut out). In contrast, to possess a capability in the sense of second act means to not only possess the capability but to also make actual use of this potential by exercising the capability. So, Socrates possesses the capability to talk in the sense of second act if he actually talks (which, of course, presupposes the possession of this capability in the sense of first act). Therefore, Aquinas states that as "principium operationis" (Sentencia De Anima, lib. 2 l. 2 n. 6), that is, as the principle of operation, "the first act is said to be in potentiality to the second act, which is operation" (I^{a} q. 76 a. 4 ad 1).

Possessing a soul as first act means to possess the capability for all operations characteristic of that soul. In the case of a rational soul, this includes the capability for all vegetative, sensitive and intellectual operations of a human being. Possessing a soul does not, however, mean that these capabilities must also be actualised; it only means that they are actualisable. But if a being can possess a soul without actualising the capabilities that come with its possession, how can we know that a being actually possesses a soul? Unfortunately, the possession of a capability as such cannot be proven. If we have never heard Socrates talk, we do not know whether he possesses the capability to talk or not. The only way to prove that Socrates

[6] Pasnau's expression "after the twentieth week after gestation" refers to the twentieth week after the beginning of the pregnant woman's last menstrual cycle. The gestational age of the unborn is always two weeks higher than their fertilisation age, that is, the age measured from the time of fertilisation. So, 20 weeks gestational age is 18 weeks or 126 days fertilisation age. The terms "fertilisation age" and "gestational age" are explained in more detail in Chap. VII, Sect. 2.2.1.

can talk is by hearing him actualise this capability to talk by actually talking. Unfortunately, the reverse is not true: Never having heard or observed Socrates talk does not logically allow for the conclusion that he cannot talk; perhaps he can talk but has made a vow of silence.

To my mind, this problem is the reason why Aquinas—being a child of his time—had to assume delayed rather than immediate ensoulment. Why is this? In the context of our discourse about rational ensoulment, the above-mentioned problem can be phrased as follows: If the absence of empirically observable rational activity does not conclusively prove the absence of a rational soul, how can we determine when the unborn possesses a rational soul?

What I want to suggest is that Aquinas did not—as modern proponents of delayed ensoulment think—answer this question by simply stipulating that rational ensoulment requires the presence of rational organs suitable to rational activity. As far as I can tell, Aquinas never explicitly set such a condition.[7] Aquinas' actual answer to the above question seems to be a bit more multifaceted and could be formulated as follows: We can be certain that the unborn possesses a rational soul if we are certain that he is human and alive.

> One does not determine whether an entity is a human being by first determining whether it has a rational soul. Rather, one first finds evidence that an entity is a human being and from that one concludes that he or she must have a rational soul. (Lee 2011: 78)

So, from the moment we can prove with certainty that the unborn is a living human being, he possesses a rational soul. The presence of suitable organs only enters into the equation because their presence was the only way by which Aquinas could actually determine whether a being is a human being.

To show that this is his actual position, let us start with Aquinas' statement that the embryo first lives the life of a plant and then the life of an animal before it is infused with a rational soul (cf. Contra Gentiles, lib. 2 cap. 89 n. 11). This is an interesting and momentous statement. For it logically implies that the pre-ensoulment embryo is not only not yet

[7] On the contrary, although he holds that, in general, ensoulability requires a suitable organic infrastructure, Aquinas also holds that the specifically intellectual operations of the human soul, that is, the intellect and the will, are not the acts of any bodily organ (cf. I[a] q. 76 a. 1 ad 1; Q. d. de anima, a. 10 co.; I[a] q. 77 a. 5 co.; Contra Gentiles, lib. 2 cap. 80 n. 6; Contra Gentiles, lib. 2 cap. 62 n. 9; Super Sent., lib. 2 d. 17 q. 2 a. 1 ad 1; I[a] q. 75 a. 2 co.).

ensouled, but is not yet human. If the pre-ensoulment embryo were human, it could not live the life of a plant or an animal; it would necessarily live the life of a human. To live the life of a plant means to be a plant; and to live the life of an animal means to be an animal; this a metaphysical matter of course. So, according to Aquinas, human beings seem to undergo some sort of micro-evolution. Before a human being has become a human being, it has briefly been a plant and then an animal.

Next, we should recall that the soul is the form of the body (cf. Q. d. de anima, a. 9 ad 18). In general, a form is "the real constituent principle by virtue of which a thing is determined to be of a specific kind" (Oderberg 2007: 68). It is characteristic of the form called "soul" that its possession makes a thing exist as a living being of a certain kind. So, it is because they possess a rational soul (rather than, e.g., a mere vegetative soul) that human beings are alive and human beings (rather than plants). But unfortunately, souls elude our efforts at direct empirical observation. It is not by seeing that it possesses a rational soul that we know that some living being is a human being. Rather, we know that a living being must possess a rational soul if it is a human being. For this living being would not be human if it did not possess a rational soul. So, once we know (a) that a being is alive, that is, exhibits any kind of activity, and (b) that it is human, we know that this being must possess a rational soul.

So, how does Aquinas suggest that we recognise that a conceptus has become a human being? As has been shown above, Aquinas thinks that human beings do not start their lives as, but develop into human beings. Every human being goes through a process of hominisation, that is, of becoming human, that is usually completed after 40 days (in case the unborn is male) or 90 days (in case the unborn is female). If we take a look at the passage that details this time period, we can see that for Aquinas hominisation seems to depend on human appearance:

Minimum autem quantitatis in uno individuo est in primo instanti suae figurationis et animationis; quae quantitas adeo parva est quod parum excedit quantitatem formicae, ut dicit philosophus, quod in quadragesima die muliere pariente abortum, inventum est corpus prolis omnia membra distincta habere, quamvis in quantitate esset sicut magna formica.	But the minimum of the size in one individual is at the first instant of his formation and ensoulment; and this quantity is so small that it hardly exceeds the size of an ant, as the Philosopher (i.e. Aristotle) says, because when a woman miscarries on the fortieth day, the body of her child is found to have all its distinct members, although in size it is as small as an ant.

(Super Sent., lib. 3 d. 3 q. 5 a. 2 ad 3)

What Aquinas is saying here is that the human conceptus can be infused with a rational, that is, human, soul on the fortieth day of its existence because it then resembles a full-grown human being.[8] Before that, it cannot receive such a soul because the matter that would be ensouled does not yet possess the "figuram humani corporis" (Super Sent., lib. 3 d. 3 q. 5 a. 2 co.), that is, the shape of a human body.[9]

This observation can be further elucidated by Aquinas' description of the conception of Jesus Christ (cf. Super Sent., lib. 3 d. 3 q. 5 a. 2 co.). There he explains the steps of the process of conception that takes 40 days for every other human being but was instantaneous in the case of Christ:

> Propter quod oportet illam conceptionem subitaneam ponere, ita quod haec in eodem instanti fuerunt, scilicet conversio sanguinis illius materialis in carnem et alias partes corporis Christi, et formatio membrorum organicorum et animatio corporis organici, et assumptio corporis animati in unitatem divinae personae. In aliis autem haec successive contingunt, ita quod maris conceptio non perficitur nisi usque ad quadragesimum diem, ut philosophus in 9 de animalibus dicit, feminae autem usque ad nonagesimum.

> For this reason it is fitting to consider that conception to have been instantaneous, so that these things existed in the same instant: the conversion of that material blood into flesh and the other parts of Christ's body; the formation of the organic members and the soul being infused in the organ-bearing body; and the assumption of the ensouled body into the unity of the divine Person. Now, in others these things occur successively, such that a male child's conception is not completed until the fortieth day, or a female's until the ninetieth, as the Philosopher says in On Animals 9.

(Super Sent., lib. 3 d. 3 q. 5 a. 2 co.)

[8] The focus on human appearance also offers an explanation for why Aristotle and Aquinas made a difference with respect to the time of ensoulment of the male and female unborn: After 40 days of embryological development, the developmental state of its external genitalia—or to be more precise: of the so-called genital tubercle—make the unborn look more like a male human being. The external genitalia of a female human being only become clearly discernible after about day 80 of her development.

[9] This position is not contrary to but actually in line with Aquinas' idea of the heart of the primary organ of the human body (cf. Chap. VIII, Sect. 6). According to Aquinas, a body is disposed to receive soul by means of heat and humidity (cf. Q. d. de anima, a. 9 ad 16). Therefore, as heat and humidity "are preserved in an animal by means of its blood" (De potentia, q. 3 a. 9 ad 8), Aquinas takes the disposition for ensoulment to consist in blood circulation, that is, a functioning circulatory system. Without a beating heart, there can be no soul. As Aquinas takes the human conceptus to resemble a full-grown human being from the fortieth day of its existence, the conceptus then also possesses a beating heart—albeit in miniature format—from that point on—and therefore also meets this condition for ensoulment.

As this passage clearly shows, Aquinas takes rational ensoulment to have taken place after 40 days in male and 90 days in female human beings. This ensoulment is possible because at the end of this period there already exists a *corpus organicus*, that is, a body furnished with organs and therefore—given Aquinas' condition for ensoulment (as introduced above)—suitable for the reception of a rational soul. So, unlike what modern proponents of delayed ensoulment think, the organs that must be present for rational ensoulment according to Aquinas' own logic are neither internal organs in general (e.g. kidney, liver, lungs) nor rational organs (i.e. brain and nervous system) in particular. Rather, the organs he has in mind are the external organs, that is, the body parts visible on the outside of a living being's body, that constitute a specifically human shape (e.g. arms, hands, legs, feet).[10] It is the discernible presence of these external organs that gives the conceptus a specifically human shape. And because what has a human shape is human (regardless of its size), this makes the conceptus clearly human. And as a living being that is human must, by definition, have a rational soul, a living conceptus with the shape of a human body—even if it is only a very little human being (homunculus)—must have a rational soul.[11]

Strange as this kind of reasoning might sound to modern ears, it is not surprising from a medieval perspective. Given the embryological and biological knowledge of his time, if not by its appearance, how else could Aquinas have determined whether a living being was a human being? He did not know, as we do today, that human beings are human not if and because they look human but if and because they possess—contained in a molecule called deoxyribonucleic acid (DNA)—a specifically human genetic code.

> Fertilization brings together the full human complement of 23 male and 23 female chromosomes in the nucleus of the zygote. Lined along the chromo-

[10] This might strike us as somewhat odd in light of our modern use of the term "organ". But Aquinas repeatedly uses the hands or feet as examples of bodily organs. He even calls the hand "organum organorum" (Sentencia De anima, lib. 3 l. 13 n. 4), that is, the organ of organs or the most perfect of organs. Aquinas does not give a list of external organs. In theory, such list could also include head, ears, nose, eyes, mouth and teeth.

[11] That this line of reasoning seems to be in line with Aquinas can be seen by his statement that whoever belongs to the human species is rational and that a child, even before leaving the womb, is human although he does not yet exhibit rational activity (cf. Contra Gentiles, lib. 2 cap. 59 n. 17).

> some bands are the genes, estimated at some 90,000 in man, whose essential genetic material consists of double helix strands of DNA. DNA itself is a chemical band composed of specific alternating arrangements of four paired nitrogenous bases: adenine, thymine, guanine, and cytosine, commonly designated by their initials. At fertilization, chromosomes are paired by natural attraction according to partner-patterns of A, T, G, and C. In themselves, these four chemicals are not unlike the same chemicals found in subhuman entities, but their paired sequence is such as to produce a specific human protein. The sequence of A, T, G, C is specific to the species in which it is found. […] The significant fact for our concern here is that the human specificity of the adult human being is now directly traceable to the original DNA bearing the ATGC pattern. In all living things, this pattern is an architectural blueprint of the future adult expressed in chemical bonds. An embryologist able to read such a code in an early embryo could predict its normal destination in terms of species and accidental characteristics such as eye color, sex, etc. (Gerber 1966: 245)

Aquinas did not know that the zygote that results from the fusion of two human gametes, that is, a male sperm and a female ovum, necessarily already possesses a specifically human genetic code. Therefore, he did not know that the zygote does not become but necessarily is human from the very first moment of its existence.[12]

What does this mean for the problem of the time of rational ensoulment? As has been argued, for Aquinas, it is not the often-quoted presence of organs as such that is relevant for rational ensoulment. Rather, what is

[12] Relying on the authority of Aristotle's "Generation of Animals", Aquinas based his reflections on rational ensoulment on a now obsolete state of embryology: "According to Aquinas the formation of a body apt for human ensoulment occurs as follows. The male semen, acting as an instrumental cause for the father (who he thought was the sole principal cause) gradually brings the menstrual blood (the material cause, that matter out of which the effect is produced) provided by the female, to the point where it is disposed for the reception of the human soul" (Haldane & Lee 2003b: 535). So, for Aquinas human beings come into existence once the father's formative power contained in the male semen has developed a homogenous lump of menstrual blood into a heterogenous miniature of a full-grown human being.

The Aristotelian theory that a woman's blood and a man's semen are involved in human generation, dominated ideas of generation until almost the late modern period. Although its existence had been hypothesised since the middle of the seventeenth century, it was only in 1827 that Karl Ernst von Baer first confirmed the presence of the mammalian egg under the microscope and only in 1876 that Oscar Hertwig discovered the process of fertilisation (cf. Cobb 2006a, 2006b, 2012; Lopata 2009).

crucial for rational ensoulment is the belonging to the human species. The organs are important only because and insofar as they are indicative for this species membership. Therefore, Aquinas takes the time of presence of those external organs that give the conceptus the "figuram humani corporis" (Super Sent., lib. 3 d. 3 q. 5 a. 2 co.), that is, the shape of a human body, as the time of rational ensoulment.

While the general argument is sound and the metaphysical premises true, Aquinas' underlying embryological beliefs are unfortunately outdated and wrong. Today we know that human beings do not come into existence from a lump of menstrual blood but start their lives from a single cell, the zygote, that results from the fusion of a sperm and ovum. Today we know that species membership is not determined by a being's appearance but by its DNA. And we also know that the human zygote is genetically specifically human from the first moment of its existence.

Once we substitute the erroneous assumptions in Aquinas' theory of ensoulment with the current state of embryological knowledge, we can see that the resulting position changes from delayed to immediate ensoulment. For if the zygote does not develop into a human being but is a living human being from the moment of conception, then it must possess a rational soul from the moment of conception.

It is true that "the human body must have a diversity of organs to be properly disposed for the *operations* of the soul, that is, for second act" (Heaney 1992: 37). But, as has been pointed out above, the rational soul is the *first act* of a human being, that is, grants it its existence as a living human being and the capacity for every kind of endogenous activity. Everything a rational soul needs to fulfil its role as the first act of a human being is present in the human zygote. With the material properties provided by the zygote the rational soul "can function as the life principle and go on to develop those organs by which the soul can eventually perform its second act operations such as thinking and choosing" (Bracken 2001: 66).

Therefore, we can agree with the assessment that in light of modern embryology the theory of delayed hominisation has to be considered "obsolete" (Nguyen 2018a: 417) and that "it is a mistake to claim Thomas Aquinas (still less Aristotle) as precedent for holding a doctrine of 'delayed animation' today" (Jones 2012: 35f). For applying Aquinas' philosophical principles to the embryological facts uncovered since his time clearly supports immediate rather than delayed ensoulment:

> Thus, the same sound philosophy that led the ancients to affirm a theory of delayed hominization now leads us to affirm that hominization is complete at fertilization when the human organism comes into being. (Austriaco 2011a: 60)

Before we close this chapter, one last but highly important thing should be pointed out: The fact that Aquinas was a proponent of delayed ensoulment does not allow for the conclusion that Aquinas considered abortion of an unanimated foetus to be ethically permissible. It is true that Aquinas made a distinction between the killing of an animated and an unanimated foetus and did not seem to take causing the miscarriage of a not yet animated foetus to be homicide (cf. II^{a}-IIae q. 64 a. 8 ad 2). For Aquinas, such an act would nevertheless be morally impermissible because the act of preventing the generation of a human nature (e.g. by sterilisation, contraception or the killing of a not yet animated foetus) is practically on par with the destruction of a human nature, that is, the killing of an animated human being of whatever age (cf. Contra Gentiles, lib. 3 cap. 122 n. 9). So, causing the death of a yet unanimated foetus "would be sinful but the sin would in effect be that of contraception" (Haldane & Lee 2003a: 262; cf. Jones 2012: 9) and as such violate the intrinsic finality of the conceptus to reach the moment of animation.

Even if delayed ensoulment were true, this would not make the abortion of a not yet animated foetus ethically permissible. Drawing such a conclusion would be a logical non sequitur—and is a position that was not held by Aquinas or, to the best of my knowledge, any other ancient or medieval advocate of delayed ensoulment who is usually advanced by modern delayed ensoulment theorists who support early-term abortions.[13]

[13]This statement can be corroborated by the "Declaration on Procured Abortion" issued by the "Sacred Congregation for the Doctrine of the Faith" (1974: n. 7): "In the course of history, the Fathers of the Church, her Pastors and her Doctors have taught the same doctrine—the various opinions on the infusion of the spiritual soul did not introduce any doubt about the illicitness of abortion. It is true that in the Middle Ages, when the opinion was generally held that the spiritual soul was not present until after the first few weeks, a distinction was made in the evaluation of the sin and the gravity of penal sanctions. Excellent authors allowed for this first period more lenient case solutions which they rejected for following periods. But it was never denied at that time that procured abortion, even during the first days, was objectively grave fault. This condemnation was in fact unanimous".

2. Human Cell or Human Being?

As has been established in the preceding chapter, a rational soul can and does animate a being (a) if the being is a distinctively human being and (b) if this being shows any kind of endogenous activity. As has also been established, the first condition is met if the being in question is a genetically human being, that is, possesses a distinctively human DNA; the second condition is met from the moment the being starts to show any kind of endogenous activity until it starts to show the early certain signs of death.

At this point, the critical reader might interject and ask whether the possession of human genetic makeup and the display of endogenous activity are sufficient to be able to determine with certainty that one is dealing with a living human being and not just a living human cell or a cluster of living human cells. Given the difference in their respective genomes, their DNA may be sufficient to distinguish human beings from animals or plants. But does DNA allow us to determine whether a single living human cell is a human being, that is, a zygote, or merely a human cell but not a human being?

2.1. A Primer on Selected Types of Cells

To find an answer to this question, let us first familiarise ourselves with several cell types that are at the centre of the question whether a particular cell is a human being or merely a human cell:

- The human *egg cell* (also: egg, ovum, oocyte) is the female gamete (i.e. reproductive cell). It is produced in the ovaries, that is, the female gonads (sex glands), of female humans.
- The *sperm cell* (also: sperm, spermatozoon) is the male gamete (i.e. reproductive cell). It is produced in the testicles, that is, the male gonads (sex glands), of male humans.
- An *aneuploid human zygote* is a zygote that does not have two complete but only incomplete chromosome sets. Such a cell's chromosome sets are incomplete because the cell possesses either more or fewer than the 46 chromosomes characteristic of the human spe-

cies.[14] Aneuploid human zygotes can have 45 (monosomy; e.g. Turner syndrome), 47 (trisomy; e.g. Klinefelter syndrome, Down syndrome, Patau syndrome, Edwards syndrome), 48 (tetrasomy; e.g. Tetrasomy X) or 49 (pentasomy; e.g. Pentasomy X) chromosomes.

- A *parthenote* is "an egg that has been activated to begin to divide and to develop in the absence of sperm" (Austriaco 2011b: 47). Such a parthenogenetically activated cell is a diploid egg cell that contains only the maternal genome; it "has the normal number of 46 human chromosomes" but these chromosomes "are all derived from the mother" (Austriaco 2011b: 51). The activation that gives rise to a parthenote can be spontaneous (due to activation by a sperm that degenerates without fusing with the egg) or artificially (i.e. mechanically, electrically or chemically) induced.
- An *androgenote* is a diploid egg cell that contains only the paternal genome; it has 46 human chromosomes but all of them are derived from the father. Such a cell comes into existence when "either two sperm fertilize an enucleated egg (approximately 20% of complete moles) or a single sperm fertilizes an enucleated egg and then undergoes a duplication of its haploid genome (approximately 80% of complete moles)" (Austriaco 2011b: 46; cf. Schoenwolf et al. 2015: 53; Moore et al. 2020: 60). After implantation, the androgenote produces an entity named "*complete hydatidiform mole*".[15]
- *Polyploid zygotes* are zygotes that do not have two but more than two (usually three or four) complete chromosome sets:
  - In what can be called a *diandric triploid zygote* or an *androgenetic triploid*, "two sperm fertilize a normal egg resulting in a conceptus with 69 chromosomes" (Austriaco 2011b: 46; cf. Schoenwolf

[14] A cell is called euploid if it has an exact multiple of the number (n) of chromosomes that is characteristic of a species and constitutes a chromosome set. Accordingly, not only haploid (also: monoploid; 1n) and diploid (2n) cells but also polyploid (i.e. triploid (3n) or tetraploid (4n)) cells are euploid. In an aneuploid cell the total number of chromosomes is not an exact multiple of the number of chromosomes that constitutes a haploid chromosome set.

[15] According to Schoenwolf et al. (2015: 53), complete hydatidiform moles form in 0.1% to 0.5% of pregnancies but often spontaneously abort in early pregnancy; with partial hydatidiform moles, spontaneous abortion usually does not occur until the second trimester. As Bynum et al. (2020: 90) state, most complete hydatidiform moles "result from fertilization of an ovum devoid of maternal genetic material by a single sperm that duplicates (monospermy; approximately 85% [...]), but a subset is due to fertilization by two sperm (dispermy [...])".

et al. 2015: 53; Moore et al. 2020: 60).[16] Such a zygote possesses three complete sets of chromosomes, with the extra set being of paternal origin. In the course of its development, this specific kind of triploid zygote produces a cluster of grape-like cysts named "*partial hydatidiform mole*".
  - A *digynic triploid zygote* or a *gynogenetic triploid* is a chromosomally anormal zygote with three instead of two complete sets of chromosomes. Such a zygote has 69 chromosomes, with the extra set of chromosomes being of maternal origin.[17]
  - A *tetraploid zygote* possesses four instead of the normal two complete sets of chromosomes. Such a cell has 92 chromosomes, half of which are of maternal and half of which are of paternal origin. In rare cases of triandric tetraploidy, a tetraploid zygote possesses one maternal and three paternal chromosome sets (cf. Bynum et al. 2020: 90).[18]
- *Stem cells* are characterised by two properties: Stem cells possess (1) the ability to self-renew, that is, proliferate while maintaining their identity as stem cell, and (2) the ability to differentiate into more

[16] While most cases (approximately 99%) of diandric triploidy result from dispermy, that is, fertilisation of an ovum by two sperm, it can—in very rare cases—also arise by monospermy, that is, fertilisation of an ovum by one sperm that duplicates (cf. Bynum et al. 2020: 90).

[17] According to Fleischer et al. (2015: 306), most triploid—both digynic and diandric—conceptions "end in first trimester spontaneous abortions". However, "there are few reports of live born infants with triploidy, with the longest reported survival of digynic triploidy at 10 ½ months" and "a girl with diandric triploidy who survived for 10 ½ weeks".

[18] The clinical consequences of tetraploidy include limited life expectancy. Although a great majority of tetraploid pregnancies end in miscarriage, "in rare cases, complete tetraploidy is compatible with life" (Bothur-Nowacka et al. 2013: 470). However, only 12 live-born cases with regular tetraploidy have been reported in the literature so far (cf. Bothur-Nowacka et al. 2013: 468). Of these cases, "most reported individuals have died between birth and the age of one year. The oldest described non-mosaics tetraploid patient was aged 26 months, another female was at least 22-months-old" (Bothur-Nowacka et al. 2013: 470).

specialised daughter cells[19] and by doing so—either directly or indirectly—form the various, differentiated cell types of the body that make up its tissues and organs. Depending on their "source", stem cells are usually distinguished into adult stem cells, embryonic stem cells and induced pluripotent stem cells:

- *Adult stem cells* (also: somatic stem cells; tissue-specific stem cells) are undifferentiated cells that reside in small numbers throughout the body among differentiated cells in a tissue or organ (e.g. bone marrow, brain, skin, gastrointestinal tract, muscles). They are not only capable of self-renewal but also give rise to a limited number of differentiated cell types that build the tissue in which the respective stem cells reside. In this way adult stem cells help substitute dying cells and regenerate damaged tissues. Compared to embryonic stem cells, adult stem cells are only multipotent.
- *Embryonic stem cells* are derived from the inner cell mass (also: embryoblast) of the blastocyst, that is, "the portion of the blastocyst of a mammalian embryo that is destined to become the structures of the postnatal body" (Condic 2013: 77). Embryonic stem cells are pluripotent, that is, "can produce only the cell types eventual found in the postnatal body" but "lack the ability to

[19] This latter property is called potency. A cell's potency signifies its ability to differentiate into other cell types. A distinction is made between the following kinds of potency:

- *Totipotent* cells are "capable of developing into a complete organism" (Condic 2020: 113; cf. 22) respectively "capable of producing all the cell types of the embryo, including those cell types that make up the extra-embryonic organs of the embryo; namely, the placenta, amnion, and chorion" (Condic & Condic 2018: 82f). The only truly totipotent cell is the zygote. However, as it is not self-renewing, the zygote is not considered a stem cell despite its totipotency.
- The zygote's daughter cells retain their mother's totipotency until the morula stage, that is, the embryo's 16-call stage. However, as Condic & Condic (2018: 83) point out, the totipotency of the zygote is different from the totipotency of their immediate daughter cells: "individual totipotent cells (other than the zygote) do not appear to be capable of producing an organized or complete body". This special kind of totipotency of a zygote's immediate daughter cells is called *plenipotency* (also: pluripotency): Plenipotent cells are totipotent cells that lack the ability to form "the complete body as an organized, functional whole" (Condic & Condic 2018: 84). Accordingly, from the 2-cell stage until the 16-cell stage, the individual cells of the embryo are plenipotent rather than totipotent.
- A *multipotent* cell (e.g. an adult somatic stem cell) has the potential to transform into all cell types within one particular germ layer lineage, that is, the mesoderm, endoderm or ectoderm.

form the extra-embryonic organs such as the placenta" (Condic & Condic 2018: 83).

- A *human induced pluripotent stem cell* (h-iPSC) is a pluripotent stem cell that has been generated from a human somatic cell by means of genetic reprogramming. h-iPSCs are differentiated cells that have been artificially de-differentiated back to a pluripotent state.[20]

- A *terminally differentiated cell* (e.g. muscle cell, sperm cell, root hair cell, cardiac cell, nerve cell, blood cell) is a specialised cell that is suited to a specific function and is normally—that is, without artificial intervention—not able to transform into another type of cell. A terminally differentiated cell is the cell furthest from its stem cell.
- A *clone* is a zygote that has been artificially produced by means of somatic cell nuclear transfer, that is, by transferring the nucleus of a somatic cell into an enucleated egg. In contrast to a cybrid (see below), both the somatic cell and egg used for cloning are—though not necessarily from the same individual—of the same species.
- A *hybrid* is "the product of breeding two different species (via normal copulation or in vitro fertilization)" (Seyfer 2006: 37). Every cell of a hybrid—such as a mule, that is, the crossing between a female horse and male donkey—contains a mixture of

[20] The reprogramming technique is an alternative to the technique called "Alternate Nuclear Transfer" (ANT). At its core, ANT is a form of cloning based on somatic cell nuclear transfer. However, in contrast to normal cloning, ANT takes the nucleus of a somatic cell but either alters the somatic nucleus or the cytoplasm of the egg prior to transfer of the nucleus into the egg. The result is not a totipotent cell but a cell that has "the characteristics and limited developmental disposition of a pluripotent stem cell" (Austriaco 2006: 523).

Compared to ANT, the h-IPSC approach is "an elegant (and far simpler) way of achieving the same goals" (Condic & Condic 2018: 224). While both produce a pluripotent stem cell, the h-IPSC approach does not require the use of a human egg cell. As h-IPSC makes ANT superfluous and as h-IPSC makes it practically superfluous to deal with the contested question whether the cell resulting from ANT is a human being or not—authors such as Walker (2004) and Schindler (2005) answer in the affirmative, others such as Austriaco (2006), Brugger (2005) and Condic & Condic (2018: 223ff) in the negative—this chapter does not focus on cells produced by ANT. For the interested reader: In light of this chapter's position, a cell resulting from ANT is a (albeit defective) human being and not merely a human cell. Despite the genetic engineering prior to transfer, ANT remains a form of cloning.

genetic material from both parents.[21] Accordingly, a ***human–animal hybrid cell*** is a cell that possesses a mixed human/animal nuclear genome.

- A *synkarion* is a special kind of hybrid that comes into existence by the fusion of two cells and, in consequence, their pre-existing nuclei. If two cells but not their nuclei fuse, the cell resulting from the fusion is called *syncytium* (also: heterokaryon).
- Another special kind of hybrid is a *transgenic zygote*. The chromosome sets of a transgenic zygote do not come from two but only one species; however, its nuclear genome has been altered by integration of one or more DNA sequences from the genome of another species. A human–animal transgenic zygote has an originally human nuclear genome with animal genetic elements admixed to it.

- A *cybrid* (also: cytoplasmic hybrid) is a cell that has been artificially produced by inserting the nucleus of a human somatic cell into the enucleated cytoplasm of a non-human animal egg. While such a cell possesses a human nuclear genome, its mitochondrial genome is of animal origin.[22]

[21] Just like hybrids, chimeras are beings that result from the combination of two species. However, while every cell of a hybrid contains DNA from both of the hybrid's parents, a chimera does not contain genetically mixed cells: "the cells in the body of a chimera are a mosaic of cells of different species" (Seyfer 2006: 38). A chimera can be produced, for example, by the melding of two zygotes or by the insertion of "an embryonic cell from one species [...] into the blastocyst of the other species" (Seyfer 2006: 38). As they cannot start their life as a unicellular entity but only a two-cell entity, chimeras are not the focus of our considerations. A human-nonhuman chimera contains human cells but is not a human being. In contrast to Seyfer, Condic (2020: 65) also considers beings that result from the combination of two specimens of the same species (as, e.g., in the natural fusion of two—same-sex or opposite-sex—embryos to form a single embryo) as chimeras.

[22] Mitochondria are membrane-bound cell organelles that serve as a cell's powerhouses by generating most of its chemical energy. The multitude of mitochondria present in a cell do not reside in its nucleus but its cytoplasm, that is, the gelatinous liquid that fills the cell's inside. Mitochondria contain their own genome consisting of 16,569 base pairs each (for comparison: a cell's nuclear genome comprises approximately 3.3×10^9 base pairs). Mutations in a cell's mitochondrial genome can result in a range of mitochondria-associated diseases, such as Leber hereditary optic neuropathy.

As Schoenwolf et al. (2015: 484) point out, "we inherit all our mitochondria from our mothers". This is due to the fact that when sperm and ovum fuse in vivo, "the head and tail of the sperm enter the cytoplasm of the oocyte, but the sperm's cell membrane [...] and mitochondria remain behind" (Moore et al. 2020: 26). And even in the case that male mitochondria manage to enter the egg, they are sought out and destroyed by the egg: "Mitochondria carried into the oocyte cytoplasm by the spermatozoon are sought out and destroyed, leaving only oocyte mitochondria to propagate their mitochondrial (mt) DNA to offspring" (Sutovsky & Song 2017: 56).

What all of the above-mentioned entities have in common is that they are unicellular entities, that is, entities comprising only one cell. But how do we know whether the above-mentioned cells are human beings or human cells? Where are the boundaries of humanity?

What we are looking for is a criterion that helps us answer these questions with respect to cells without knowing their developmental trajectory, that is, how they will develop. Waiting and seeing is not a helpful approach when we are trying to find out whether a cell is a human being at a certain point in time.

2.2. The Status of Selected Types of Cells

To address the above-mentioned questions, let us compare these cells with the one cell that we know is a human being, namely the zygote, that is, a human being in his one-cell embryonic stage. In order to be able to make this comparison it is helpful to first describe the essential characteristics of a human zygote and find out why we consider it a human being rather than merely a human cell. So, what characterises a human zygote essentially? I propose that there are four such essential characteristics:

- First and most obviously, a human zygote is a unicellular being. The zygote is a complete, individual being consisting of one cell only.
- Second, a human zygote possesses a specifically human genome. As the genome of human beings not only consists of a nuclear genome (nDNA) but also a mitochondrial genome (mtDNA), this characterisation can be rephrased as follows: A human zygote possesses both a specifically human nuclear and a mitochondrial genome.
- Third, while its mitochondrial genome is of maternal origin only, the zygote's diploid nuclear genome is of both maternal and paternal origin. The zygote's diploid nuclear genome is a combination of a human haploid paternal and a human haploid maternal chromosome set.
- Fourth, while the zygote is genetically of both maternal and paternal origin, that is, while its nuclear genome contains a paternal and maternal chromosome set, materially it is virtually exclusively of maternal origin only. You could say that the zygote comes into exis-

> tence when a sperm—or to be more precise, the sperm's haploid nucleus—moves in with an egg cell. On that occasion, the sperm contributes a haploid chromosome set to the newly formed zygote, but the egg cell not only also provides a haploid chromosome set but in addition the actual "housing" of the zygote, that is, the cell membranes, the organelles and the cytoplasm. So, it is characteristic of the human zygote that its material-cellular basis originates from a human egg cell (rather than, e.g., a stem cell or a somatic cell). From the perspective of its material cause, that is, neglecting the origin of its nuclear genome, the zygote only has a maternal parent cell.

Some readers might correctly note that the above list lacks—what others might consider—important characteristics. The list does not contain the zygote's totipotency or developmental potential, that is, its "active potential" (Austriaco 2011b: 46) to generate—"in an ordered and sequential way" (Austriaco 2011b: 52)—all the cell types and tissues necessary to develop into a mature human being.[23] And neither does the list contain "organismal behavior" (Condic 2013: 48; cf. Condic & Condic 2018: 231), that is, behaviour that preserves and promotes the life, health and continued development of the whole. These are important characteristics of the zygote, indeed. However, I do not think that the question whether a cell is a human being should be made dependent on these behavioural or functional characteristics.

It is true that "agere sequitur esse" or—as the original statement reads—"agere sequitur ad esse in actu" (Contra Gentiles, lib. 3 cap. 69 n. 20), that is, that acting follows upon being in act.[24] We must, however, be careful to not misunderstand this statement. The statement does not say that an entity's behaviour suffices to account for the nature of the entity; an entity cannot simply be defined by what it does. What this statement does say is that being in act is causally anterior to acting and that acting is,

[23] As a sidenote: Others characterise the active potential characteristic of a zygote not as the potential to develop into a mature human specimen. Rather, it is "the intrinsic active potential to develop to the stage where it can (if not impeded) perform operations specific to human organisms" (Lee 2011: 82) which is equated with "the intrinsic capacity to develop a brain" (Lee 2011: 82).

[24] In a parallel passage, Aquinas states that "agere sequitur ad esse perfectum, cum unumquodque agat secundum quod est in actu" (Super Sent., lib. 3 d. 3 q. 2 a. 1 co.; cf. I^{a}-IIae q. 55 a. 2 ad 1), that is, that acting follows upon perfect existence, because each thing acts insofar as it is in act.

therefore, consequent upon being in act. An entity can only act insofar as it is in act.

So, while we can learn about what an entity is and get to understand its nature by observing its behaviour, an entity is not what it is because of its behaviour. An entity's behaviour is the effect of its nature; its nature is not the effect of its behaviour. In short, behaviour reveals but is not constitutive of nature. A being is what it is not because of what it does or can do but because it is what it is.

A being is what it is because it possesses a substantial form as first act. It is the possession of its substantial form that grants a being a specific range of capacities or active potentials. However, a being's substantial form can actualise its capacities only by means of and based on the matter it informs. But if a material defect hinders a being to actualise one or more of the capacities granted by its substantial form, this does not allow the conclusion that it does not possess the substantial form. It simply means that it is a being with a defective material infrastructure, that is, an infrastructure that does not allow the being's substantial form to move from first to second act to a lesser or greater extent.

Having said this, let us summarise the non-behavioural or non-functional essential characteristics of a human zygote. A human zygote

- is a unicellular being
- with both a human nuclear (nDNA) and a human mitochondrial (mtDNA) genome
- whose nuclear genome is composed of a human haploid maternal and a human haploid paternal chromosome set and
- whose material-cellular basis originates from a human egg cell.

It is because of these four characteristics that we should consider a human zygote a human being rather than merely a human cell. This means that the question whether a cell is a human being or merely a human cell can be answered with reference to these criteria, too. If the cell types listed in the previous chapter meet all four of these criteria, the respective cell type should be considered a human being rather than merely a human cell.

Figure 1 shows how the individual cell of the cell types introduced above fare with respect to the above-mentioned criteria. As can be seen,

Type of Cell	Is the cell's nuclear genome (nDNA) human?	Ploidy of human nuclear genome (nDNA)			Is nDNA composed of both maternal and paternal human chromosome set?	Is the cell's mito-chondrial genome (mtDNA) human?	Does the cell have a human ovum as parent cell?	Is the cell a human being?
		Does nDNA consist of one or more complete chromosome sets?	# of complete chromosome sets in nDNA	# of chromo-somes in nDNA				
Human male gamete (also: sperm or spermatozoon)	✓	✓ (euploid cell)	1 (haploid cell)	23	✗ (paternal genome only)	✓	✗	✗
Human female gamete (also: egg cell, ovum or oocyte)	✓	✓ (euploid cell)	1 (haploid cell)	23	✗ (maternal genome only)	✓	✗	✗
Human zygote	✓	✓ (euploid cell)	2 (diploid cell)	46	✓	✓	✓	✓
Aneuploid human zygote	✓	✗ (aneuploid cell)	n/a	45, 47, 48 or 49	✓	✓	✓	✓ (albeit defective)
Tetraploid human zygote	✓	✓ (euploid cell)	4 (tetraploid cell)	92	✓ (+ extra maternal and paternal set <u>or</u> + two extra paternal or maternal sets)	✓	✓	✓ (albeit defective)
Digynic triploid human zygote	✓	✓ (euploid cell)	3 (triploid cell)	69	✓ (+ extra maternal set)	✓	✓	✓ (albeit defective)
Diandric triploid human zygote (develops into "partial hydatidiform mole")	✓	✓ (euploid cell)	3 (triploid cell)	69	✓ (+ extra paternal set)	✓	✓	✓ (albeit defective)
Human androgenote (develops into "complete hydatidiform mole")	✓	✓ (euploid cell)	2 (diploid cell)	46	✗ (paternal genome only)	✓	✓	✗
Human parthenote	✓	✓ (euploid cell)	2 (diploid cell)	46	✗ (maternal genome only)	✓	✓	✗
Human embryonic stem cell (h-ESC)	✓	✓ (euploid cell)	2 (diploid cell)	46	✓	✓	✗	✗
Human induced pluripotent stem cell (h-IPSC)	✓	✓ (euploid cell)	2 (diploid cell)	46	✓	✓	✗	✗
Adult human stem cell	✓	✓ (euploid cell)	2 (diploid cell)	46	✓	✓	✗	✗
Terminally differentiated human somatic cell	✓	✓ (euploid cell)	2 (diploid cell)	46	✓	✓	✗	✗
Clone cell (i.e. cloned human zygote)	✓	✓ (euploid cell)	2 (diploid cell)	46	✓	✓	✓	✓
Cybrid cell	✓	✓ (euploid cell)	2 (diploid cell)	46	✓	✗	✗	✗
Human-animal hybrid cell	✗	n/a	n/a	n/a	n/a	n/a	n/a	✗

Fig. 1 Which cell is a human being? Status of selected types of cells

normal human zygotes but also aneuploid human zygotes, tetraploid human zygotes, digynic triploid human zygotes, diandric triploid human zygotes or clones are human beings while human gametes, androgenotes, parthenotes, human embryonic stem cells, human induced pluripotent stem cells, adult human stem cells, terminally differentiated human somatic cells and cybrids are merely human cells. Human–animal hybrid cells are not human cells.

The status of the above-mentioned unicellular beings transfers to the multicellular beings into which these cells develop. A human zygote is a human being. As it is already a human being a zygote can and does not develop into a human being; given the right circumstances, the only thing it develops into is a mature human being. The same holds true for all cells that are human beings; they develop as human beings but not into human beings.

> No living being can become anything other than what it already essentially is. [...] Living beings come into existence all at once and then gradually unfold to themselves and to the world what they already but only incipiently are. (Joyce, 1978: 103/ 106)

The same holds true for cells that are not a human being but merely or—as in case of a human–animal hybrid cell—not even a human cell. Such cells can and do not develop into a human being. They develop as whatever else they may be.

The fact that a cell is not a human being has implications for what we are allowed and not allowed to do with it. As we are not allowed to kill human beings, we are also not allowed to kill a cell that is a human being. If a cell is not a human being, this prohibition does not apply. We have more degrees of freedom in dealing with this cell. This does not, however, automatically mean that it is morally permissible to create human–animal hybrids or chimeras.

The above list of cell types does not cover all possible types of cells that exist—or do not yet exist. Accordingly, there might be situations when there is reasonable doubt whether a certain type of cell is a human being or not. For reasons of prudence, in these cases, we ought to treat such cell as a human being—which also implies that (and be it for research purposes) we ought not to kill it and especially not bring it into existence in order to kill it.

3. Delayed Desoulment: Do We Die Twice?

Given the nature of human life and the human soul outlined above (cf. Chap. IV, Sect. 4.2), human beings cannot lose their rational life while keeping their vegetative life. When a human being loses his rational life, he necessarily also loses the capacity to exhibit any signs of life or any endogenous activity. If the rational soul is gone, all life and all capacities are gone. Or to put it succinctly, human beings only die once.

However, there are those who hold the contradictory view, namely that human beings die twice: "We need to distinguish two concepts of death: the death of the person, or self, and the death of the physical organism" (McMahan 2010: 110; also cf. McMahan 2002: 439; Madea 2014: 51; Vij 2014: 76). According to this view, human beings first lose their rational life and then lose whatever rest of life they have in their organs, tissues and cells; so, they die a "personal death" (also: individual death, death of the person) before they die a "biological death" (also: cellular death, molecular death, death of the organism). As biological death is supposedly preceded by personal death, proponents of this view call for a separate criterion for the diagnosis of personal death, namely brain death, that is, the complete and irreversible loss of brain activity.

In an article defending the total brain death criterion, Lee and Grisez (2012) couch McMahan's two-death hypothesis in the language of hylomorphism. As they stipulate, "if the living remains (in a brain-dead body) themselves have a soul, then it is a vegetative soul, not a rational or animal soul" (Lee and Grisez 2012: 282). So, according to Lee and Grisez, with brain death a person loses her rational (i.e. human) as well as sensitive (i.e. animal) soul. Having died this first death, her remains are only animated by a vegetative soul. These remains die a second death when this vegetative soul separates from them. As a human being cannot remain a human being if his rational soul is substituted by a vegetative soul, this position logically implies that only the being who dies the first death is a human being; the being that dies the second death is—to put it pointedly—a vegetable.

What Lee and Grisez seem to stipulate is a position that could be termed "delayed desoulment". This would make the death of human beings a two-stage process: In a first step, a human being's rational soul would be substituted with a vegetative soul; in consequence, the being that has been a human being would turn into some kind of merely living plant. Sometime after this first death and in a second step, this plant would then die a second death when its vegetative soul separates from it and turns it into a lifeless corpse.

Unfortunately, Lee and Grisez do not provide an argument or cite any sources that back their position. Van Dyke (2012: 252) tries to fill this gap by referencing the following passage from Aquinas' "Commentary on the Book of the Causes":

Rursumque in via corruptionis primo amittit usum rationis et remanet vivum et spirans, secundo amittit vitam et remanet ipsum ens, quia non corrumpitur in nihilum. [...] Nam priora sunt existentia viventibus et viventia hominibus, quia remoto homine non removetur animal secundum continentiam, sed e converso quia, si non est animal, non est homo.	Again, in the process of corruption he first loses the use of reason, while living and breathing remain. Second, he loses life, while being itself remains, because he does not corrupt into nothing. [...] For existing things are prior to living things and living things to men, because when 'man' is removed what 'animal' entails is not removed. Instead, the opposite is the case, because, if there is no animal, there is no man.

(Thomas Aquinas, Super De causis, l. 1)

In this passage, Aquinas seems to parallel the process of the corruption of human beings with the process of their generation. As, for Aquinas, the latter is characterised by delayed ensoulment (cf. Chap. VIII, Sect. 1), he therefore seems to advance some sort of delayed desoulment with respect to the former: "what the Liber de causis commentary passage suggests is that the substantial form 'human being' is sometimes (or, perhaps, always) replaced by at least one intermediate form before that being becomes a corpse" (van Dyke 2012: 254).

Unfortunately, van Dyke's reading of the passage from Aquinas' "Commentary on the Book of the Causes" is somewhat problematic for several reasons. As Eberl (2005b: 37) argues, in this passage

> Aquinas is referring to logical distinctions among species and genera. Thus, when Aquinas asserts that an animal remains if rationality is removed from a human being, he is not offering an ontological description of what happens in the process of human death. Rather, he is claiming that, if one mentally abstracts the concept of rationality from the definition of a human being, the concept of animality will yet remain.

Acknowledging Eberl's argument, van Dyke (2012: 253), however, thinks that his position is "unconvincing". For her, Aquinas does not make a statement about mere logical priority between being an animal and being a human being: As Aquinas is clearly contrasting the process of corruption with the process of generation and as he describes "an actual ontological process in the case of generation, it would be an extremely forced reading that denied he was also describing an actual, ontological process in the case of corruption immediately following" (van Dyke 2012: 253).

Even if we were to grant van Dyke's interpretation of the passage and agree that Aquinas proposes some sort of delayed desoulment, there remains a serious problem: As has been shown in a previous chapter (cf. Chap. VIII, Sect. 1), Aquinas' theory of delayed ensoulment rests on the erroneous and outdated embryological knowledge of his time and must therefore be considered obsolete. But, if we think that—although his metaphysical principles are sound—Aquinas' theory of delayed hominisation is mistaken, why should we follow him in transferring this mistaken theory to the end of life? As the theory of delayed ensoulment is wrong, the process of the corruption of human beings cannot be modelled by it.[25] So, unfortunately, the passage from Aquinas' "Super De causis" cannot be cited as a convincing argument for delayed desoulment.

Moreover, in light of his position that the rational soul is the substantial soul of man, it is hard to see how Aquinas—despite the fact that he advocated a theory of delayed *en*soulment—could actually have been a proponent of delayed *de*soulment. Why this is so becomes clear when we consider the following two things: For Aquinas, "there is no other substantial form in man besides the intellectual soul" (I^a^ q. 76 a. 4 co.). In addition, Aquinas holds that the intellectual soul "virtually contains the sensitive and nutritive soul" (I^a^ q. 76 a. 4 co.). To put the latter point more generally, while a less noble kind of soul (e.g. the vegetative soul) does not contain and cannot evolve into a nobler kind of soul (e.g. into a sensitive or rational soul), the nobler kind of soul (e.g. the rational soul) virtually contains the less noble kinds of soul (e.g. the sensitive and vegetative soul). So, by virtue of possessing a rational soul, man is already an animal and a living being:

Sic ergo dicimus quod in hoc homine non est alia forma substantialis quam anima rationalis; et quod per eam homo non solum est homo, sed animal et vivum et corpus et substantia et ens.	Thus, therefore, we say that in 'this man' there is no other substantial form than the rational soul, and that by it man is not only man, but animal, and living being, and body, and substance, and being.

(Thomas Aquinas, De spiritualibus creaturis, a. 1 ad 24)

[25] All the less so because Aquinas' statement that "in the process of corruption he first loses the use of reason, while living and breathing remain" (Super De causis, l. 1) cannot claim absolute empirical validity. There are many cases of human beings who die without having first lost the use of reason.

As his substantial form man's rational soul not only grants him the capacity for rational activity but also all activities characteristic of the sensitive and nutritive soul. According to the theory of delayed ensoulment, the lower-level souls of the growing conceptus are destroyed and replaced with a higher-level soul; with each step of ensoulment, the conceptus gains a range of additional capacities—until it finally possesses a rational soul and with it the capacity for all the activities characteristic of a rational, sensitive and vegetative soul.

However, this process cannot simply be mirrored onto the end of life. Even if delayed ensoulment were true, this does not logically imply that the expansive process of ensoulment takes place in reverse order at the end of life. Why is this? To be able to die a personal death (i.e. brain death), a human being must possess a rational soul; if there is no rational soul to replace, there can be no personal death. As the rational soul virtually contains the sensitive and vegetative soul, it can do what these souls can do. In contrast to delayed ensoulment, where a less powerful soul is replaced by a more powerful soul, in delayed desoulment a more powerful soul would be replaced by a less powerful soul. But why does a rational soul have to be substituted by a sensitive or vegetative soul if it can already do what these souls can do? From a metaphysical perspective, a regressive substitution of souls would be unnecessary and superfluous—and therefore in violation of Ockham's razor, that is, the problem-solving principle according to which the simplest explanation is usually the right one.[26]

The only reason that would make this substitution logically compelling would be if a brain-dead body were no longer suited to host a rational soul. However, as has been argued in a previous chapter (cf. Chap. IV, Sect. 5), a rational soul can animate a being

- (a) if the being is a distinctively human being and
- (b) as soon as and as long as this being shows any kind of endogenous activity

The first of the above conditions is met if the being in question is (i) a unicellular being or (ii) a being that has developed from a unicellular being

[26] As a sidenote: Although it is called Ockham's razor and therefore attributed to William of Ockham, a fourteenth-century English Franciscan monk and philosopher, the principle of simplicity as such is already found in earlier texts, for example in Aristotle's Posterior Analytics (86 a 33–35; book 1, chapter 25) and Aquinas' Contra Gentiles (lib. 3 cap. 70 n. 3).

- that possesses both a human nuclear (nDNA) and a human mitochondrial (mtDNA) genome,
- whose nuclear genome is composed of a human haploid maternal and a human haploid paternal chromosome set and
- whose material-cellular basis originates from a human egg cell.

Such a human being meets the second of the above conditions from the moment it exhibits any kind of endogenous activity until the moment when its physical-material infrastructure, that is, its body, makes it impossible to exhibit any kind of endogenous activity. In other words, such a human being possesses a rational soul from the moment of conception until it exhibits the early certain signs of death.

Accordingly, as a brain-dead body is clearly suited to host a rational soul, the regressive substitution of souls implied by delayed desoulment is unnecessary and superfluous.

4. The Twinning Charge to Immediate Ensoulment

Proponents of delayed ensoulment sometimes base their position on the so-called twinning charge (also: individuality charge). According to this charge, the possibility of twinning which exists approximately 14 days following conception precludes rational ensoulment until after this period.[27]

> Identical twins (unlike fraternal twins, who derive from two distinct ova, fecundated by two distinct spermatozoa) start life as one ovum, fecundated by one spermatozoon. For the proponents of immediate hominization, this fecundated ovum is one human person. Very early in pregnancy this ovum splits into two (or more) parts, each one of which develops into an adult. This fact is difficult to reconcile with immediate hominization. A human person does not split into two or more human persons. (Donceel 1970: 98)[28]

[27] It should be noted that the twinning charge is only applicable to cases of monozygotic twinning, that is, cases in which a zygote splits sometime during the period from conception to implantation (which coincides with the formation of the primitive streak). As Condic (2020: 42) points out "monozygotic twinning is a rare event, occurring in approximately 0.35 percent of all live births". In contrast to monozygotic (also: identical) twins, dizygotic twins (approximately 1% of all pregnancies) always and only come about if two different ova are fertilised by two different spermatozoa at the same time.

[28] This argument by Donceel is also echoed by, for example, Ford (1988).

According to this position, the undoubtedly existing possibility of twinning[29] makes immediate ensoulment impossible because rational ensoulment presupposes ontological individuality or identity, that is, continuous existence "as the same continuing individual living body, being or entity" (Ford 1988: xv). In the mind of those persuaded by the twinning charge, the requirement of ontological individuality is violated as long as twinning remains possible:

> How could a zygote be one distinct human individual whilst it still had the capacity to become more than one distinct individual? (Ford 1988: xvi)

> (W)hile fertilization establishes genetic individuality, it does not establish a second aspect of individuality, namely oneness or singleness. (Grobstein cf. 1979: 66f; 1988: 346)

It is only after the formation of the so-called primitive streak, which is the precursor of the spinal cord and takes place approximately 14 days after fertilisation, that embryonic fission, which produces identical twins, can no longer occur. Given this biological fact, proponents of the twinning charge stipulate that prior to the formation of the primitive streak the embryo is not an individual but merely a cluster of individual cells or—as Grobstein (1979: 66; 1988: 346) calls it—a "preembryo":

> This is when the human body is first formed with a definite body plan and definitive axis of symmetry. Prior to this stage it would seem to be quite unreal to speak of the presence of a distinct human individual. (Ford 1988: xviii)

Authors such as Ford, therefore, take the appearance of the primitive streak at about 14 days of development as the time of rational ensoulment:

> A new human individual begins once the matter of the epiblastic cells becomes one living body, informed or actuated by a human form, life-principle or soul that arises through the creative power of God. The appearance of one primitive streak signals that only one embryo proper and human individual has been formed and begun to exist. (Ford 1988: 172; cf. xviii)

[29] The early embryo's potential for twinning is rooted in the quality of the cells of the early embryo, which are of a special and different quality than the cells of his later developmental stages. The early embryo's cells are totipotent, that is, they have the capacity to develop into any cell type (and consequently into any tissue and organ) and form an entire organism. This totipotency is lost after the formation of the primitive streak which occurs about the time of the embryo's implantation into the uterine wall (between 13 and 16 days after conception).

So, what are we to make of the twinning charge? Unfortunately, pointing to the possibility of twinning is not a cogent argument against immediate ensoulment. The twinning charge does not establish why we should equate indivisibility with individuality. On the contrary, divisibility is perfectly consistent with individuality. The ability to divide, that is, the potential to become two or many, does not as such preclude the individuality of the holder of that potential.

> It is an old insight of Greek philosophy that if something is potentially two (or more) does not logically follow that it is not actually one. A stick in my hand is potentially two sticks. I can break it into two parts, but before I do so, it is just one stick. The fact that out of a single morula two blastocysts could arise, does not entail that the morula is not actually one. (Damschen et al. 2006: 172)

Rather than being indivisible, being an individual means being a countably distinct instance of a kind that can be distinguished from other instances of that kind: "For an individual is definable as an entity which is a countably distinct instance of the kind of which it is a member. And an embryo younger than fourteen days, even if we know it is going to divide, is clearly a countable instance of the kind 'human embryo'" (Oderberg, 1997: 274). If the embryo is not an individual prior to twinning, then—as Napier (2010: 790) points out—"what is it that undergoes twinning?" As the embryo even prior to twinning clearly is a countably distinct human being and therefore an individual, the twinning charge falls flat. Although it looks like a powerful charge at first glance, science itself has provided us with the argument to refute the twinning charge, namely cloning. It is the very possibility of cloning that serves as a rebuttal to Ford's stipulated ontological individuality as prerequisite for ensoulment.

Twinning and cloning are very similar processes. In both processes, a new human being is generated from an already existing one. There are only two differences:

- While twinning is a natural (albeit seldom) occurrence, cloning is an artificial one.
- Whereas the age difference of identical twins can only be up to two weeks, the age difference between a clone and its original can be much more than two weeks.

In fact, "twinning is a form of cloning that is not artificial but results from an embryological accident" (Ashley & Moraczewski 2001: 194). The younger identical twin can be considered the clone of an earlier embryo. In the end, both clone and identical twin are copies—the former an artificial, the latter a natural one.

What makes cloning a rebuttal of the twinning charge is the fact that we do not hold that cloning negates the ontological individuality of the being that is cloned. For if we consistently apply the logic of the twinning charge to the process of cloning, human beings could only be rationally ensouled once cloning has become impossible (or once they have been cloned at least once). But because cloning remains possible until (and perhaps some time after) death, no living human being would and could be rationally ensouled. Given the fact that we do not consider living human beings rationally unensouled in general or until they have been cloned, the possibility of twinning—for reasons of logical consistency—does and cannot be advanced as an argument against immediate ensoulment.[30]

Immediate ensoulment is perfectly possible despite the possibility that the rationally ensouled zygote can twin. Before twinning, there is an individual human being with a human soul. In twinning, one or more totipotent cells separate from this individual and form a second (or possibly even a third, fourth, etc.) individual human being that is animated by another human soul.

> If by a developmental accident twinning occurs [...] the only reasonable explanation is that a clone of that individual has been formed by its loss of a part. By reason of its totipotentiality, that part can begin an independent development. Thus organism A that was developing normally up to the point of accidental twinning continues its development alongside its somewhat younger clone, its monozygotic twin (organism B). (Ashley & Moraczewski 2001: 198)

It is—as Seifert (2004: 117) puts it—perfectly possible that "twinning may just play the same role as conception and give rise to a new human

[30] Napier (2010) offers an additional—albeit somewhat similar—line of reasoning against the twinning charge. As he points out, identical twins do not come into being at conception; rather, one twin come into being at conception, the other comes "into being at the point of twinning through a form of asexual reproduction" (Napier 2010: 792). For reasons of consistency, proponents of the twinning charge must also adhere to the position that no living being capable of asexual reproduction is an individual. But such position would be unrealistic: "But numerous organisms are capable of asexual reproduction at some point in their development, and yet we consider such organisms individual organisms prior to the reproductive event. After all, it makes sense to ask, 'What is it that reproduces?' Any answer will refer to an organism of some sort such as 'that organism reproduced another organism of the same sort'" (Napier 2010: 792).

being that receives a new soul". Unfortunately, the possibility of twinning provides no evidence at all against immediate ensoulment. Proponents of the twinning charge neglect the fact that in twinning a rationally ensouled human being is kept in existence while another rationally ensouled human being is being brought into existence.[31]

So, immediate ensoulment theory can also be upheld in light of the fact that the zygote can twin up until approximately 14 days after fertilisation. For just like every cloneable human being, the twinnable embryo is just that: a rationally ensouled individual human being with the potential to become two.

5. Asymmetry at the Margins of Life?

The life that ends in death must have come into being beforehand. Our lives have not only one but two boundaries. And these boundaries are logically interdependent: The way we draw the boundary that marks the end of life predetermines the way we have to draw the boundary that marks its beginning—and vice versa. Our definition of life is not only decisive for our answer to the question "When has life ended?" but also for our answer to the question "When does life begin?" In other words, if we want to be logically consistent, the criteria we use to determine the point in time when life ends must also be used to determine the point in time when life begins.

Despite real differences in the biological processes that typically govern the beginning and the end of life, a proponent of brain death has to not only hold that human beings are alive as long but also only as soon as they exhibit brain activity.[32] This is not a matter of biology but of logical consistency.[33]

[31] As Condic (2020: 57) points out "monozygotic twinning most closely resembles regeneration or wound healing of an injured human organism". Accordingly, "one of the two twins produced by splitting is ontologically identical to the original zygote".

[32] Such a position is, for example, held by Goldenring (1985). Advancing what could be called the "brain life theory", Goldenring (1985: 200) holds that "whenever a functioning brain is present, a human being is alive".

[33] According to Eberl & Brown (2011: 45), those who formulate this condition do so based on the assumption that "the ascent of development and growth is mirrored by the degenerative processes that wear away our vitality". It should be noted here that this book does not hold this assumption. From a biological perspective, there is no symmetry at the margins of life; the biological processes that typically govern the beginning and the end of life are obviously different as are the processes that govern the formation and disintegration of the brain. But although these biological processes differ, the nature of the soul and with it the metaphysical principles governing its possessability do not change over time.

According to this requirement, human beings would not, however, be alive during the early stages of their development. Why this is the case becomes obvious once we ask ourselves at what point in time during their development human beings become alive according to the brain-based definition of death.

Answering this question amounts to finding an answer to the question of when human beings start to exhibit brain activity. The earliest point in time when it is not impossible for them to do so is once at least some basic brain-like structures have developed and start to function. As embryology tells us, the development of the structural precursors of a human being's brain starts with a process called "primary neurulation". This process is normally initiated 19 days after conception (Carnegie stage 7)[34] with the formation of the so-called neural plate, that is, the key developmental structure that serves as the basis for the entire nervous system.

It is, however, questionable whether the neural plate can be said to exhibit brain activity that can be detected and measured by means of an electroencephalography (EEG). Such electrocerebral activity can usually be measured only around week six or seven fertilisation age[35] of an embryo's development (Carnegie stage 19).

> The earliest neurons are born at approximately four weeks, with neural activity beginning at approximately seven weeks and connections forming between the spinal cord and lower brain centers at approximately eighteen weeks. Circuitry capable of supporting 'higher' functions, such as consciousness, develops between twenty-two to twenty-four weeks. (Condic 2013: 71f)[36]

[34] The Carnegie stages (also: embryonic stages) are a standardised system comprising 23 stages for the description of human embryonic development during the eight weeks of the embryonic period, that is, from fertilisation to the beginning of the foetal period.

[35] Fertilisation age is a measure for the age of an unborn human being and the duration of a pregnancy. The fertilisation age is indicated in completed development weeks and/or days. Week six fertilisation age (also: six weeks post conceptionem (p.c.)) means that the conception is assumed to have taken place six weeks ago.

An alternative measure for the duration of a pregnancy is gestational age, which is measured from the beginning of the pregnant woman's last menstrual period. The calculation of the gestational age assumes that a menstrual cycle lasts 28 days and that fertilisation has occurred approximately on the fourteenth day of the cycle. Accordingly, the gestational age is always two weeks higher than the fertilisation age. So, week six fertilisation age is tantamount to week eight gestational age (also: eight weeks post menstruationem (p.m.)).

[36] While the last point in time mentioned by Condic might be relevant for proponents of higher brain definitions of death, it is, however, irrelevant for the whole brain definition of death that defines death as electrocerebral silence, that is, the cessation of *all* electrocerebral activity (as established by an isoelectric EEG (also: zero-line EEG; flat EEG)) and not just activity in the form of regular brain waves.

If it is only at week six or seven fertilisation age that the brain of the embryo first generates electrical waves (brain waves) discernible by an EEG, we are led to conclude that proponents of a brain-based definition of death have to accept that human beings are definitely not alive during the first six weeks of their existence.

If this were true, this would mean that a pregnant woman carries and nourishes a dead being in her womb for the first two months of the pregnancy, that human embryonic stem cell researchers hope to engineer tissue from dead cells, that the human beings who come into existence by means of in vitro fertilisation are "conceived" dead, that the embryos transferred into the mother's uterus after an in vitro fertilisation are dead embryos and that pre-implantation (genetic) diagnosis is performed on dead human beings. Moreover, if all this were true, this would mean that human embryos cannot be killed (at least not during the first six weeks of their existence) because it is impossible to kill a being that is not alive. The utter strangeness of this position speaks for itself and needs no further comment.

Unlike Eberl & Brown (2011: 47), I do not see why it should be a "logical absurdity" to accept that logic requires proponents of brain death "to admit that a person is not truly alive—at least in the same way that a brain dead individual is truly dead—until sometime after birth". Although this admission is absurd indeed, it is not *logically* absurd. It is simply the logically correct deduction of the whole brain definition of death. If this deduction sounds absurd to anyone, it is not the fault of logic but the deduction's premises. If they want to avoid the absurdity contained in this deduction, proponents of brain death should not point to obvious but irrelevant biological differences between the formation and disintegration of the brain; rather, proponents of brain death should either (a) question the premise that brain death is death (and by doing so discover that this premise is the cause of the absurdity) or (b) just bite the bullet and accept that defining death—and therefore life—by neurological criteria leads to absurd conclusions when applied to the beginning of life.

6. Brain as Intermediary?

Some authors—such as Battro et al. (2008: 18f), Eberl (2005b: 31; 2015: 238), Moschella (2016) and Hanley (2020: 147)—might interject that Aquinas states that as motor the soul is united to the body through an intermediary (cf. Q. d. de anima, a. 9 co.; Q. d. de anima, a. 11 ad 16). For Aquinas, this intermediary is the heart, as he states that it is "mediante corde" (Q. d. de anima, a. 9 co.), that is, by means of the heart, that the soul moves the other parts of the body to perform their operations.[37] So, the soul does not move the body directly but indirectly via the heart as the body's primary organ.

While adopting Aquinas' idea of a primary organ, the above-mentioned authors and proponents of brain death substitute the heart for the brain. For them, the brain serves as the human body's primary motor-organ "which has the role of directing, coordinating and integrating the activities of the whole body" (Battro et al. 2008: 18).

> The brain is [...] the 'primary organ,' not only because it is necessary for the exercise of our rational capacities, but also because it is through the brain that the soul operates to regulate and coordinate the other parts, enabling them to function integrally. (Moschella 2016: 292)

Accordingly, they stipulate that when the brain dies, the "intermediary of the soul in its dynamic and operative function (as a motor) within the body has been removed" (Battro et al. 2008: 19).

However, the question is whether it is permissible to replace the heart with the brain. To answer this question, it is important to first understand that Aquinas explicitly teaches that the existence of a primary organ does not imply that the soul is in this organ only. Rather, the whole soul is present in the whole human body and in each part thereof (cf. I^a q. 76 a. 8;

[37] On another occasion, Aquinas states that "spiritus" is the first instrument of motion (cf. I^a q. 76 a. 7 ad 2). This term is usually translated as "breath", a meaning that the Latin word "spiritus" can have, indeed. But doing so would lead to a contradiction: As heart and lungs, that is, the organs doing the breathing, are two distinct organs, which of them is actually the primary organ? So, does Aquinas really think of "breath" when he uses the term "spiritus" in the cited passage?

Jones (2020: 41ff) helps us out by pointing to the fact that Aquinas' thinking was underpinned by the anatomical knowledge produced by his teacher, Albert the Great. And it is in Albert's treatise "On Animals" ("De Animalibus") where we can find the solution to our question. There, Albert defines "spiritus" as "that by which the powers of the soul are distributed throughout the body" (Jones 2020: 42). Albert takes "spiritus" to be some sort of vapor that the heart sends through the arteries. It is this understanding that Aquinas seems to refer to in the passage cited above. So, the spirit Aquinas is speaking of is not breath as in the common sense of the term, that is, the inhalation and exhalation of air performed by the lungs. Rather, it is that which Albert and Aquinas took the heart to "breathe" into the body through the arteries in order to move the body. Understood like this, the above passage only reiterates Aquinas' idea of the heart as the primary organ of the human body.

Q. d. de anima, a. 10).[38] The only thing that Aquinas takes to reside in the primary organ is "the motive power of the soul" (I^a q. 76 a. 8 ad 1). It is by its motive power that the soul moves the animated body.

Having said this, we can now turn to the question whether one can retain the idea of a primary organ but substitute the heart for the brain. As Jones' (2020) analysis of Aquinas' notion of the motion of the heart suggests, this substitution is problematic. Why this is the case becomes obvious once we understand why Aquinas takes the heart to be the primary organ.[39] The key to answering this question lies in understanding what Aquinas means when he says that the heart moves the body. As Jones (2020) shows, it is because of its being in continuous local motion, that is, in motion with regard to place composed of push and pull, that Aquinas takes the heart to be the primary organ:

> Because, for Thomas, to move is to be in act, and the heart is the only organ in the body that is continuously and involuntarily moving as long as the animal is alive, it is the only animal organ that is continuously in act. Additionally, because to be in act for a material thing is to be receptive of a form, the heart is the organ of the body that is continuously receptive of the form of the soul as a mover. Thus the motion of the heart is the first motion of the body and the vital motion of the animal. (Jones 2020: 111)

So, for Aquinas the primacy of the heart is grounded in its being the only organ that is in constant local motion. It is because the heart is constantly and unconsciously moving that it serves the role of an intermediary. The heart's primacy is grounded in its local push-and-pull-motion. It is the local motion of the heart that indicates that the soul is still acting on the body.

[38] As Nguyen (cf. 2017: 180 (FN 41), 2018b: 277) points out "this is referred to in modern language as the non-localization of the soul. [...] That the soul is wholly present in each part of the body does not refer to quantitative wholeness, however, since the soul is not a magnitude. Rather, it refers to wholeness of the soul with respect to essence and perfection, that is, each part of the body is alive because it is animated by the soul. Moreover, in this context, 'whole' does not mean wholeness according to the totality of the soul's powers; rather, the soul is whole in a particular part with respect to that power which corresponds to the function(s) of the particular part".

[39] As Spencer (2010: 863) points out, Aquinas insists on the heart being the primary organ despite knowing that "the brain is necessary for the preservation of the whole organism, and that the brain controls our sleep-wake cycles" and "that an injury to the brain causes the soul to lose the ability to understand and to direct itself in relation to the body".

> The heart is primary because it provides the force which moves blood to the organs of the body, providing them with what they need to stay alive. The other organs depend on the primary organ not to rule them in a conscious or neurological sense, but to provide them with the material they need to stay alive. Oxygenated and nutrient-rich blood, or a functionally equivalent substitute, is a necessary material condition for biological life, and it is thus the matter through which the soul provides its motive power. (Spencer 2010: 864f)

Unfortunately, this reasoning cannot be extended to the brain. Aquinas supports his argument of the primacy of the heart not by its critical or integrative role in the body as such; rather, he provides a metaphysical argument that links local motion with act and form. The intermediary role of the heart is grounded in its constant local motion. If the heart did not constantly contract and expand, it would not be the primary organ of the body. So, according to Aquinas, an organ can only be a primary organ if it is in constant motion.[40]

> In his writings, Aquinas has always insisted that that the primary organ is an instrument of motion and not an instrument of integration or an instrument to cause the union of the soul and body. (Nguyen 2018b: 340)

However, the brain is not in constant local motion; it does not constantly contract and expand. It actually does not move at all in a way comparable to the local motion of the heart. Of course, there is movement in the brain, but there is also movement in every other organ of the human body. Therefore, proponents of brain death cannot point to Aquinas' idea of an intermediary and simply switch organs. There cannot be a Thomistic defence of brain death that takes recourse to Aquinas' idea of a primary organ. For according to Aquinas, the only organ of a human being that can serve as primary organ is the heart.

[40] Ashley & Moraczewski (2001: 194) stipulate that "by 'a body proportionate to ensoulment as a human person,' Aquinas meant a body with a principal organ "capable of being the efficient cause of the activities specific to a human person, given the organism's state of development". For them, the primary organ must be "capable of producing a brain with the capacity for intellectual cognition in the body at some appropriate phase of the human life cycle" (Ashley & Moraczewski 2001: 200). Unfortunately, they do not reference any of Aquinas' works or give any quote that corroborates their statement. It is therefore not possible to address and examine this claim in a meaningful way.

Does this mean that human beings can only be rationally ensouled as soon and as long as they have a beating heart? Not necessarily. In his reply to the objection that "the soul is united to the body by means of the heart" (Q. d. de anima, a. 9 arg. 13), Aquinas explicitly states that the soul is united to the other parts of the body through the medium of the heart only "ut motor" (Q. d. de anima, a. 9 ad 13), that is, as their mover. However, "ut forma" (Q. d. de anima, a. 9 ad 13), that is, as form, the soul is united to every part of the body essentially and directly ("per se et immediate" (Q. d. de anima, a. 9 ad 13)). A few paragraphs before, Aquinas expresses this even more clearly:

[...] anima unitur corpori ut forma sine medio, ut motor autem per medium.	[...] as a form the soul is united to the body without an intermediary, but as a mover through an intermediary.

(Thomas Aquinas, Q. d. de anima, a. 9 co.; also cf. Super Sent., lib. 3 d. 1 q. 2 a. 1 co.; Contra Gentiles, lib. 2 cap. 57 n. 11 & 12; De spiritualibus creaturis, a. 3 co.)

What Aquinas is saying here is that the soul is joined to the body in two ways, namely as its form and motor. Expressed differently, the soul is both formal and efficient cause; it makes a being the kind of being it is and it is the source of its activity. As it is joined to the body in two ways, the soul can disjoin from the body in two ways, too, namely as its form and motor.

As Aquinas' intermediary, the heart is the transmission belt by which the soul exercises its motive power (vis movens), that is, its power "directed to local movement" (Q. d. de anima, a. 13 co.). When the heart fails, this transmission belt ruptures and the soul loses its means to set the body in motion. However, the fact that the soul is disjoined from the body as its motor does not logically imply that it is simultaneously disjoined from the body as its substantial form. The former does not necessitate the latter.

The concept of an intermediary organ is only relevant with respect to the union of the soul to the body as its mover. But it is not relevant for the union of the soul to the body as its substantial form. It is, however, the dissolution of the latter union, that is, the separation of the soul from the body as its substantial form, that is relevant for the determination of death. So, the presence of a functioning intermediary organ is not decisive for a human being's ensoulment with a rational soul as his substantial form. On the contrary, "precisely because the soul is united to the body as its substantial form and not as its motor, the body-soul union involves no intermediary element" (Nguyen 2018b. 276).

7. Artificial Life?

In contrast to what, for example, Battro et al. (2008: 13) hold, there cannot be such a thing as "artificial life". For them, the body of a brain-dead human being is a corpse that only seems alive because of, for example, a ventilator that camouflages or masks death: "In essence, an artificial instrument has become the principal cause of such a non-natural 'life'. In this way, death is camouflaged or masked by the use of the artificial instrument" (Battro et al. 2008: 13).

If ventilators actually had the ability to grant life to a dead human being, we could plug any corpse into a ventilator that would then make the respective corpse artificially alive. If life support machines could actually grant life, we could raise the dead. Unfortunately, this "miracle" only works with brain-dead bodies or—to put it more broadly—bodies that do not show the certain signs of death. A body that shows the certain signs of death cannot be plugged into a ventilator and respirated.

Given the account of life and death developed above, the reason for this is rather straightforward: As has been stated above, a being is alive as soon and as long as it shows any kind of endogenous activity. This definition does not, however, preclude cooperation with an external principle, such as a ventilator. According to Aquinas, "those things are without life which can be moved by an exterior principle only" (Sentencia de Anima, lib. 2 l. 1 n. 9). The important word for our purposes is the word "only".

While in the case of mechanical ventilation the first impetus might come from an exterior principle, the living body is capable of endogenous activity that processes the oxygen pumped into it by the ventilator. The ventilator can and does not process the oxygen; it can and does not circulate oxygenated blood; it can and does not perform cellular respiration, that is, put the oxygen into the cells of the body; it only offers oxygen to the body, which then processes it by endogenous activity. A brain-dead person on a ventilator is not moved by an external principle only. The fact that it can—in contrast to a corpse, that is, a body showing the certain signs of death—cooperate with the ventilator shows that it does not draw its life from the ventilator but—the name says it already—merely is on life support. A ventilator does not grant life to an otherwise lifeless corpse; it supports the workings of the rational soul present in a perhaps dying but still living human being by moving breathable air into and out of his lungs. If we were to believe differently, we would also have to hold that an infant who cannot yet eat by herself and is therefore fed by her parents is not alive and that the parents are the principal cause of her life.

Against the claim that ventilators can mask or camouflage death, we can therefore agree with Nguyen (2018b: 94) that "the ventilator and other technological tools are instruments of life support, which implies that they can only work if there is still some life present in the individual. Hence technological prowess cannot mask death. It can only retard the moment of death".

8. Roma locuta, causa finita?

Some Catholic readers might object that—while its positions happen to coincide with Catholic doctrine on virtually all other aspects—this book's rejection of a neurological definition of death is not in line with the magisterium of the Catholic Church. Such an impression might be due to authors who claim that the whole-brain criterion of death has been "affirmed by the Vatican" (Eberl 2015: 244 (FN 1)), that the brain-death standard has been "given the stamp of approval of the Roman Catholic Church" (Lock 2004: 137) or that Pope John Paul II has given "definitive approval to the use of neurological criteria for the determination of death" (Furton 2002: 455).

The above and other similar statements are usually made with reference to the address of Pope John Paul II to the 18th International Congress of the Transplantation Society in August 2000 (John Paul II 2000) and/or the results of three working groups convened by the Pontifical Academy of Sciences (cf. Chagas 1986; White et al. 1992; Sánchez Sorondo 2007).

As far as the latter reports are concerned, we must be careful to not confuse statements by Catholic authors or consultative but non-magisterial bodies of the Catholic Church (e.g. the Pontifical Academy of Sciences or its working groups) with the magisterium of the Catholic Church. If a report issued by the academy favours the acceptance of a neurological definition of death this is far from tantamount to an affirmation of this definition by the Catholic magisterium. The reason is rather straightforward: The Pontifical Academy of Sciences has no magisterial authority.

As far as the above-mentioned address of Pope John Paul II is concerned, one has to be careful to not misinterpret its actual content and magisterial weight. With respect to its content, the address does not allow for the conclusion that the Pope and the magisterium definitively endorse or approve a neurological definition of death. Rather, the Pope's approval is only a conditional and cautious approval "pending the fulfillment of several specific presuppositions or conditions" (Nguyen 2017: 156; 2018b:

458). As the critical analysis of this address and other relevant magisterial statements shows, “from the perspective of the Magisterium, the ‘brain-death’ issue still remains unsettled” (Nguyen 2017: 174; 2018b: 463ff).

At most, the address of John Paul II allows us to conclude that the magisterium has morally allowed the application of the whole-brain criterion of death. This means that the magisterium holds that the whole-brain criterion of death provides health-care workers with the necessary and sufficient level of certitude to justify action, namely moral certitude (also: moral certainty).[41] A health-care worker can be morally certain to not commit a sin when she uses the whole-brain criterion of death to declare another person dead or when she surgically explants vital organic material from a brain-dead person. However, this moral permission does not logically imply that the whole-brain death criterion cannot be inconsistent with the Catholic Church’s anthropology. The lack of absolute certitude on this matter is also admitted by the Pope himself:

> The criterion adopted in more recent times for ascertaining the fact of death, namely the complete and irreversible cessation of all brain activity, if rigorously applied, does not seem to conflict with the essential elements of a sound anthropology. (Pope John Paul II 2000: 5)[42]

For the Pope, it only *seems* but is not necessarily true that the whole-brain death criterion is not inconsistent with the Catholic Church’s anthropology. This does not sound—as claimed by the above-cited authors—like a definitive approval of the whole-brain criterion of death. The option that the criterion is inconsistent with the Catholic Church’s anthropology is

[41] In contrast to absolute certainty, moral certainty does not exclude the possibility for error. A judgement is morally certain “when error is impossible according to what is customary among mankind” (McHugh & Callan 1958: §643). For example: We can be morally certain that what a reputedly truthful and competent person relates to us is true. So, although we do not possess absolute certitude that what such person tells us is true, we are allowed to consider her testimony as true and act upon this assessment.

The permissibility to act on moral certainty is enshrined by a decree of the Holy Office issued in 1690. This decree condemned various errors of the Jansenists, among them the proposition that “it is not licit to follow a [probable] opinion, even the most probable among the probable ones” (DH 2303). The decree therefore condemns the position that acting requires absolute certainty and allows to act on moral certainty.

[42] As we can read in the same document, it is—amongst others—characteristic of such sound anthropology that the human body stands in “substantial union with a spiritual soul” (John Paul II 2000: 3) and that death “results from the separation of the life-principle (or soul) from the corporal reality of the person” (John Paul II 2000: 4).

deliberately left open by the Pope. Rome—to borrow and slightly adapt a saying initially coined by Augustine of Hippo (Sermones 131, 10)—has spoken but in a way that has not definitely decided the matter. Accordingly, one cannot be said to contradict the magisterium by holding that the brain-death criterion of death is wrong or by conscientiously objecting to participate in the determination of death by neurological criteria and the explantation of organs from merely brain-dead persons. As long as the Catholic magisterium has not given definitive teaching on a certain matter, there cannot—as such—be a duty to accept and not contradict a non-definitive teaching.[43]

To sum up, a closer analysis of the actual content and weight of Catholic magisterial teaching on brain death allows for the following conclusions:

> First, although there seems to be a preponderance in favor of accepting the whole brain definition of death, there is ongoing debate both among Catholic scholars and within the groups of advisors assembled by the Vatican. Even though a poll of these individuals would likely show that most of them approve of the neurological criteria, the presence of dissenting voices demonstrates a lack of solid consensus. Second, even if such consensus existed, the magisterial statements produced to date have not been definitive. As of now, we lack a "manifestly evident" statement of the Church's teaching in a document such as a papal encyclical or a statement of a general synod of bishops, either of which would generally be binding on Catholics. (Ostertag & Karches 2019: 339f)

This book's rejection of a neurological definition of death cannot be said to be not in line with the definitive Catholic magisterium. On the contrary, its rejection of the brain-death criterion of death is based on its coherence with Catholic doctrine. Although this has not been its aim or premise, the positions outlined in this book are in line with the Catholic Church's anthropology, especially its definitive teachings on the human soul and the soul's union to the body. Without claiming completeness, the essentials of these teachings are expressed in the following magisterial statements:

[43] When assessing the magisterial weight of the Pope's address, one has to also consider that "not all magisterial documents bear the same authoritative weight. Speeches are directed at specific audiences, in circumstances decidedly conditioned by history and culture—for example, by the current state of scientific understanding concerning a specific matter of inquiry—and thus, carry little authority in the grand sweep of the Tradition" (Bulzacchelli 2013: 51 (FN 20); cf. Ostertag & Karches 2019: 337ff).

- In the constitution "Fidei Catholicae", the Council of Vienne—which was the 15th ecumenical council of the Catholic Church and was held between October 1311 and May 1312—declared that "whoever presumes to assert, defend, or obstinately hold that the rational and intellectual soul is not of itself and essentially the form of the human body is to be censured as heretic" (DH 902).
- In the bull "Apostolici regiminis", the 5th Lateran Council—which was the 18th ecumenical council of the Catholic Church and was held between May 1512 and March 1517—reaffirmed this teaching and also asserted the immortality of the soul (cf. DH 1440).
- In 1914, the Sacred Congregation of Studies—at the command of Pope Pius X—approved of the Thomistic theses as safe directional norms (a) that the rational soul "is united with a body in such a way that it is the unique substantial form of the same body" and (b) that it is by means of the soul that man is living (DH 3616).
- In his Apostolic letter "Dolore haud mediocri" (April 30, 1860) Pope Pius IX declared that "the judgement that assumes that there is one principle of life in man—namely, the rational soul, from which the body also receives its movement and all its life and feeling—is the most universal within the Church of God" (DH 2833).
- The 4th Lateran Council—which was the 12th ecumenical council of the Catholic Church and was held in November 1215—teaches that man is composed of a spiritual soul and a body (DH 800).[44]
- The teaching that man is composed of a rational soul and body is also affirmed by the 1st Vatican Council—which was the 20th ecumenical council of the Catholic Church and was held between December 1869 and October 1870—in its dogmatic constitution "Dei Filius" (cf. DH 3002).
- The 4th Council of Constantinople—which was the 8th ecumenical council of the Catholic Church and was held between October 869 and February 870—defined as Catholic dogma the statement that man has not two or more souls but only "one rational and intellectual soul" (cf. DH 657f).
- With approval by Pope John Paul II, the Congregation for the Doctrine of the Faith recalls the Catholic teaching that after death the

[44] To be precise, the original text states that man is composed "ex spiritu et corpore", that is, of spirit and body. However, as the Congregation for the Doctrine of the Faith points out, the word "soul" is used to designate the spiritual element of man (cf. DH 4653). It is therefore licit to translate the Latin word "spiritus" with "soul".

soul exists "without the complement of its body" (DH 4653), thereby implicitly affirming death as the loss of the union of body and soul.
- The Catechism of the Catholic Church defines death as "animae et corporis separatione" (CCC 997), that is, as the separation of the soul from the body.

None of these definitive teachings of the Catholic magisterium is contradicted in the book. As this book's position on the soul as the principle of life and its substantial union to the body is in line with definitive Catholic teaching on the soul in general and the magisterial statements above in particular, its definition of death cannot not be in line with Catholic teaching—especially in the absence of a clear and unconditional magisterial acceptance of brain death.

9. Brain Death and Decapitation

In an effort to strengthen their case, it is not uncommon for proponents of a brain-based definition of death to compare brain death with decapitation. The following statement by Christopher Pallis, a proponent of brainstem death, is an example of such effort:

> All human death [...] is (and always has been) brain death. [...] The identity of brain death with death has been very perceptively realised by people with little or no knowledge of physiology: we have been hanging and decapitating for centuries. We have only to think 'whole-brain death' as physiological decapitation for the relationship to become crystal clear. (Pallis 1983: 33f)

What Pallis seems to suggest is—to use a clearer formulation—that "if we are willing to accept decapitation as death, we should also be willing to accept physiological decapitation (total brain failure) as death" (Lizza 2009a: 5). The argument contained in these statements can be expressed formally by means of the following three-step categorical syllogism:

> *1. Brain death constitutes physiological decapitation. 2. Decapitation is an infallible sign and sufficient condition of death. 3. Hence, individuals diagnosed as brain dead are necessarily dead.* (Miller & Truog 2010: 632)[45]

[45] A similar but syllogistically less stringently formulated version of this deductive argument has been offered by Shewmon (2007. 299). "(1) A decapitated person is dead. (2) Brain death is physiologically equivalent to decapitation. (3) Therefore, a brain-dead person is dead".

The above syllogism is undoubtedly valid as its conclusion is the necessary consequence of its premises, that is, as its conclusion is logically entailed by its premises with certainty. However, validity alone does not yet make a sound deductive argument; a deductive argument is only sound if it is valid and if all of its premises are true (cf. Erk 2019: 302ff). So, what about the above argument's soundness? Let us grant the argument's major premise that brain death is actually physiologically equivalent to decapitation.[46] But what about the argument's minor premise? Is it actually safe to stipulate that a decapitated human being is dead? Only if it is true that a previously living human being is dead from the moment his head is severed can decapitation be a sufficient condition of death. So, to find out whether the above syllogism is actually sound, we need to address the following question: Does a human being die the moment his head is severed? Depending on the answer, decapitation is not a certain diagnosis but merely a certain prognosis of death.

For example, Lizza (2009b: 393) thinks that the view that decapitation does not "necessarily entail death" is "absurd". For him, "if anything entails one's death, decapitation certainly does" (Lizza 2009a: 5). In contrast, Shewmon (2007: 303) holds that it "cannot simply be assumed without question" that a headless body is dead. In a similar fashion, Miller & Truog (2010: 633) doubt that a decapitated human being is "invariably dead at the moment when decapitation occurs". So, which is it? Do living human beings die the moment their head is severed or not?

9.1. Decapitation and Death

The question of whether a living human being dies the moment his head is severed can only be answered by reference to a definition and criterion of death. Luckily, Chap. IV has already provided us with such a definition and criterion. There we have learned that

[46] Shewmon (2007: 303) doubts such equivalence as decapitations, unlike brain death, are accompanied by severe mutilation and exsanguination. If at all, one can only speak of an equivalence if the decapitation in question features "the hypothetical details of immediate suturing of severed neck vessels and cauterization of tissues to prevent bleeding, plus mechanical ventilation through an endotracheal tube placed in the tracheal stump" (Shewmon 2007: 303). Only such a ventilated, non-bleeding headless body would actually be physiologically identical to a brain-dead body. For Shewmon (2007: 302), thought experiments involving brain-body separation are not only "highly speculative" but "in the final analysis [...] irrelevant to understanding clinical brain death, in which no such separation is involved".

> A previously living human being is dead, that is, has completely lost his life and can therefore no longer show any signs of life or endogenous activity, if he shows no empirically observable signs of life and if he shows the early certain signs of death.

Given this criterion of death, can human beings be said to die the moment their head is severed? The answer is rather clear: As neither the decapitated body nor the severed head meet the above criterion of death the moment they are separated, we are forced to conclude that neither the former nor the latter die the moment they are separated.

Surprising as it may be, this conclusion seems to be not only theoretically but also practically sound. For both the severed head and the headless body have been—although by means of gruesome animal experiments—shown to exhibit empirically observable signs of life. In this context, Miller & Truog (2010) point to an experiment conducted and reported by Steinberg & Hersch (1995):

> A pregnant near-term sheep was anaesthetised and connected by intubation to a respirator. Dissection of the head exposed the carotid arteries, which were tied and cannulated. The sheep's head was fully cut-off and the endotracheal tube placed in the tracheal opening. Pharmacological measures were introduced to maintain blood pressure and heart rate. Thirty minutes after decapitation, a healthy lamb was delivered by caesarian section. (Miller & Truog 2010: 633)

As this experiment shows, the body of a decapitated animal can be kept alive with the aid of technological intervention. Because it continues to exhibit empirically observable signs of life after its head has been severed, the headless body cannot have died at the moment of the decapitation. But what about the head?

Having previously performed the first successful total isolation of the monkey and the canine brain with retention of biological activity (cf. White et al. 1963, 1964, 1965), in 1970 Robert J. White and his team performed the first successful cephalic exchange transplantation in primates (cf. White et al. 1971). Lang et al. (2019) describe the experiment and its results as follows:

> On March 14, 1970, White, along with his team of approximately 30 doctors, nurses, and technicians, performed cephalic exchange on 4 monkeys. They isolated the head from the donor monkey by severing the spinal cord

> and ligating all the vasculature except for the carotid arteries and jugular veins, which were directly sutured to the carotid and jugular veins of the recipient monkey heads. Continuous heparin infusion was required due to the possibility of clot formation from the direct vascular anastomoses. Three to four hours postoperatively, the recipients' heads demonstrated the ability to chew, swallow, track movement with eyes, and bite. The heads also performed learned responses from prior to the procedure, suggesting intact higher cortical function. Recording electroencephalograms showed awake activity patterns. The monkeys, however, were unable to move the transplanted body, as the spinal cords were not connected. With subsequent blood loss and the need for continuous mechanical respiratory support due to severing of the spinal cord, the monkeys survived for 6 to 36 h. (Lang et al. 2019: E369; cf. White et al. 1971)[47]

As this experiment shows, not only a headless animal body but also the severed head of a decapitated animal can be kept alive with the aid of technological intervention—while even maintaining the brain's higher cortical function.

Why are the results of the experiments conducted by White et al. (1971) as well as Steinberg & Hersch's (1995) relevant to the question at hand? They are relevant because they show that both a severed head and a decapitated body survive their separation. For if both a severed head and a decapitated body can be kept alive by artificial means beyond the time of decapitation, then they cannot have died at the moment of the decapitation.

Fortunately, the above-mentioned findings gained by means of animal experiments cannot and hopefully never will be repeated and verified by means of human experiments. However, history has already provided us with a valuable piece of information that proves the continued existence of life in a severed human head beyond the time of decapitation. In 1884, the *Revue Scientifique* published the report of a medical doctor who happened

[47] For an overview of the general history of head transplantation—or cephalosomatic anastomosis (CSA) as it is called in technical terms—cf. Lamba et al. (2016). As Lamba et al. (2016: 2246) point out, what is commonly called a head transplantation truly is "a body transplantation, in which a person's brain is receiving a new body".

An article by Ren et al. (2017) even reported the rehearsal of a head transplantation on two recently deceased human cadavers—confirming that "CSA is feasible and without compelling impediments to abandon the operation" and ending with the gruesome suggestion (or announcement?) that "the next stage would be a full rehearsal on heart-beating brain-dead organ donors to replicate the actual operation in a live patient with a medical condition appropriate for a CSA".

to witness the decapitation of the leader of a band of Annamite (i.e. Vietnamese) pirates in Saigon in 1875. The relevant excerpt from his report reads as follows:

> Les préparatifs terminés, je me tins à 2 mètres de lui; il s'était agenouillé, et, avant de baisser la tête, il avait encore échangé avec moi un rapide regard. La tête tomba à 1m, 20 de moi, sans rouler, comme il arrive d'ordinaire; mais, la surface de section s'appliquant immédiatement sur le sable, l'hémorragie se trouva ainsi accidentellement réduite au minimum. A ce moment, je fus effrayé de voir les yeux du supplicié fixés franchement sur les miens. N'osant croire à une manifestation consciente, je décrivis vivement un quart de cercle autour de la tête gisant à mes pieds, et je dus constater que les yeux me suivaient pendant ce mouvement. Je revins alors à ma position première, mais plus lentement cette fois: les yeux me suivirent encore pendant un instant fort court, puis me quittèrent subitement. La face exprimait à ce moment une angoisse manifeste, l'angoisse poignante d'une personne en état d'asphyxie aiguë. La bouche s'ouvrit violement, comme pour un dernier appel d'air respirable; et la tête, ainsi déplacée de sa position d'équilibre, roula sur le côté. Cette contraction des muscles maxillaires fut la dernière manifestation de la vie. Depuis le moment de la décollation, il s'était écoulé de 15 à 20 secondes.

> When the preparations were completed, I stood two metres away from him; he had knelt down, and before lowering his head, he had again exchanged a quick glance with me. The head fell 1.20 metres from me, without rolling, as it usually does; but, as the surface of the cut immediately applied itself to the sand, the haemorrhage was thus accidentally reduced to a minimum. At that moment I was startled to see the eyes of the victim fixed squarely upon mine. I dared not believe that this was a conscious manifestation, so I made a quarter circle around the head lying at my feet, and I could see that the eyes were following me during this movement. I then returned to my original position, but more slowly this time: the eyes followed me for a short time, and then suddenly left me. At this moment the face expressed a manifest anguish, the poignant anguish of a person in a state of acute asphyxia. The mouth opened violently, as if for a last gasp of breathable air; and the head, thus displaced from its position of equilibrium, rolled to one side. This contraction of the jaw muscles was the last manifestation of life. From the moment of decapitation, 15 to 20 seconds had elapsed.

(Petitgand 1884: 11)

As all this shows, decapitation is a severe form of mutilation, injury or physical trauma that is obviously incompatible with life, but it does not kill at the moment of the decapitation.

If both head and headless body are technologically kept alive through attaching the body to a ventilator and keeping the head perfused with oxygenated blood by attaching its major vessels to a cardiopulmonary bypass

> machine, then we can legitimately question whether even complete physical separation is per se death or rather a condition that would quickly lead to death if heroic medical intervention had not taken place. (Shewmon 2007: 312)

So, decapitation as such is merely a prognosis but not a certain diagnosis of death—and cannot, accordingly, be invoked as a defence of brain death.

9.2. Ensoulment after Decapitation

However, this conclusion immediately raises a follow-up question: Decapitation necessarily results in the division of a human body into two parts, a severed head and a headless body. If neither die at the moment of the decapitation, what happens to the human soul the moment a living human being's head is severed? Does it continue to inform the severed head, the headless body or both? To answer these questions, it is helpful to consider an important characteristic of the human soul:

> The soul is first and foremost the form of a body—in its natural condition it informs a body—and it will naturally tend to inform a body until material conditions deteriorate to the point where it simply no longer can do so. (Spencer 2010: 856)

This statement aptly sums up the position held by Aquinas, namely that the human soul has "an aptitude and a natural inclination to be united to the body" (I^a q. 76 a. 1 ad 6), that "the state of the soul in the body is more perfect than outside the body" (Suppl. IIIae q. 75 a. 1 ad 4), that "the soul does not have the perfection of its own nature, except in union with the body" (De spiritualibus creaturis, a. 2 ad 5) and that "the rational soul naturally desires to be united to a body" (De potentia, q. 3 a. 10 co.). In his commentary on St Paul's first letter to the Corinthians (1 Cor 15: 12–19), Aquinas summarises his position as follows:

Constat enim quod anima naturaliter unitur corpori, separatur autem ab eo contra suam naturam, et per accidens. Unde anima exuta a corpore, quamdiu est sine corpore, est imperfecta.	For it is clear that the soul is naturally united to the body but is separated from it contrary to its nature and per accidens. Hence the soul devoid of its body is imperfect, as long as it is without the body.

(Thomas Aquinas, Super I Cor., cap. 15 l. 2; also cf. Suppl. IIIae q. 75 a. 2 co.; De potentia, q. 5 a. 10 ad 5)

To put it even more succinctly, "The soul, as long as it is separated from the body, is imperfect" (Super I Cor., cap. 15 l. 9). By its very nature, the human soul is meant to be the substantial form of a human body.[48] Once embodied, the human soul will therefore continue to inform its body as long as somehow possible and will not separate from its body unless absolutely necessary.

However, this absolute necessity is not already given when the soul's physical-material infrastructure makes the actualisation of its highest, namely its rational, powers impossible: "The soul seeks to implement its rational capacities, but seeks more strongly to continue informing matter" (Spencer 2010: 869). Given its natural inclination to be united to its body, the soul will rather content itself to only realise its lower, namely its sensible and vegetative, powers than separate from its body:

> The human soul is a rational soul and so will implement these powers in relation to matter as long as possible. However, if the implementation of its rational capacities is not possible, the same soul will continue to implement its lower powers rather than separate entirely from matter and take on a separated existence in which it can only implement some powers unnaturally without matter. (Spencer 2010: 856f)

Why is this important? It is important because it helps us to answer the question what happens to the human soul at the moment of the decapitation of a living human being.

If—as has been argued above—the soul has the natural inclination "both to continue informing matter and to implement its highest potencies" (Spencer 2010: 857), the decapitation riddle can be solved as follows: When—as, for example, in the case of a decapitation—the brain, that is, the physical-material infrastructure necessary for the actualisation of the human soul's rational faculties, is separated from the rest of the body, "the soul goes with it and continues to inform this small piece of matter" (Spencer 2010: 857). The head of a decapitated human being would, therefore, be animated by the soul that has animated the whole human being before his decapitation: "The presence of the soul in the brain or in the body is only dependent on where it can best implement its powers at

[48]Accordingly, Aquinas holds that the human soul must resurrect bodily, that is, must be reunited with its body after death (cf. Suppl. IIIae q. 75), and that it does not pre-exist its body, that is, is not created before and apart from its body (cf. De potentia, q. 3 a. 10 co.).

the moment of separation" (Spencer 2010: 858f).[49] In short, at the moment of the decapitation, the rational soul continues to inform the severed head.

So far, we have dealt with the ensoulment status of the severed head of a decapitated human being at the moment right after the decapitation. Let us now turn to the decapitated human being's headless body. The fact that it shows signs of life even after the decapitation and does not show the early certain signs of death logically suggests that the body is still alive. But is it also rationally ensouled? If what has just been said is true, we must conclude that this is not the case. The soul is unable to be divided or duplicated among different material substances. As the decapitated human being's rational soul has "decided" at the moment of separation to continue to inform that part of the body in which it can best implement its powers, namely the severed head, the soul cannot also be in the headless body.

However, because the headless body is still alive, there must be some soul in it. But what kind of soul? It cannot be a rational soul. As it is lacking the necessary physical-material infrastructure and is not geared towards the development of such infrastructure, the headless body cannot be said to be disposed to receive a rational form. Accordingly, the soul that begins to ensoul the headless body at the moment of decapitation must be a separate, non-rational soul:

> In case of cerebrum removal, a substantial change would occur in the body: a new soul would emerge in the cerebrumless body and the organs of that body would act for the sake of a new end, the functioning of the new life form. In terms of observable vital processes, no change would seem to have occurred aside from the removal of the cerebrum. However, if substances are defined in terms of forms and ends, the body which has had its cerebrum removed cannot be informed by the same soul, because this soul goes with the cerebrum. (Spencer 2010: 859)

Put differently, the decapitated body has the metaphysical status of a severed limb that is animated by a non-rational soul beyond the time of separation. And as "vegetative souls are generated in those cells which naturally

[49] In contrast, when the whole brain or parts of the brain are destroyed in the body, "the soul remains with the body, implementing only its lower potencies but remaining connected to matter. Only when the matter to which it is connected becomes completely unfit for being informed by the soul does the soul depart entirely" (Spencer 2010: 857).

divide off from the body, viz. the reproductive cells" (Phillips 1934: 189), a vegetative soul "could be generated in the excised part by the power of the rational soul which gives vegetative life to the whole organism".

With this, we now have all the elements in place to answer the question of what happens to the human soul the moment a living human being's head is severed. So, does it continue to inform the severed head, the headless body or both? The answer is as follows: At the moment of decapitation the human soul continues to inform and ensoul the severed head; the headless body becomes informed and ensouled by a non-rational soul. Both the severed head and the headless body continue to be informed by their respective souls until their physical-material infrastructure makes it impossible to remain ensouled.[50] The time span between decapitation and death of the severed head and/or the headless body may be prolonged by artificial means. But we know that death has occurred when the severed head or the headless body shows no empirically observable signs of life and shows the early certain signs of death.

Of course, all of this is somewhat speculative as the soul cannot be empirically observed and as we do not know how a rational soul differs in appearance from a sensitive or vegetative soul. Nevertheless, the above conclusion is in line with the general theory of soul, life and death on which this book is based.

Bibliography

Ashley, Benedict. "A Critique of the Theory of Delayed Animation." In: McCarthy, Donald G. & Albert S. Moraczewski. *An Ethical Evaluation of Fetal Experimentation.* St Louis, MO: Pope John XXIII Center, 1976. pp. 113–137.

Ashley, Benedict & Albert Moraczewski. "Cloning, Aquinas, and the Embryonic Person." *The National Catholic Bioethics Quarterly* 1.2 (2001): 189–201.

Austriaco, Nicanor Pier Giorgio. "The Moral Case for ANT-Derived Pluripotent Stem Cell Lines." *The National Catholic Bioethics Quarterly* 6.3 (2006): 517–537.

[50] As Spencer (2010: 859) points out that if the head could be re-attached to a headless but living body, "the soul informing the cerebrum would then "re-inform" the whole body, again implementing all of its nonrational powers, since the matter would be adequate to receive this higher sort of soul, and this higher soul could take over the functions of any lower soul".

Austriaco, Nicanor Pier Giorgio. *Biomedicine and Beatitude: An Introduction to Catholic Bioethics*. Washington, DC: The Catholic University of America Press, 2011a.

Austriaco, Nicanor Pier Giorgio. "Complete Moles and Parthenotes Are Not Organisms." In: Suarez, Antoine & Joachim Huarte (Eds.). *Is this Cell a Human Being? Exploring the Status of Embryos, Stem Cells and Human-Animal Hybrids*. Heidelberg et al.: Springer, 2011b. pp. 45–54. This article is an updated version of the author's "Are Teratomas Embryos or Non-embryos? A Criterion for Oocyte-Assisted Reprogramming." *The National Catholic Bioethics Quarterly* 5.4 (2005): 697–706.

Battro, Antonio M., James L. Bernat, Marie-Germaine Bousser, Nicola Cabibbo, Card Georges M. M. Cottier, Robert B. Daroff, Stephen Davis, Lüder Deecke, Conrado J. Estol, Werner Hacke, Michael G. Hennerici, Johannes C. Huber, Card Alfonso López Trujillo, Card Carlo M. Martini, José C. Masdeu, Heinrich Mattle, Jerome B. Posner, Louis Puybasset, Marcus E. Raichle, Allan H. Ropper, Paolo M. Rossini, Marcelo Sánchez Sorondo, Herbert Schambeck, Elio Sgreccia, Prakash Narain Tandon, Rafael Vicuña, Eelco F. M. Wijdicks & Antonino Zichichi. *Why the Concept of Death is Valid as a Definition of Brain Death. Statement by the Pontifical Academy of Sciences and Responses to Objections*. Vatican City: Pontificia Academia Scientiarum, 2008. (The Pontifical Academy of Sciences, Extra Series 31) Available at: https://www.pas.va/en/publications/extra-series/es31pas.html Accessed: March 28, 2022. Previously published as: "Why the Concept of Death is Valid as a Definition of Brain Death. Statement by Neurologists and Others." In: Sánchez Sorondo, Marcelo (Ed.). *Working Group on The Signs of Death. The Proceedings of the Working Group 11–12 September 2006*. Vatican City: Pontificia Academia Scientiarum, 2007. pp. XXI–XXIX. (The Pontifical Academy of Sciences, Scripta Varia 110) Available at: https://www.pas.va/en/publications/scripta-varia/sv110pas.html Accessed: March 28, 2022.

Bothur-Nowacka, Joanna, Aleksandra Jezela-Stanek, Katarzyna Zaniuk, Bożenna Goryluk-Kozakiewicz, Małgorzata Krajewska-Walasek & Anna Dobrzańska. "Tetraploidy in the era of molecular karyotyping—What we need to remember." *Pediatra Polska* 88.5 (2013): 467–471.

Bracken, W. Jerome. "Is the Early Embryo a Person?" *The Linacre Quarterly* 68.1 (2001): 49–70.

Brugger, E. Christian. "ANT-OAR: A Morally Acceptable Means for Deriving Pluripotent Stem Cells. A Reply to Criticisms." *Communio: International Catholic Review* 32.4 (2005): 753–769.

Bulzacchelli, Richard H.. "The diagnosis of death and the irreducibility of the human person." *The Linacre Quarterly* 80.1 (2013): 39–55.

Bynum, Jennifer, Denise Batista, Rena Xian, Deyin Xing, James R. Eshleman, Brigitte M. Ronnett & Gang Zheng. "Tetraploid Partial Hydatidiform Moles.

Molecular Genotyping and Determination of Parental Contributions." *The Journal of Molecular Diagnostics* 22.1 (2020): 90–100.
Chagas, Carlos (Ed.). *Working Group on the Artificial Prolongation of Life and the Determination of the Exact Moment of Death. 19–21 October 1985.* Vatican City: Pontificia Academia Scientiarum, 1986. (The Pontifical Academy of Sciences, Scripta Varia 60) Available at: https://www.pas.va/en/publications/scripta-varia/sv60pas.html Accessed: March 28, 2022.
Cobb, Matthew. *The Egg & Sperm Race. The Seventeenth-Century Scientists Who Unravelled the Secrets of Sex, Life and Growth.* London: Free Press, 2006a.
Cobb, Matthew. "Heredity before Genetics: A History." *Nature Reviews Genetics* 7.12 (2006b): 953–958.
Cobb, Matthew. "An Amazing 10 Years: The Discovery of Egg and Sperm in the 17th Century." *Reproduction in Domestic Animals* 47.s4 (2012): 2–6.
Condic, Maureen L. "When Does Human Life Begin? The Scientific Evidence and Terminology Revisited." *The National Catholic Bioethics Quarterly* 9.1 (2009): 129–149. Originally published in 2008 by The Westchester Institute for Ethics & the Human Person as part of their White Paper Series (Volume 1, Number 1).
Condic, Maureen L.. "When Does Human Life Begin? The Scientific Evidence and Terminology Revisited." *University of St. Thomas Journal of Law and Public Policy* VIII.1 (2013): 44–81.
Condic, Maureen L.. *Untangling Twinning: What Science Tells Us about the Nature of Human Embryos.* Notre Dame, IN: University of Notre Dame Press, 2020. (Notre Dame Studies in Medical Ethics and Bioethics)
Condic, Samuel B. & Maureen L. Condic. *Human Embryos, Human Beings: A Scientific & Philosophical Approach.* Washington, DC: The Catholic University of America Press, 2018.
Damschen, Gregor, Alfonso Gómez-Lobo & Dieter Schönecker. "Sixteen Days? A Reply to B. Smith and B. Brogaard on the Beginning of Human Individuals." *Journal of Medicine and Philosophy* 31.2 (2006): 165–175.
Donceel, Joseph F.. "Immediate Animation and Delayed Hominization." *Theological Studies* 31.1 (1970): 71–105.
Dombrowski, Daniel A. & Robert Deltete. *A Brief, Liberal, Catholic Defense of Abortion.* Urbana, IL: University of Illinois Press, 2000.
Eberl, Jason T.. "Aquinas's Account of Human Embryogenesis and Recent Interpretations." *Journal of Medicine and Philosophy* 30.4 (2005a): 379–94.
Eberl, Jason T.. "A Thomistic understanding of human death." *Bioethics* 19.1 (2005b): 29–48.
Eberl, Jason T.. "A Thomistic defense of whole-brain death." *The Linacre Quarterly* 82.3 (2015): 235–250.
Eberl, Jason T. & Brandon P. Brown. "Brain Life and the Argument from Potential: Affirming the Ontological Status of Human Embryos and Fetuses." In: Napier, Stephen (Ed.). *Persons, Moral Worth, and Embryos: A Critical Analysis of Pro-*

Choice Arguments. Dordrecht, Heidelberg, London & New York: Springer, 2011. pp. 43–66. (Series "Philosophy and Medicine", Volume 111)

Erk, Christian. *Moral Philosophy. Fundamental Concepts*. Second, revised and expanded Edition. Heusenstamm: editiones scholasticae, 2019.

Fleischer, Julie, Archana Shenoy, Katherine Goetzinger, Catherine E. Cottrell, Dustin Baldridge, Frances V. White & Marwan Shinawi. "Digynic triploidy: utility and challenges of noninvasive prenatal testing." *Clinical Case Reports* 3.6 (2015): 406–410.

Ford, Norman M.. *When Did I Begin? Conception of the Human Individual in History, Philosophy and Science*. Cambridge: Cambridge University Press, 1988.

Furton, Edward J.. "Brain Death, the Soul, and Organic Life." *The National Catholic Bioethics Quarterly* 2.3 (2002): 455–470.

Gerber, Rudolph Joseph. "When is the human soul infused?" *Laval théologique et philosophique* 22.2 (1966): 234–247.

Goldenring, John M.. "The brain-life theory: towards a consistent biological definition of humanness." *Journal of Medical Ethics* 11.4 (1985): 198–204.

Grobstein, Clifford. "External Human Fertilization." *Scientific American* 240.6 (1979): 57–67.

Grobstein, Clifford. "Biological Characteristics of the Preembryo." *Annals of the New York Academy of Sciences* 541.1 (1988): 346–348.

Haldane, John & Patrick Lee. "Aquinas on Human Ensoulment, Abortion and the Value of Life." *Philosophy* 78.2 (2003a): 255–278.

Haldane, John & Patrick Lee. "Rational Souls and the Beginning of Life (A Reply to Robert Pasnau)." *Philosophy* 78.4 (2003b): 532–540.

Hanley, Matthew. *Determining Death by Neurological Criteria. Current Practice and Ethics*. Washington, D.C.: The Catholic University of America Press, 2020.

Heaney, Stephen J.. "Aquinas and the Presence of the human rational Soul in the early Embryo." *The Thomist* 56.1 (1992): 19–48.

John Paul II. Pope. *Address to the 18th International Congress of the Transplantation Society, August 29, 2000*. Available at: http://www.vatican.va/holy_father/john_paul_ii/speeches/2000/jul-sep/documents/hf_jp-ii_spe_20000829_transplants_en.html Accessed: March 28, 2022.

Johnson, Mark. "Delayed Hominization: Reflections on some recent Catholic Claims for Delayed Hominization." *Theological Studies* 56.4 (1995): 743–763.

Jones, David Albert. "Thomas Aquinas, Augustine, and Aristotle on "Delayed Animation"." *The Thomist: A Speculative Quarterly Review* 76.1 (2012): 1–36.

Jones, David Albert. "Aquinas as an advocate of abortion? The appeal to 'delayed animation' in contemporary Christian ethical debates on the human embryo." *Studies in Christian Ethics* 26.1 (2013): 97–124.

Jones, Jordan P. *The Function of the Human Heart as It Relates to the Soul According to the Thought of Thomas Aquinas*. Thesis Presented to the Graduate Faculty of Saint Louis University in Partial Fulfillment of the Requirements of the Degree of Master of Arts. Saint Louis, MO: Saint Louis University, 2020. Available at: https://www.heartmath.org/research/research-library/dissertations/function-of-the-human-heart-as-it-relates-to-the-soul and https://slu.academia.edu/JordanJones Accessed: March 28, 2022.

Joyce, Robert E. "Personhood and the Conception Event." In: Goodman, Michael F. (Ed.). *What is a Person?* Clifton, NJ: The Humana Press, 1988. pp. 199–212. Originally published in: *The New Scholasticism* 52 (1978): 92–109.

Lamba, Nayan, Daniel Holsgrove & Marike L. Broekman. "The history of head transplantation: a review." *Acta Neurochirurgica* 158 (2016): 2239–2247.

Lang, Min, John Tsiang, Nina Z. Moore, Mark D. Bain & Michael P. Steinmetz. "A Tribute to Dr Robert J. White." *Neurosurgery* 85.2 (2019): E366–E373.

Lee, Patrick. "Distinguishing Embryos from Non-embryos." In: Suarez, Antoine & Joachim Huarte (Eds.). *Is this Cell a Human Being? Exploring the Status of Embryos, Stem Cells and Human-Animal Hybrids.* Heidelberg et al.: Springer, 2011. pp. 77–95.

Lee, Patrick & Germain Grisez. "Total Brain Death: A Reply to Alan Shewmon." *Bioethics* 26.5 (2012): 275–284.

Lizza, John P.. "And She's Not Only Merely Dead, She's Really Most Sincerely Dead." *The Hastings Center Report* 39.5 (2009a): 5–6.

Lizza, John P.. "Commentary on 'The incoherence of determining death by neurological criteria.'" *Kennedy Institute of Ethics Journal* 19.4 (2009b): 393–395.

Lock, Margaret. "Living Cadavers and the Calculation of Death." *Body & Society* 10.2–3 (2004): 135–52.

Lopata, Alex. "History of the Egg in Embryology." *Journal of Mammalian Ova Research* 26.1 (2009): 2–9.

Madea, Burkhard. "Nature and Definition of Death." In: Madea, Burkhard (Ed.). *Handbook of Forensic Medicine.* Hoboken, NJ: Wiley Blackwell, 2014. pp. 49–56.

McHugh, John A. & Charles J. Callan. *Moral Theology: A Complete Course Based on St. Thomas Aquinas and the Best Modern Authorities.* Revised and enlarged by Edward P. Farrell. New York City: Joseph F. Wagner; London: B. Herder, 1958. Available at: http://www.gutenberg.org/ebooks/35354 Accessed: March 28, 2022.

McMahan, Jeff. *The Ethics of Killing: Problems at the Margins of Life.* Oxford: Oxford University Press, 2002.

McMahan, Jeff. "The Metaphysics of Brain Death." In: Brennan, Samantha & Robert J. Stainton (Eds.) *Philosophy and Death: Introductory Readings.* Ontario: Broadview Press, 2010. pp. 109–142. Originally published in: *Bioethics* 9.2 (1995): 91–126.

Mercier, Désiré-Joseph. *A Manual of Modern Scholastic Philosophy. Volume I: Cosmology, Psychology, Epistemology (Criteriology), General Metaphysics*

(Ontology). Second Edition (Revised). London: Kegan Paul, Trench, Trubner & Co., 1921.

Miller, Franklin G. & Robert D. Truog. "Decapitation and the Definition of Death." *Journal of Medical Ethics* 36.10 (2010): 632–634.

Moore, Keith L., T. V. N. Persaud & Mark G. Torchia. *The Developing Human: Clinically Oriented Embryology.* Eleventh Edition. Edinburgh et al.: Elsevier, 2020.

Moschella, Melissa. "Deconstructing the Brain Disconnection–Brain Death Analogy and Clarifying the Rationale for the Neurological Criterion of Death." *Journal of Medicine and Philosophy* 41.3 (2016): 279–299.

Napier, Stephen. "Vulnerable Embryos: A Critical Analysis of Twinning, Rescue, and Natural-Loss Arguments." *American Catholic Philosophical Quarterly* 84.4 (2010): 783–812.

Nguyen, Doyen. "Pope John Paul II and the neurological standard for the determination of death: A critical analysis of his address to the Transplantation Society." *The Linacre Quarterly* 84.2 (2017): 155–186.

Nguyen, Doyen. "Why the Thomistic Defense of 'Brain Death' Is Not Thomistic: An Analysis from the Perspectives of Classical Philosophy and Contemporary Biophilosophy." *The Thomist: A Speculative Quarterly Review* 82.3 (2018a): 407–446.

Nguyen, Doyen. *The New Definitions of Death for Organ Donation. A Multidisciplinary Analysis from the Perspective of Christian Ethics.* Bern et al.: Peter Lang AG, 2018b.

Noldin, Hieronymus. *Summa Theologiae Moralis: II. De Praeceptis.* Editio XXIII. Oeniponte: F. Rauch, 1935.

Oderberg, David S.. "Modal Properties, Moral Status, and Identity." *Philosophy & Public Affairs* 26.3 (1997): 259–298.

Oderberg, David S.. *Real Essentialism.* London: Routledge, 2007.

Ostertag, Christopher & Kyle Karches. "Brain Death and the Formation of Moral Conscience." *The Linacre Quarterly* 86.4 (2019): 335–346.

Pallis, Christopher. "Whole-Brain Death Reconsidered: Physiological Facts and Philosophy." *Journal of Medical Ethics* 9.1 (1983): 32–37.

Pasnau, Robert. *Thomas Aquinas on Human Nature: A Philosophical Study of Summa Theologiae, 1a 75–89.* Cambridge: Cambridge University Press, 2002.

Petitgand, M.. "Observations sur un décapité annamite." *Revue Scientifique* (1884): 10–12. (Troisième Série—Tome VIII. Tome XXXIV de la Collection. 21me Année—2e Semestre. Juillet 1884 a Janvier 1885) Available at: https://gallica.bnf.fr/ark:/12148/bpt6k215103f?rk=42918;4 Accessed: March 28, 2022.

Phillips, Richard Percival. *Modern Thomistic Philosophy: An Explanation for Students. Volume 1: The Philosophy of Nature.* Westminster, MD: The Newman Bookshop, 1934.

Prümmer, Dominicus M.. *Manuale Theologiae Moralis Secundum Principia S. Thomae Aquinatis in Usum Scholarum.* Tomus II. Editio Quinta Decima. Barcinone, Friburgi Brisgoviae & Romae: Herder, 1960.

Ren, Xiaoping, Ming Li, Xin Zhao, Zehan Liu, Shuai Ren, Yafang Zhang, Shide Zhang & Sergio Canavero. "First cephalosomatic anastomosis in a human model." *Surgical Neurology International* 8.276 (2017): https://surgicalneurologyint.com/surgicalint-articles/first-cephalosomatic-anastomosis-in-a-human-model

Sacred Congregation for the Doctrine of the Faith. *Declaration on Procured Abortion.* Vatican: Sacred Congregation for the Doctrine of the Faith, 1974. Available at: https://www.vatican.va/roman_curia/congregations/cfaith/documents/rc_con_cfaith_doc_19741118_declaration-abortion_en.html Accessed: March 28, 2022.

Sánchez Sorondo, Marcelo (Ed.). *Working Group on The Signs of Death. The Proceedings of the Working Group 11–12 September 2006.* Vatican City: Pontificia Academia Scientiarum, 2007. (The Pontifical Academy of Sciences, Scripta Varia 110) Available at: https://www.pas.va/en/publications/scripta-varia/sv110pas.html. Accessed: March 28, 2022.

Schindler, David L.. "A Response to the Joint Statement, 'Production of Pluripotent Stem Cells by Oocyte Assisted Reprogramming'." *Communio: International Catholic Review* 32.2 (2005): 369–380.

Schoenwolf, Gary C., Steven B. Bleyl, Philip R. Brauer & Philippa H. Francis-West. *Larsen's Human Embryology.* Fifth Edition. Edinburgh et al.: Elsevier, 2015.

Seifert, Josef. *The Philosophical Diseases of Medicine and Their Cure. Philosophy and Ethics of Medicine, Vol. 1: Foundations.* Dordrecht: Springer, 2004. (Philosophy and Medicine, Volume 82)

Seyfer, Tara L.. "An Overview of Chimeras and Hybrids." *The National Catholic Bioethics Quarterly* 6.1 (2006): 37–49.

Shewmon, D. Alan. "Mental Disconnect: 'Physiological Decapitation' as a Heuristic for Understanding Brain Death." In: Sánchez Sorondo, Marcelo (Ed.). *Working Group on The Signs of Death. The Proceedings of the Working Group 11–12 September 2006.* Vatican City: Pontificia Academia Scientiarum, 2007. pp. 292–333. (Pontificiae Academiae Scientiarum Scripta Varia 110)

Spencer, Mark K.. "A Reexamination of the Hylomorphic Theory of Death." *The Review of Metaphysics* 63.4 (2010): 843–870.

Steinberg, Avraham & M. Hersch. "Decapitation of a pregnant sheep: a contribution to the brain death controversy." *Transplantation Proceedings* 27.2 (1995): 1886–1887.

Sutovsky, Steve Peter & Won-Hee Song. "Post-fertilisation sperm mitophagy: the tale of Mitochondrial Eve." *Reproduction, Fertility and Development* 30.1 (2017): 56–63.

van Dyke, Christina. "The End of (Human) Life as We Know It: Thomas Aquinas on Persons, Bodies and Death." *The Modern Schoolman* 89.3–4 (2012): 243–257.

Vij, Krishan. *Textbook of Forensic Medicine and Toxicology: Principles and Practice*. Fifth Edition. London: Elsevier, 2014.

Wallace, William A.. "Aquinas's Legacy on Individuation, Cogitation, and Hominization." In: Gallagher, David M. (Ed.). Thomas Aquinas and His Legacy. Washington, D.C.: The Catholic University of America Press, 1994, pp. 173–193. Gallagher's book has been republished 2018 as volume 28 of the series "Studies in Philosophy and The History of Philosophy" edited by The Catholic University of America Press.

Walker, Adrian J.. "Altered Nuclear Transfer: A Philosophical Critique." *Communio: International Catholic Review* 31.4 (2004): 649–684.

White Robert J., Heinz Angstwurm & Ignacio Carrasco de Paula (Eds.). *Working Group on the Determination of Brain Death and its Relationship to Human Death. 10–14 December 1989*. Vatican City: Pontificia Academia Scientiarum, 1992. (The Pontifical Academy of Sciences, Scripta Varia 83) Available at: https://www.pas.va/en/publications/scripta-varia/sv83pas.html Accessed: March 28, 2022.

White, Robert J., Maurice S. Albin & Javier Verdura. "Isolation of the Monkey Brain: In vitro Preparation and Maintenance." *Science* 141.3585 (1963): 1060–1061.

White, Robert J., Maurice S. Albin & Javier Verdura. "Preservation of Viability in the Isolated Monkey Brain Utilizing a Mechanical Extracorporeal Circulation." *Nature* 202.4937 (1964): 1082–1083.

White, Robert J., Maurice S. Albin, George E. Locke & Eugene Davidson. "Brain Transplantation: Prolonged Survival of. Brain after Carotid-Jugular Interposition." *Science* 150.3697 (1965): 779–781.

White, Robert J., Lee R. Wolin, Leo C. Massopust, Norman Taslitz, & Javier Verdura. "Cephalic exchange transplantation in the monkey." *Surgery* 70.1 (1971): 135–139.

Literature Cited in Abbreviation

(a) St. Thomas Aquinas

Unless indicated otherwise, the original (Latin) version of the following works by St. Thomas Aquinas is—as of March 28, 2022—available at http://www.corpusthomisticum.org/iopera.html and/ or https://isidore.co/aquinas.

I^{a}: *Summa Theologiae, Prima pars.* English translation by the Fathers of the English Dominican Province available at: https://isidore.co/aquinas and http://newadvent.org/summa Accessed: March 28, 2022.

I^{a}-IIae: *Summa Theologiae, Prima pars secundae partis (Prima secundae).* English translation by the Fathers of the English Dominican Province available at: https://isidore.co/aquinas and http://newadvent.org/summa Accessed: March 28, 2022.

II^{a}-IIae: *Summa Theologiae, Secunda pars secundae partis (Secunda secundae).* English translation by the Fathers of the English Dominican Province available at: https://

C. Erk, *The Ethics of Killing*,
https://doi.org/10.1007/978-3-031-07183-6

	isidore.co/aquinas and http://newadvent.org/summa Accessed: March 28, 2022.
III[a]:	*Summa Theologiae, Tertia pars.* English translation by the Fathers of the English Dominican Province available at: https://isidore.co/aquinas and http://newadvent.org/summa Accessed: March 28, 2022.
Contra Gentiles:	*Summa contra Gentiles (On the Truth of the Catholic Faith).* English translation (several translators) available at: https://isidore.co/aquinas and http://www.documentacatholicaomnia.eu/20_50_1225-1274-_Thomas_Aquinas,_Sanctus.html Accessed: March 28, 2022.
De Malo:	*Quaestiones Disputatae de Malo (Disputed Questions on Evil).*
De motu cordis:	*De motu cordis ad magistrum Philippum de Castro Caeli (On the Motion of the Heart).* English translation by Gregory Froelich available at: https://isidore.co/aquinas Accessed: March 28, 2022.
De potentia:	*Quaestiones disputatae de potentia (Disputed Questions on the Power of God).* English translation by Gregory Froelich available at: https://isidore.co/aquinas Accessed: March 28, 2022.
De spiritualibus creaturis:	*Quaestio disputata de spiritualibus creaturis (Disputed Question on Spiritual Creatures).* English translation by Mary C. Fitzpatrick and John J. Wellmuth available at: https://isidore.co/aquinas Accessed: March 28, 2022.
De unitate intellectus:	*De unitate intellectus contra Averroistas (On the Unity of the Intellect Against the Averroists).*
De Veritate:	*Quaestiones Disputatae de Veritate (Disputed Questions on Truth).* English translation (several translators) available at: https://isidore.co/aquinas Accessed: September 10th, 2021.

De Virtutibus: *Quaestiones Disputatae de Virtutibus (Disputed Questions on the Virtues).*

Q. d. de Anima: *Quaestiones Disputatae de Anima (Disputed Questions on the Soul).* English translation by John Patrick Rowan available at: https://isidore.co/aquinas Accessed: March 28, 2022.

Sentencia de Anima: *Sentencia Libri De Anima (Commentary on Aristotle's De Anima).* English translation by John Patrick Rowan available at: https://isidore.co/aquinas Accessed: March 28, 2022.

Sententia Ethic.: *Sententia Libri Ethicorum (Commentary on Aristotle's Nicomachaean Ethics).* English translation by C. I. Litzinger available at: https://isidore.co/aquinas Accessed: March 28, 2022.

Super De causis: *Super librum De causis expositio (Commentary on the Book of the Causes).* English translation by Vincent A. Guagliardo, Charles R. Hess and Richard C. Taylor available at: https://isidore.co/aquinas Accessed: March 28, 2022.

Super I Cor.: *Super I Epistolam B. Pauli ad Corinthios lectura (Commentary on the First Epistle to the Corinthians).* English translation by Fabian Larcher available at: https://isidore.co/aquinas Accessed: March 28, 2022.

Super Sent.: *Scriptum super Sententiis (Magistri Petri Lombardi)* (Commentary on the "Libri Quattuor Sententiarum" by Petrus Lombardus).

Suppl. IIIae: *Supplementum Tertiae Partis Summa Theologiae* (Supplement to the Third Part of the Summa Theologiae). English translation by the Fathers of the English Dominican Province available at: http://newadvent.org/summa Accessed: March 28, 2022.

(b) Aristotle

De Anima: Περί Ψυχῆς (Perí Psychēs; On the Soul; Bekker Pagination 402a ff). Original (Greek) version available at: http://www.isnature.org/Files/Aristotle (Bekker Edition 1837), https://isidore.co/aquinas and http://www.swisseduc.ch/altphilo/antike/alinks/texte.html and https://sites.google.com/site/ancienttexts/gk-a3 English translation available at: https://isidore.co/aquinas, http://classics.mit.edu/Aristotle/soul.html and http://www.documentacatholicaomnia.eu/04z/z_-384_-322__Aristoteles__On_The_Soul__EN.pdf.html Accessed: March 28, 2022.

Metaphysics: Τὰ *μετὰ* τὰ φυσικά (tà metà tà physiká; Metaphysics; Bekker Pagination 980a ff). Original (Greek) version available at: https://archive.org/details/aristotelisopera02arisrich (Bekker Edition 1831) and http://www.isnature.org/Files/Aristotle (Bekker Edition 1837) and http://www.perseus.tufts.edu/hopper/collection?collection=Perseus:collection:Greco-Roman, https://isidore.co/aquinas and http://www.swisseduc.ch/altphilo/antike/alinks/texte.html Accessed: March 28, 2022.

Posterior Analytics: 'Αναλυτικὰ ὕστερα (Analytica hýstera; Posterior Analytics; Bekker Pagination 71a ff) Original (Greek) version available at: http://www.hs-augsburg.de/~harsch/graeca/Chronologia/S_ante04/Aristoteles/ari_a200.html English translation by Richard McKeon available at: https://archive.org/details/basicworksofaris00aris Accessed: March 28, 2022.

(c) Plato

Alcibiades I: Ἀλκιβιάδης α (Alcibiades I; Stephanus Pagination 103a–135e (Volume 2)). Original (Greek) version as well as English translation by W. R. M. Lamb available at: http://www.perseus.tufts.edu/hopper/collection?collection=Perseus:collection:Greco-Roman Accessed: March 28, 2022.

Phaedo: Φαίδων (Phaidon; Stephanus Pagination 57a–118a (Volume 1)). Original (Greek) version as well as English translation by Harold North Fowler available at: http://www.perseus.tufts.edu/hopper/collection?collection=Perseus:collection:Greco-Roman Accessed: March 28, 2022.

(d) Augustine (also: Augustine of Hippo; Aurelius Augustinus Hipponensis)

In Psalmos: *Enarrationes in Psalmos* (Expositions on the Psalms). Original (Latin) version available at: http://www.augustinus.it/latino/esposizioni_salmi/index.htm and http://www.documentacatholicaomnia.eu/20_40_0354-0430-_Augustinus,_Sanctus.html English translation by J. E. Tweed available at: https://www.newadvent.org/fathers/1801.htm and http://www.documentacatholicaomnia.eu/20_40_0354-0430-_Augustinus,_Sanctus.html Accessed: March 28, 2022.

Sermones: *Sermones de Novo Testamento* (Sermons on Selected Lessons of the New Testament). Original (Latin) version available at: http://www.augustinus.it/latino/discorsi/index.htm and http://www.documentacatholicaomnia.eu/20_40_0354-0430-_Augustinus,_Sanctus.html English translation by R. G. MacMullen available at: https://www.newadvent.org/fathers/1603.htm and http://www.documentacatholicaomnia.eu/20_40_0354-0430-_Augustinus,_Sanctus.html Accessed: March 28, 2022.

(e) Dionysius Areopagita (Dionysius the Areopagite)

De Divinis Nominibus: *On the Divine Names* (Περί θείων Ονομάτων). Original (Greek) version available at http://www.documentacatholicaomnia.eu/30_20_0531-0533-_Dionysius_Areopagita.html English translation available at: http://www.tertullian.org/fathers/areopagite_03_divine_names.htm German translation available at: https://www.unifr.ch/bkv/awerk.htm Accessed: March 28, 2022.

(f) Enchiridion Symbolorum

DH: *Denzinger-Hünermann*. The number following the abbreviation "DH" refers to the recital in the "Enchiridion Symbolorum", the compendium including all basic texts of Catholic teachings, first edited by Heinrich Denzinger in 1854 and from 1991 by Peter Hünermann who also added a German translation. The quotes in this book are taken from the English translation of the 43rd edition of the Denzinger-Hünermann: Denzinger, Heinrich. *Enchiridion symbolorum, definitionum et declarationum de rebus fidei et morum. / Compendium of Creeds, Definitions, and Declarations of the Catholic Church*. Latin—English. 43rd Edition. Edited by Peter Hünermann. San Francisco: Ignatius Press, 2012.

(g) Catechism of the Catholic Church

CCC: *Catechism of the Catholic Church*. Latin and English version available at: https://www.vatican.va/archive/ccc/index.htm Accessed: March 28, 2022.

Bibliography

Accad, Michel. "Of wholes and parts: A Thomistic refutation of 'brain death'." *Linacre Quarterly* 82.3 (2015): 217–234.

Anderson, Marie A., Robert L. Fastiggi, David E. Hargroder, Joseph C. Howard Jr. & C. Ward Kischer. "Ectopic Pregnancy and Catholic Morality. A Response to Recent Arguments in Favor of Salpingostomy and Methotrexate." *The National Catholic Bioethics Quarterly* 11.1 (2011): 65–82. 667–684.

Anscombe, Gertrude Elizabeth Margaret. "War and Murder." In: Stein, Walter (Ed.). *Nuclear Weapons: A Catholic Response.* London: Sheed & Ward, 1961. pp. 44–52.

Ashley, Benedict. "A Critique of the Theory of Delayed Animation." In: McCarthy, Donald G. & Albert S. Moraczewski. *An Ethical Evaluation of Fetal Experimentation.* St Louis, MO: Pope John XXIII Center, 1976. pp. 113–137.

Ashley, Benedict & Albert Moraczewski. "Cloning, Aquinas, and the Embryonic Person." *The National Catholic Bioethics Quarterly* 1.2 (2001): 189–201.

Atkins, E. Margarete & Thomas Williams. "Introduction." In: Atkins, E. Margarete & Thomas Williams (Eds.). *Thomas Aquinas: Disputed Questions on the Virtues.* Translated by E. Margarete Atkins. Cambridge: Cambridge University Press, 2005. IX–XXX.

Austriaco, Nicanor Pier Giorgio. "The Moral Case for ANT-Derived Pluripotent Stem Cell Lines." *The National Catholic Bioethics Quarterly* 6.3 (2006): 517–537.

Austriaco, Nicanor Pier Giorgio. *Biomedicine and Beatitude: An Introduction to Catholic Bioethics.* Washington, DC: The Catholic University of America Press, 2011a.

C. Erk, *The Ethics of Killing*,
https://doi.org/10.1007/978-3-031-07183-6

Austriaco, Nicanor Pier Giorgio. "Complete Moles and Parthenotes Are Not Organisms." In: Suarez, Antoine & Joachim Huarte (Eds.). *Is this Cell a Human Being? Exploring the Status of Embryos, Stem Cells and Human-Animal Hybrids.* Heidelberg et al.: Springer, 2011b. pp. 45–54. This article is an updated version of the author's "Are Teratomas Embryos or Non-embryos? A Criterion for Oocyte-Assisted Reprogramming." *The National Catholic Bioethics Quarterly* 5.4 (2005): 697–706.

Balch, Thomas J. "Catholic Treatment Ethics and Secular Law: How Can They Cohere?" *Solidarity: The Journal of Catholic Social Thought and Secular Ethics* 6.1 (2016): 1–15 (Article 4). Available at: https://researchonline.nd.edu.au/solidarity/vol6/iss1 Accessed: March 28, 2022.

Baschab, Charles Reinhard. *A Manual of Neo-Scholastic Philosophy. Fourth revised Edition.* St. Louis & London: B. Herder Book Co., 1937.

Battro, Antonio M., James L. Bernat, Marie-Germaine Bousser, Nicola Cabibbo, Card Georges M. M. Cottier, Robert B. Daroff, Stephen Davis, Lüder Deecke, Conrado J. Estol, Werner Hacke, Michael G. Hennerici, Johannes C. Huber, Card Alfonso López Trujillo, Card Carlo M. Martini, José C. Masdeu, Heinrich Mattle, Jerome B. Posner, Louis Puybasset, Marcus E. Raichle, Allan H. Ropper, Paolo M. Rossini, Marcelo Sánchez Sorondo, Herbert Schambeck, Elio Sgreccia, Prakash Narain Tandon, Rafael Vicuña, Eelco F. M. Wijdicks & Antonino Zichichi. *Why the Concept of Death is Valid as a Definition of Brain Death. Statement by the Pontifical Academy of Sciences and Responses to Objections.* Vatican City: Pontificia Academia Scientiarum, 2008. (The Pontifical Academy of Sciences, Extra Series 31) Available at: https://www.pas.va/en/publications/extra-series/es31pas.html Accessed: March 28, 2022. Previously published as: "Why the Concept of Death is Valid as a Definition of Brain Death. Statement by Neurologists and Others." In: Sánchez Sorondo, Marcelo (Ed.). *Working Group on The Signs of Death. The Proceedings of the Working Group 11–12 September 2006.* Vatican City: Pontificia Academia Scientiarum, 2007. pp. XXI–XXIX. (The Pontifical Academy of Sciences, Scripta Varia 110) Available at: https://www.pas.va/en/publications/scripta-varia/sv110pas.html Accessed: March 28, 2022.

Beckwith, Francis J. "The Human Being, a Person of Substance: A Response to Dean Stretton." In: Napier, Stephen (Ed.). *Persons, Moral Worth, and Embryos: A Critical Analysis of Pro-Choice Arguments.* Dordrecht, Heidelberg, London & New York: Springer, 2011. pp. 67–83. (Series "Philosophy and Medicine", Volume 111)

Bedau, Mark A.. "The Nature of Life." In: Boden, Margaret A. (Ed.). *The Philosophy of Artificial Life.* Oxford: Oxford University Press, 1996. pp. 332–357.

Bender, Ludwig. "Occisio directa et indirecta." *Angelicum* 28.3 (1951): 224–253.

Bernat, James L.. "A Defense of the Whole-Brain Concept of Death." *The Hastings Center Report* 28.2 (1998): 14–23.

Bernat, James L. "On Irreversibility as a Prerequisite for Brain Death Determination." In: Machado, Calixto & D. Alan Shewmon. *Brain Death and Disorders of Consciousness.* New York: Springer, 2004. pp. 161–167. (Advances in Experimental Medicine and Biology, Volume 550)

Bernat, James L.. "The Whole-Brain Concept of Death Remains Optimum Public Policy." *Journal of Law, Medicine & Ethics* 34.1 (2006): 35–43.

Bishop, Jeffrey P.. "When is somebody just some body? Ethics as first philosophy and the brain death debate." *Theoretical Medicine and Bioethics* 40.5 (2019):419–436.

Blechschmidt, Erich. *The Beginnings of Human Life.* New York: Springer-Verlag, 1977.

Blechschmidt, Erich. *The Ontogenetic Basis of Human Anatomy: A Biodynamic Approach to Development from Conception to Birth.* Revised, edited, and translated by Brian Freeman. Berkeley, CA: Atlantic Books, 2004.

Boëthius, Anicius Manlius Severinus. "Liber contra Eutychen et Nestorium." In: Boëthius, Anicius Manlius Severinus. *The Theological Tractates and the Consolation of Philosophy.* Latin/ English Edition. Translated and edited by Hugh Fraser Stewart & Edward Kennard Rand. London: The Loeb Classical Library, 1918. Available at: http://www.gutenberg.org/etext/13316 Accessed: March 28, 2022.

Bothur-Nowacka, Joanna, Aleksandra Jezela-Stanek, Katarzyna Zaniuk, Bożenna Goryluk-Kozakiewicz, Małgorzata Krajewska-Walasek & Anna Dobrzańska. "Tetraploidy in the era of molecular karyotyping—What we need to remember." *Pediatra Polska* 88.5 (2013): 467–471.

Bracken, W. Jerome. "Is the Early Embryo a Person?" *The Linacre Quarterly* 68.1 (2001): 49–70.

Brenner, Steven A.. "Defining Life." *Astrobiology* 10.10 (2010): 1021–1030.

Brock, Dan Willets. "Voluntary Active Euthanasia." *The Hastings Center Report* 22.2 (1992): 10–22.

Brugger, E. Christian. "ANT-OAR: A Morally Acceptable Means for Deriving Pluripotent Stem Cells. A Reply to Criticisms." *Communio: International Catholic Review* 32.4 (2005): 753–769.

Bulzacchelli, Richard H.. "The diagnosis of death and the irreducibility of the human person." *The Linacre Quarterly* 80.1 (2013): 39–55.

Buris, László. *Forensic Medicine.* Budapest et al.: Springer-Verlag, 1993.

Bynum, Jennifer, Denise Batista, Rena Xian, Deyin Xing, James R. Eshleman, Brigitte M. Ronnett & Gang Zheng. "Tetraploid Partial Hydatidiform Moles. Molecular Genotyping and Determination of Parental Contributions." *The Journal of Molecular Diagnostics* 22.1 (2020): 90–100.

Campbell, Neil A., Lisa A. Urry, Michael L. Cain, Steven A. Wasserman, Peter V. Minorsky & Jane B. Reece. *Biology: A Global Approach.* Eleventh Edition. New York. Pearson, 2017.

Carlson, Bruce M.. *Human Embryology and Developmental Biology*. Sixth edition. St. Louis, MO: Elsevier, 2019.

Cathrein, Victor. *Philosophia Moralis in Usum Scholarum*. Editio Quinta ab Auctore Recognita. Friburgi Brisgoviae: B. Herder, 1905.

Chabot, Boudewijn & Christian Walther. *Ausweg am Lebensende Sterbefasten—Selbstbestimmtes Sterben durch Verzicht auf Essen und Trinken*. 5., aktualisierte und erweiterte Auflage. München: Ernst Reinhardt Verlag, 2017.

Chagas, Carlos (Ed.). *Working Group on the Artificial Prolongation of Life and the Determination of the Exact Moment of Death. 19–21 October 1985*. Vatican City: Pontificia Academia Scientiarum, 1986. (The Pontifical Academy of Sciences, Scripta Varia 60) Available at: https://www.pas.va/en/publications/scripta-varia/sv60pas.html Accessed: March 28, 2022.

Cobb, Matthew. *The Egg & Sperm Race. The Seventeenth-Century Scientists Who Unravelled the Secrets of Sex, Life and Growth*. London: Free Press, 2006a.

Cobb, Matthew. "Heredity before Genetics: A History." *Nature Reviews Genetics* 7.12 (2006b): 953–958.

Cobb, Matthew. "An Amazing 10 Years: The Discovery of Egg and Sperm in the 17th Century." *Reproduction in Domestic Animals* 47.s4 (2012): 2–6.

Coffey, Peter. *Ontology or the Theory of Being: An Introduction to General Metaphysics*. London, New York, Bombay, Calcutta & Madras: Longmans, Green and Co., 1918.

Commer, Ernst. *System der Philosophie. Zweite Abtheilung: Naturphilosophie und Psychologie*. Münster: Nasse'sche Verlagsbuchhandlung, 1884.

Commer, Ernst. *System der Philosophie. Vierte (Schluss-) Abtheilung: Ethik*. Münster & Paderborn: Ferdinand Schöningh, 1886.

Condic, Maureen L. "When Does Human Life Begin? The Scientific Evidence and Terminology Revisited." *The National Catholic Bioethics Quarterly* 9.1 (2009): 129–149. Originally published in 2008 by The Westchester Institute for Ethics & the Human Person as part of their White Paper Series (Volume 1, Number 1).

Condic, Maureen L. "Preimplantation Stages of Human Development: The Biological and Moral Status of Early Embryos." In: Suarez, Antoine & Joachim Huarte (Eds.). *Is this Cell a Human Being? Exploring the Status of Embryos, Stem Cells and Human-Animal Hybrids*. Heidelberg et al.: Springer, 2011. pp. 25–43.

Condic, Maureen L.. "When Does Human Life Begin? The Scientific Evidence and Terminology Revisited." *University of St. Thomas Journal of Law and Public Policy* VIII.1 (2013): 44–81.

Condic, Maureen L.. *Untangling Twinning: What Science Tells Us about the Nature of Human Embryos*. Notre Dame, IN: University of Notre Dame Press, 2020. (Notre Dame Studies in Medical Ethics and Bioethics)

Condic, Samuel B. & Maureen L. Condic. *Human Embryos, Human Beings: A Scientific & Philosophical Approach*. Washington, DC: The Catholic University of America Press, 2018.

Cronin, Michael. *The Science of Ethics. Volume II: Special Ethics.* New York et al.: Benziger Brothers, 1922.

Cronin, Michael. *The Science of Ethics. Volume I: General Ethics.* Third Impression. Dublin: M.H. Gill and Son, 1930.

Cronin, Daniel Anthony. "The Moral Law in Regard to the Ordinary and Extraordinary Means of Conserving Life." In: Smith, Russell Edward (Ed.). *Conserving Human Life.* Braintree, MA: Pope John XXIII Medical-Moral Research and Education Center, 1989. pp. 3–116. Originally published in 1958 as Doctoral Dissertation at the Pontifical Gregorian University (Rome).

Damschen, Gregor, Alfonso Gómez-Lobo & Dieter Schönecker. "Sixteen Days? A Reply to B. Smith and B. Brogaard on the Beginning of Human Individuals." *Journal of Medicine and Philosophy* 31.2 (2006): 165–175.

Dan-Cohen, Meir. "The Value of Ownership." In: Dan-Cohen, Meir. *Harmful Thoughts: Essays on Law, Self, and Morality.* Princeton, NJ: Princeton University Press, 2002. pp. 264–301. Originally published as: Dan-Cohen, Meir. "The Value of Ownership." *Journal of Political Philosophy* 9.4 (2001): 404–434.

Davis, Henry. *Moral and Pastoral Theology in Four Volumes. Volume One: Human Acts, Law, Sin, Virtue.* Eighth Edition, revised and enlarged. Edited by L. W. Geddes. London & New York: Sheed and Ward, 1959.

de Locht, Pierre. "Discussion." In: *L'Avortement: Actes du Xeme Colloque International de Sexologie. 2ème partie.* Louvain: Centre International Cardinal Suenens, 1968.

de Wulf, Maurice. *Mediaeval Philosophy Illustrated from the System of Thomas Aquinas.* Cambridge, MA: Harvard University Press, 1922.

Dettmeyer, Reinhard B., Harald F. Schütz & Marcel A. Verhoff. *Forensic Medicine—Fundamentals and Perspectives.* Berlin & Heidelberg: Springer-Verlag, 2014.

Dillon, Robin S. "Respect." *The Stanford Encyclopedia of Philosophy.* February 18, 2018. Available at: http://plato.stanford.edu/entries/respect Accessed: March 28, 2022.

Donceel, Joseph F.. "Immediate Animation and Delayed Hominization." *Theological Studies* 31.1 (1970): 71–105.

Dombrowski, Daniel A. & Robert Deltete. *A Brief, Liberal, Catholic Defense of Abortion.* Urbana, IL: University of Illinois Press, 2000.

Dunne, Kathryn & Pamela Doherty. "Donation after circulatory death." *Continuing Education in Anaesthesia Critical Care & Pain* 11.3 (2011): 82–86.

Eberl, Jason T.. "Aquinas's Account of Human Embryogenesis and Recent Interpretations." *Journal of Medicine and Philosophy* 30.4 (2005a): 379–94.

Eberl, Jason T.. "A Thomistic understanding of human death." *Bioethics* 19.1 (2005b): 29–48.

Eberl, Jason T.. "A Thomistic defense of whole-brain death." *The Linacre Quarterly* 82.3 (2015): 235–250.

Eberl, Jason T. & Brandon P. Brown. "Brain Life and the Argument from Potential: Affirming the Ontological Status of Human Embryos and Fetuses." In: Napier, Stephen (Ed.). *Persons, Moral Worth, and Embryos: A Critical Analysis of Pro-Choice Arguments.* Dordrecht, Heidelberg, London & New York: Springer, 2011. pp. 43–66. (Series "Philosophy and Medicine", Volume 111)

Edwards, John M., Richard D. Hass, Jr. & Virginia M. Robertson. "Non–Heart-Beating Organ Donation: Process and Review." *AACN Advanced Critical Care* 10.2 (1999): 293–300.

Elders, Leo. *The Metaphysics of Being of St. Thomas Aquinas in a Historical Perspective.* Leiden, New York & Köln: E. J. Brill, 1993.

Emanuel, Ezekiel J.. "The Painful Truth About Euthanasia." *Wall Street Journal.* January 7, 1997. p. A16.

Erk, Christian. *Health, Rights and Dignity. Philosophical Reflections on an Alleged Human Right.* Frankfurt, Paris, Lancaster & New Brunswick: ontos Verlag, 2011.

Erk, Christian. "Das Eigentliche des Todes. Ein Beitrag zur Be-Lebung der Debatte über Hirntod und Transplantation." *Ethik in der Medizin* 26.2 (2014a): 121–135.

Erk, Christian. "The Diagnosis of the Absence of Life." *Bioethica Forum* 7.3 (2014b): 105–108.

Erk, Christian. *Rationierung im Gesundheitswesen. Eine wirtschafts- und sozialethische Analyse der Rationierung nach Selbstverschulden.* Berlin & Boston, MA: de Gruyter, 2015.

Erk, Christian. "(Brain) Death, (Brain) Life and the Value of Life." *International Journal of Bioethics and Health Policy* 1.1 (2016): 27–34.

Erk, Christian. *Moral Philosophy. Fundamental Concepts.* Second, revised and expanded Edition. Heusenstamm: editiones scholasticae, 2019.

Fagothey, Austin. *Right and Reason. Ethics in Theory and Practice.* Third edition. Saint Louis: The C.V. Mosby Company, 1963.

Feinberg, Joel. "Some Conjectures on the Concept of Respect." *Journal of Social Philosophy* 4.2 (1973): 1–3.

Feser, Edward. *Philosophy of Mind. A Beginner's Guide.* London: Oneworld Publications, 2006.

Feser, Edward. *Aquinas.* London: Oneworld Publications, 2009.

Feser, Edward. *Scholastic Metaphysics. A Contemporary Introduction.* Heusenstamm: editiones scholasticae, 2014a.

Feser, Edward. "*Remarks on God, Reason and Reality.*" Paper presented at the Symposium "God, Reason and Reality." November 8, 2014b. Dominican School of Philosophy and Theology, Berkeley, CA. Available via: https://web.archive.org/web/20161010002912/https://www.dspt.edu/files/FeseronGodReasonReality.pdf Accessed: March 28, 2022.

Feser, Edward. "In Defense of the Perverted Faculty Argument." In: Feser, Edward. *Neo-Scholastic Essays*. South Bend, Indiana: St. Augustine's Press, 2015. pp. 378–415.

Feser, Edward & Joseph Bessette. *By Man Shall His Blood Be Shed: A Catholic Defense of Capital Punishment*. San Francisco: Ignatius Press, 2017.

Finnis, John. "The Fragile Case for Euthanasia: A Reply to John Harris." In: Keown, John (Ed.). *Euthanasia Examined: Ethical, Clinical, and Legal Perspectives*. Cam-bridge: Cambridge University Press, 1995. pp. 46–55.

Fleischer, Julie, Archana Shenoy, Katherine Goetzinger, Catherine E. Cottrell, Dustin Baldridge, Frances V. White & Marwan Shinawi. "Digynic triploidy: utility and challenges of noninvasive prenatal testing." *Clinical Case Reports* 3.6 (2015): 406–410.

Foot, Philippa. "Killing and Letting Die." In: Steinbock, Bonnie & Alastair Norcross (Eds.). *Killing and Letting Die*. New York: Fordham University Press, 1994. pp. 280–289.

Ford, Norman M.. *When Did I Begin? Conception of the Human Individual in History, Philosophy and Science*. Cambridge: Cambridge University Press, 1988.

Frey, Jennifer A.. "Happiness as the constitutive principle of action in Thomas Aquinas." *Philosophical Explorations* 22.2 (2019): 208–221.

Furton, Edward J.. "Brain Death, the Soul, and Organic Life." *The National Catholic Bioethics Quarterly* 2.3 (2002): 455–470.

Gerber, Rudolph Joseph. "When is the human soul infused?" *Laval théologique et philosophique* 22.2 (1966): 234–247.

Giubilini, Alberto & Francesca Minerva. "After-birth abortion: why should the baby live?" *Journal of Medical Ethics* 39.5 (2013): 261–263.

Giacomini, Mita. "A change of heart and a change of mind? Technology and the redefinition of death in 1968." *Social Science and Medicine* 44.10 (1997): 1465–1482.

Goldenring, John M.. "The brain-life theory: towards a consistent biological definition of humanness." *Journal of Medical Ethics* 11.4 (1985): 198–204.

Grisez, Germain Gabriel. "The First Principle of Practical Reason: A Commentary on the Summa Theologiae, 1–2 Question 94, Article 2." *Natural Law Forum* 10 (1965): 168–201.

Grisez, Germain Gabriel. "Toward a Consistent Natural-Law Ethics of Killing." *American Journal of Jurisprudence* 15.1 (1970): 64–96.

Grisez, Germain. "When do People Begin?" *Proceedings of the American Catholic Philosophical Association* 63 (1989): 27–47.

Grisez, Germain & Patrick Lee. "Total Brain Death: A Reply to Alan Shewmon." *Bioethics* 26.5 (2012): 275–284.

Grobstein, Clifford. "External Human Fertilization." *Scientific American* 240.6 (1979): 57–67.

Grobstein, Clifford. "Biological Characteristics of the Preembryo." *Annals of the New York Academy of Sciences* 541.1 (1988): 346–348.

Haldane, John & Patrick Lee. "Aquinas on Human Ensoulment, Abortion and the Value of Life." *Philosophy* 78.2 (2003a): 255–278.

Haldane, John & Patrick Lee. "Rational Souls and the Beginning of Life (A Reply to Robert Pasnau)." *Philosophy* 78.4 (2003b): 532–540.

Hanley, Matthew. *Determining Death by Neurological Criteria. Current Practice and Ethics.* Washington, D.C.: The Catholic University of America Press, 2020.

Harris, John. *The Value of Life: An Introduction to Medical Ethics.* London: Routledge, 1985.

Heaney, Stephen J.. "Aquinas and the Presence of the human rational Soul in the early Embryo." *The Thomist* 56.1 (1992): 19–48.

Heinberg, Eric M.. "Ectopic Pregnancy." In: Sokol, Andrew I. & Eric R. Sokol (Eds.). *General Gynecology. The Requisites in Obstetrics and Gynecology.* Philadelphia, PA: Mosby, 2007. pp. 257–285.

Henke, Donald E.. "A History of Ordinary and Extraordinary Means." *The National Catholic Bioethics Quarterly* 5.3 (2005): 555–575.

Henßge, Claus & Burkhard Madea. "Leichenerscheinungen und Todeszeitbestimmung." In: Brinkmann, Bernd & Burkhard Madea (Hrsg.). *Handbuch Gerichtliche Medizin, Band I.* Berlin & Heidelberg: Springer, 2004. pp. 79–225.

Higgins, Thomas J.. *Man as Man: The Science and Art of Ethics.* Milwaukee: The Bruce Publishing Company, 1949.

Hillis, David M., Mary V. Price, Richard W. Hill, David W. Hall & Marta J. Laskowski. *Principles of Life.* Third Edition. New York: Sinauer Associates, 2019.

Hittinger, Russell. *A Critique of the New Natural Law Theory.* Notre Dame, IN: University of Notre Dame Press, 1987.

Hooker, Brad. "Rule Consequentialism." *The Stanford Encyclopedia of Philosophy.* November 18, 2015. Available at: https://plato.stanford.edu/entries/consequentialism-rule Accessed: March 28, 2022.

Hüntelmann, Rafael. *Grundkurs Philosophie VI: Natürliche Ethik.* Neunkirchen-Seelscheid: editiones scholasticae, 2017. (Reihe Philosophie des gesunden Menschenverstandes, Band 6).

Hursthouse, Rosalind. *Beginning Lives.* Oxford: Oxford University Press, 1987.

Iglesias, Teresa. "In vitro fertilisation: the major issues." *Journal of Medical Ethics* 10 (1984): 32–37.

Jensen, Steven J.. *Good & Evil Actions. A Journey through Saint Thomas Aquinas.* Washington, D.C.: The Catholic University of America Press, 2010.

Jensen, Steven J.. "Killing and Letting Die." In: Jensen, Steven J. (Ed.). *The Ethics of Organ Transplantation.* Washington, D.C.: The Catholic University of America Press, 2011. pp. 170–191.

Jensen, Steven J.. *Living the Good Life. A Beginner's Thomistic Ethics.* Washington, D.C.: The Catholic University of America Press, 2013.

Jensen, Steven J.. *The Human Person. A Beginner's Thomistic Psychology.* Washington, D.C.: Catholic University of America Press, 2018.

Jessop, Nancy M. *Theory and Problems of Zoology.* New York: McGraw-Hill, 1988. (Schaum's Outline Series)

John Paul II. Pope. *Address to the 18th International Congress of the Transplantation Society, August 29, 2000.* Available at: http://www.vatican.va/holy_father/john_paul_ii/speeches/2000/jul-sep/documents/hf_jp-ii_spe_20000829_transplants_en.html Accessed: March 28, 2022.

Johnson, Mark. "Delayed Hominization: Reflections on some recent Catholic Claims for Delayed Hominization." *Theological Studies* 56.4 (1995): 743–763.

Jones, David Albert. "Thomas Aquinas, Augustine, and Aristotle on "Delayed Animation"." *The Thomist: A Speculative Quarterly Review* 76.1 (2012): 1–36.

Jones, David Albert. "Aquinas as an advocate of abortion? The appeal to 'delayed animation' in contemporary Christian ethical debates on the human embryo." *Studies in Christian Ethics* 26.1 (2013): 97–124.

Jones, Jordan P. *The Function of the Human Heart as It Relates to the Soul According to the Thought of Thomas Aquinas.* Thesis Presented to the Graduate Faculty of Saint Louis University in Partial Fulfillment of the Requirements of the Degree of Master of Arts. Saint Louis, MO: Saint Louis University, 2020. Available at: https://www.heartmath.org/research/research-library/dissertations/function-of-the-human-heart-as-it-relates-to-the-soul and https://slu.academia.edu/JordanJones Accessed: March 28, 2022.

Joyce, Robert E.. "Personhood and the Conception Event." *New Scholasticism* 52.1 (1978): 97–109.

Joyce, Robert E. "Personhood and the Conception Event." In: Goodman, Michael F. (Ed.). *What is a Person?* Clifton, NJ: The Humana Press, 1988. pp. 199–212. Originally published in: *The New Scholasticism* 52 (1978): 92–109.

Kaczor, Christopher. *The Edge of Life. Human Dignity and Contemporary Bioethics.* Dordrecht: Springer, 2005. (Series "Medicine and Philosophy", Volume 85; Subseries "Catholic Studies in Bioethics")

Kaczor, Christopher. "Artificial Wombs and Embryo Adoption." In: Brakman, Sarah-Vaughan & Darlene Fozard Weaver (Eds.). *The Ethics of Embryo Adoption and the Catholic Tradition: Moral Arguments, Economic Reality and Social Analysis.* Dordrecht: Springer, 2007. pp. 307–322. (Series "Medicine and Philosophy", Volume 95; Subseries "Catholic Studies in Bioethics")

Kaczor, Christopher. *The Ethics of Abortion: Women's Rights, Human Life, and the Question of Justice.* Second Edition. New York & London: Routledge, 2015.

Kaczor, Christopher. "Abortion as Human Rights Violation." In: Greasley, Kate & Christopher Kaczor. *Abortion Rights. For and Against.* Cambridge: Cambridge University Press, 2018. pp. 86–164.

Kaczor, Christopher. *Disputes in Bioethics. Abortion, Euthanasia, and other Controversies.* Notre Dame, IN: University of Notre Dame Press, 2020.

Kälin, Bernard. *Lehrbuch der Philosophie. Band II: Einführung in die Ethik.* Zweite Auflage; umgearbeitet von Raphael Fäh. Sarnen: Selbstverlag Benediktiner kollegium Sarnen, 1954.

Kant, Immanuel. "Grundlegung zur Metaphysik der Sitten (1785)." In: Kant, Immanuel. *Akademieausgabe von Immanuel Kants Gesammelten Werken. Band 4 (Kritik der reinen Vernunft, Prolegomena, Grundlegung zur Metaphysik der Sitten, Metaphysische Anfangsgründe der Naturwissenschaft).* pp. 385–464. Available at: http://www.korpora.org/Kant/verzeichnisse-gesamt.html and http://www.korpora.org/Kant/aa04 Accessed: March 28, 2022.

Kass, Leon R. "Defending Human Dignity." In: The President's Council on Bioethics (Ed.). *Human Dignity and Bioethics. Essays Commissioned by the President's Council on Bioethics.* Washington, D.C.: The President's Council on Bioethics, March 2008. pp. 297–331. Available at: https://bioethicsarchive.georgetown.edu/pcbe/reports/human_dignity Accessed: March 28, 2022.

Kaveny, M. Cathleen. "Appropriation of Evil: Cooperation's Mirror Image." *Theological Studies* 61.2 (2000): 280–313.

Kelly, Gerald A.. *Medico-Moral Problems.* St. Louis, MO: The Catholic Hospital Association of the United States and Canada, 1958.

Kitwood, Tom M.. *Dementia Reconsidered: The Person Comes First.* Buckingham: Open University Press, 1997.

Klubertanz, George Peter. *The Philosophy of Human Nature.* Heusenstamm: editiones scholasticae, 2014. (Reprint of the second edition, 1951)

Koch, Antony & Arthur Preuss. *A Handbook of Moral Theology. Volume III: Man's Duties to Himself.* Third, revised edition. St. Louis & London: B. Herder Book Co., 1926.

Koshland, Daniel E. Jr.. "The Seven Pillars of Life." *Science* 295.5563 (2002): 2215–2216.

Kuhse, Helga & Peter Singer. *Should the Baby live? The Problem of Handicapped Infants.* Oxford: Oxford University Press, 1985.

Laitinen, Arto. "Interpersonal Recognition: A Response to Value or a Precondition of Personhood?" *Inquiry* 45 (2002): 463–478.

Laitinen, Arto. "Sorting Out Aspects of Personhood Capacities, Normativity and Recognition." *Journal of Consciousness Studies* 14.5–6 (2007): 248–270.

Lamba, Nayan, Daniel Holsgrove & Marike L. Broekman. "The history of head transplantation: a review." *Acta Neurochirurgica* 158 (2016): 2239–2247.

Lang, Min, John Tsiang, Nina Z. Moore, Mark D. Bain & Michael P. Steinmetz. "A Tribute to Dr Robert J. White." *Neurosurgery* 85.2 (2019): E366–E373.

Larmer, Robert. "Abortion, Personhood and the Potential for Consciousness." *Journal of Applied Philosophy* 12.3 (1995): 241–251.

Lee, Patrick. "Substantial Identity and the Right to Life: A Rejoinder to Dean Stretton." *Bioethics* 21.2 (2007): 93–97.

Lee, Patrick. "Distinguishing Embryos from Non-embryos." In: Suarez, Antoine & Joachim Huarte (Eds.). *Is this Cell a Human Being? Exploring the Status of Embryos, Stem Cells and Human-Animal Hybrids.* Heidelberg et al.: Springer, 2011. pp. 77–95.

Lee, Patrick. "The New Natural Law Theory." In: Angier, Tom (Ed.). *The Cambridge Companion to Natural Law Ethics.* Cambridge: Cambridge University Press, 2019. pp. 73–91.

Lee, Patrick & Germain Grisez. "Total Brain Death: A Reply to Alan Shewmon." *Bioethics* 26.5 (2012): 275–284.

Lee, Patrick & Robert P. George. *Body-Self Dualism in Contemporary Ethics and Politics.* Cambridge: Cambridge University Press, 2008a.

Lee, Patrick & Robert P. George. "The Nature and Basis of Human Dignity." In: The President's Council on Bioethics (Ed.). *Human Dignity and Bioethics. Essays Commissioned by the President's Council on Bioethics.* Washington, DC: The President's Council on Bioethics, 2008b. pp. 409–433. Available at: https://bioethicsarchive.georgetown.edu/pcbe/reports/human_dignity Accessed: March 28, 2022.

Lehmen, Alfons. *Lehrbuch der Philosophie auf aristotelisch-scholastischer Grundlage zum Gebrauch an höheren Lehranstalten und zum Selbstunterricht. Zweiter Band: Kosmologie und Psychologie.* Dritte, verbesserte und vermehrte Auflage. Herausgegeben von Peter Beck. Freiburg im Breisgau: Herdersche Verlagsbuchhandlung, 1911.

Ligouri, Alphonsus. *Theologia Moralis S. Alphonsi de Ligorio. Tomus Secundus, Continens Tractatus de Actibus Humanis, de Peccatis, de Praeceptis Virtutum Theol. et 1. et 2. Praecepto Decalogi.* Edited by Mich. Heilig. Mechliniae: P. J. Hanicq, 1852.

Lisska, Anthony J.. *Aquinas's Theory of Natural Law: An Analytic Reconstruction.* Oxford: Clarendon Press, 1996.

Lizza, John P.. "And She's Not Only Merely Dead, She's Really Most Sincerely Dead." *The Hastings Center Report* 39.5 (2009a): 5–6.

Lizza, John P.. "Commentary on 'The incoherence of determining death by neurological criteria.'" *Kennedy Institute of Ethics Journal* 19.4 (2009b): 393–395.

Lock, Margaret. "Living Cadavers and the Calculation of Death." *Body & Society* 10.2–3 (2004): 135–52.

Locke, John. *An Essay Concerning Human Understanding.* London: 1690. Available at: http://oll.libertyfund.org/title/762 Accessed: March 28, 2022.

Lockwood, Michael. "When does a life begin?" In: Lockwood, Michael (Ed.). *Moral Dilemmas in Modern Medicine.* Oxford & New York: Oxford University Press, 1985. pp. 9–31.

Long, Steven. "Fundamental Errors of the New Natural Law Theory." *The National Catholic Bioethics Quarterly* 13.1 (2013): 105–131.

Lopata, Alex. "History of the Egg in Embryology." *Journal of Mammalian Ova Research* 26.1 (2009): 2–9.

Lübbe, Andreas S. "Palliativmedizin als Angebot gegen eine Normalisierung des Tötens." In: Hoffmann, Thomas Sören & Marcus Knaup (Hrsg.). *Was heißt: In Würde sterben? Wider die Normalisierung des Tötens.* Wiesbaden: Springer Fachmedien, 2015. pp. 151–170.

Madea, Burkhard. "Nature and Definition of Death." In: Madea, Burkhard (Ed.). *Handbook of Forensic Medicine.* Hoboken, NJ: Wiley Blackwell, 2014. pp. 49–56.

Madea, Burkhard & Gerhard Kernbach-Wighton. "Early and Late Postmortem Changes." In: Houck, Max M. (Ed.). *Forensic Pathology.* London: Academic Press, 2017. pp. 41–54.

Madea, Burkhard, Clauss Henssge, Saskia Reibe, Michael Tsokos & Gerhard Kernbach-Wighton. "Postmortem Changes and Time since Death." In: Madea, Burkhard (Ed.). *Handbook of Forensic Medicine.* Hoboken, NJ: Wiley Blackwell, 2014. pp. 75–133.

Madea, Burkhard, Katja Jachau, Saskia Reibe, Peter Schmidt, Gerhard Kernbach-Wighton, Oliver Peschel, Véronique Henn, Christoph Meissner, Manfred Oehmichen, Michael Thali, Rüdiger Lessig, Stefan Pollak & Ulrich Zollinger. "Thanatologie." In: Madea, Burkhard (Hrsg.). *Rechtsmedizin: Befunderhebung, Rekonstruktion, Begutachtung.* Berlin & Heidelberg: Springer-Verlag, 2015. pp. 33–170.

Mader, Sylvia S. & Michael Windelspecht. *Biology.* Thirteenth Edition. New York: McGraw-Hill Education, 2019.

Maguire, Marjorie Reiley. "Symbiosis, Biology, and Personalization." In: Doerr, Edd & James W. Prescott. *Abortion Rights and Fetal 'Personhood'.* Long Beach, CA: Centerline Press, 1989. pp. 11–19.

Maltoni, M., C. Pittureri, E. Scarpi, L. Piccinini, F. Martini, P. Turci, L. Montanari, O. Nanni & D. Amadori. "Palliative sedation therapy does not hasten death: Results from a prospective multicenter study." *Annals of Oncology* 20.7 (2009): 1163–1169.

Marquis, Don. "Why Abortion is Immoral." *The Journal of Philosophy* 86.4 (1989): 183–202.

Mason, Kenneth A., Tod Duncan, George B. Johnson, Jonathan B. Losos & Susan R. Singer. *Biology.* Eleventh Edition. New York: McGraw-Hill, 2017.

McHugh, John A. & Charles J. Callan. *Moral Theology: A Complete Course Based on St. Thomas Aquinas and the Best Modern Authorities.* Revised and enlarged by Edward P. Farrell. New York City: Joseph F. Wagner; London: B. Herder, 1958. Available at: http://www.gutenberg.org/ebooks/35354 Accessed: March 28, 2022.

McMahan, Jeff. *The Ethics of Killing: Problems at the Margins of Life.* Oxford: Oxford University Press, 2002.

McMahan, Jeff. "Infanticide." *Utilitas* 19.2 (2007): 131–159.

McMahan, Jeff. "The Metaphysics of Brain Death." In: Brennan, Samantha & Robert J. Stainton (Eds.) *Philosophy and Death: Introductory Readings.* Ontario: Broadview Press, 2010. pp. 109–142. Originally published in: *Bioethics* 9.2 (1995): 91–126.

Melé, Domènec. *Business Ethics in Action: Seeking Human Excellence in Organizations*. Basingstoke, UK: Palgrave Macmillan, 2009.

Mercier, Désiré-Joseph. *A Manual of Modern Scholastic Philosophy. Volume I: Cosmology, Psychology, Epistemology (Criteriology), General Metaphysics (Ontology). Second Edition (Revised)*. London: Kegan Paul, Trench, Trubner & Co., 1921.

Merkelbach, Benedictus Henricus. *Summa Theologiae Moralis ad Mentem D. Thomae et ad Normam Iuris Novi: I. De Principiis*. Parisiis: Typis Desclée de Brouwer et Soc., 1935.

Messner, Johannes. *Ethik. Kompendium der Gesamtethik*. Innsbruck, Wien & München: Tyrolia Verlag, 1955.

Miech, Ralph P.. "Physician Cooperation in Patient Suicide." *Ethics & Medics* 24.7 (1999): 1–4.

Miller, Franklin G. & Robert D. Truog. "Decapitation and the Definition of Death." *Journal of Medical Ethics* 36.10 (2010): 632–634.

Monteverde, Settimio & Annette Rid. "Controversies in the Determination of Death." *Swiss Medical Weekly* 142 (2012): w13667.

Moore, Keith L., T. V. N. Persaud & Mark G. Torchia. *The Developing Human: Clinically Oriented Embryology*. Eleventh Edition. Edinburgh et al.: Elsevier, 2020.

Morris, James, Daniel Hartl, Andrew Knoll, Robert Lue, Melissa Michael, Andrew Berry, Andrew Biewener, Brian Farrell, N. Michele Holbrook, Jean Heitz, Mark Hens, Elena R. Lozovsky, John Merrill, Randall Phillis & Debra Pires. *Biology. How Life Works*. Third Edition. New York: W. H. Freeman, 2019.

Moschella, Melissa. "Deconstructing the Brain Disconnection–Brain Death Analogy and Clarifying the Rationale for the Neurological Criterion of Death." *Journal of Medicine and Philosophy* 41.3 (2016): 279–299.

Murphy, Mark C. "The Natural Law Tradition in Ethics." *The Stanford Encyclopedia of Philosophy*. May 26, 2019. Available at: http://plato.stanford.edu/entries/natural-law-ethics Accessed: March 28, 2022.

Napier, Stephen. "Vulnerable Embryos: A Critical Analysis of Twinning, Rescue, and Natural-Loss Arguments." *American Catholic Philosophical Quarterly* 84.4 (2010): 783–812.

Newman, John Henry. "Discipline of Mind." In: Newman, John Henry. *The Idea of a University*. Edited, with an Introduction and Notes by Martin J. Svaglic. Notre Dame, IN: University of Notre Dame Press, 1982. pp. 361–380.

Nguyen, Doyen. "Pope John Paul II and the neurological standard for the determination of death: A critical analysis of his address to the Transplantation Society." *The Linacre Quarterly* 84.2 (2017): 155–186.

Nguyen, Doyen. "Why the Thomistic Defense of 'Brain Death' Is Not Thomistic: An Analysis from the Perspectives of Classical Philosophy and Contemporary Biophilosophy." *The Thomist. A Speculative Quarterly Review* 82.3 (2018a): 407–446.

Nguyen, Doyen. *The New Definitions of Death for Organ Donation. A Multidisciplinary Analysis from the Perspective of Christian Ethics.* Bern et al.: Peter Lang AG, 2018b.

Noldin, Hieronymus. *Summa Theologiae Moralis: II. De Praeceptis.* Editio XXIII. Oeniponte: F. Rauch, 1935.

Noldin, Hieronymus. *Summa Theologiae Moralis: I. De Principiis.* Editio XXIV. Oeniponte: F. Rauch, 1936.

O'Donovan, Oliver. "Again: Who is a Person." In: Channer, John Hugh (Ed.). *Abortion and the Sanctity of Human Life.* Exeter: The Paternoster Press, 1985, pp. 125–137.

Oderberg, David S.. "Modal Properties, Moral Status, and Identity." *Philosophy & Public Affairs* 26.3 (1997): 259–298.

Oderberg, David S.. *Applied Ethics. A Non-Consequentialist Approach.* Oxford: Blackwell, 2000.

Oderberg, David S. "The Ethics of Co-operation in Wrongdoing." In: O'Hear, Anthony (Ed.). *Modern Moral Philosophy.* Cambridge: Cambridge University Press, 2004a. pp. 203–227. (Royal Institute of Philosophy Supplement 54; Royal Institute of Philosophy Annual Lecture Series 2002–2003).

Oderberg, David. "Ordinary Duties, Extraordinary Means." *Human Life Review* 30.1 (2004b): 50–56.

Oderberg, David S.. *Real Essentialism.* London: Routledge, 2007.

Oderberg, David S.. "The Doctrine of Double Effect." In: O'Connor, Timothy & Constantine Sandis (Eds.). *A Companion to the Philosophy of Action.* Oxford: Wiley-Blackwell, 2010. pp. 324–330.

Oderberg, David S. "Synthetic Life and the Bruteness of Immanent Causation." In: Feser, Edward (Ed.). *Aristotle on Method and Metaphysics.* Basingstoke: Palgrave Macmillan, 2013. pp. 206–235.

Oderberg, David S. "The Metaphysics of Privation." In: Hüntelmann, Rafael & Johannes Hattler (Eds.). *New Scholasticism Meets Analytic Philosophy.* Heusenstamm: editiones scholasticae, 2014. pp. 63–88.

Oderberg, David S. "Further Clarity on Cooperation and Morality." *Journal of Medical Ethics* 43.4 (2017a): 192–200.

Oderberg, David S. "Cooperation in the Age of *Hobby Lobby*: When Sincerity is not Enough." *Expositions* 11.1 (2017b): 15–30.

Oderberg, David S. *Opting Out. Conscience and Cooperation in a Pluralistic Society.* London: The Institute of Economic Affairs, 2018.

Oderberg, David S.. "Death, Unity, and the Brain." *Theoretical Medicine and Bioethics* 40.5 (2019): 359–379.

Oderberg, David S.. *The Metaphysics of Good and Evil.* New York & London: Routledge, 2020.

Ostertag, Christopher & Kyle Karches. "Brain Death and the Formation of Moral Conscience." *The Linacre Quarterly* 86.4 (2019): 335–346.

Pakaluk, Michael. "Natural Inclinations in Aquinas' Account of Natural Law." In: Wolfe, Christopher & Steven Brust (Ed.) *Natural Law Today. The Present State of the Perennial Philosophy*. Lanham: Lexington Books, 2018. pp. 19–32.

Pallis, Christopher. "Whole-Brain Death Reconsidered: Physiological Facts and Philosophy." *Journal of Medical Ethics* 9.1 (1983): 32–37.

Patel, Chirag, Paul Kleinig, Michael Bakker & Paul Tait. "Palliative sedation: A safety net for the relief of refractory and intolerable symptoms at the end of life." *Australian Journal of General Practice* 48.12 (2019): 838–845.

Parfit, Derek. *Reasons and Persons*. Oxford: Oxford University Press, 1984.

Pasnau, Robert. *Thomas Aquinas on Human Nature: A Philosophical Study of Summa Theologiae, 1a 75–89*. Cambridge: Cambridge University Press, 2002.

Payne-James, Jason, Richard Jones, Steven Karch & John Manlove. *Simpson's Forensic Medicine*. 13th Edition. London: Hodder Arnold, 2011.

Penrod, Janice, Fang Yu, Ann Kolanowski, Donna M. Fick, Susan J. Loeb & Judith E. Hupcey. "Reframing Person-Centered Nursing Care for Persons with Dementia." *Research and Theory for Nursing Practice* 21.1 (2007): 57–72.

Petitgand, M.. "Observations sur un décapité annamite." *Revue Scientifique* (1884): 10–12. (Troisième Série—Tome VIII. Tome XXXIV de la Collection. 21me Année—2e Semestre. Juillet 1884 a Janvier 1885) Available at: https://gallica.bnf.fr/ark:/12148/bpt6k215103f?rk=42918;4 Accessed: March 28, 2022.

Phelan, Jay. *What is Life? A Guide to Biology*. Fourth Edition. New York: W. H. Freeman, 2018.

Phillips, Richard Percival. *Modern Thomistic Philosophy: An Explanation for Students. Volume 1: The Philosophy of Nature*. Westminster, MD: The Newman Bookshop, 1934.

Pieper, Josef. *Death and Immortality*. Translation by Richard and Clara Winston. South Bend, IN: St. Augustine's Press, 2000. Originally published in German as: *Tod und Unsterblichkeit*. München: Kösel-Verlag, 1968.

Pojman, Louis P. & Jeffrey Reiman. *The Death Penalty: For and Against*. Lanham et al.: Rowman & Littlefield Publishers, 1998.

Prahlow, Joseph A. & Roger W. Byard. *Atlas of Forensic Pathology*. New York: Springer, 2012.

Prümmer, Dominicus M.. *Manuale Theologiae Moralis Secundum Principia S. Thomae Aquinatis in Usum Scholarum*. Tomus II. Editio Quinta Decima. Barcinone, Friburgi Brisgoviae & Romae: Herder, 1960.

Prümmer, Dominicus M.. *Manuale Theologiae Moralis Secundum Principia S. Thomae Aquinatis in Usum Scholarum*. Tomus I. Editio Quinta Decima. Barcinone, Friburgi Brisgoviae & Romae: Herder, 1961.

Rachels, James. "Active and Passive Euthanasia." *The New England Journal of Medicine* 292.2 (1975): 78–80.

Raven, Peter H., George B. Johnson, Kenneth A. Mason, Jonathan B. Losos & Susan R. Singer. *Biology*. Eleventh Edition. New York: McGraw-Hill, 2017.

Regatillo, Eduardo F. & Marcellino Zalba Erro. *Theologiae Moralis Summa*. Vol. I: De Principiis. Madrid: Biblioteca de Autores Cristianos, 1953.

Regan, Richard J.. *Just War: Principles and Cases. Second Edition*. Washington, D.C.: Catholic University of America Press, 2013.

Reichlin, Massimo. "The Argument from Potential: A Reappraisal." *Bioethics* 11.1 (1997): 1–23.

Ren, Xiaoping, Ming Li, Xin Zhao, Zehan Liu, Shuai Ren, Yafang Zhang, Shide Zhang & Sergio Canavero. "First cephalosomatic anastomosis in a human model." *Surgical Neurology International* 8.276 (2017): https://surgicalneurologyint.com/surgicalint-articles/first-cephalosomatic-anastomosis-in-a-human-model

Renard, Henri. *The Philosophy of Being*. Second edition, revised and enlarged. Milwaukee: The Bruce Publishing Co., 1948.

Rickaby, Joseph. *Moral Philosophy. Ethics, Deontology and Natural Law*. Fourth Edition. London: Longman, Green and Co., 1918.

Ross, John Elliot. *Christian Ethics. A Textbook of Right Living*. New York: The Devin-Adair Company, 1919.

Ross, John Elliot. *Ethics. From the Standpoint of Scholastic Philosophy*. Heusenstamm: editiones scholasticae, 2012. (Reprint of the third edition, New York: The Devin-Adair Company, 1938.)

Sacred Congregation for the Doctrine of the Faith. *Declaration on Procured Abortion*. Vatican: Sacred Congregation for the Doctrine of the Faith, 1974. Available at: https://www.vatican.va/roman_curia/congregations/cfaith/documents/rc_con_cfaith_doc_19741118_declaration-abortion_en.html Accessed: March 28, 2022.

Sadava, David E., David M. Hillis, H. Craig Heller & Sally D. Hacker. *Life: The Science of Biology*. Eleventh Edition. Sunderland, MA: Sinauer, 2016.

Sade, Robert M.. "Consequences of the Dead Donor Rule." *The Annals of Thoracic Surgery* 97.4 (2014): 1131–1132.

Sadler, T.W.. *Langman's Medical Embryology*. Fourteenth Edition. Philadelphia: Wolters Kluwer, 2019.

Sánchez Sorondo, Marcelo (Ed.). *Working Group on The Signs of Death. The Proceedings of the Working Group 11–12 September 2006*. Vatican City: Pontificia Academia Scientiarum, 2007. (The Pontifical Academy of Sciences, Scripta Varia 110) Available at: https://www.pas.va/en/publications/scripta-varia/sv110pas.html. Accessed: March 28, 2022.

Scanlon, Thomas. *What We Owe to Each Other*. Cambridge, MA: Harvard University Press, 1998.

Scarfone, Matthew. "Incoherent Abortion Exceptions." *Journal of Social Philosophy* 53.1 (2022): 127–140.

Schindler, David L.. "A Response to the Joint Statement, 'Production of Pluripotent Stem Cells by Oocyte Assisted Reprogramming'." *Communio: International Catholic Review* 32.2 (2005): 369–380.

Schoenwolf, Gary C., Steven B. Bleyl, Philip R. Brauer & Philippa H. Francis-West. *Larsen's Human Embryology.* Fifth Edition. Edinburgh et al.: Elsevier, 2015.

Seifert, Josef. *What is Life? The Originality, Irreducibility and Value of Life.* Amsterdam and Atlanta, GA: Rodopi, 1997. (Value Inquiry Book Series (VIBS), Volume 51)

Seifert, Josef. "The right to life and the fourfold root of human dignity." In: de Dios Vial Correa, Juan & Elio Sgreccia (Eds.). *The Nature and Dignity of the Human Person as the Foundation of the Right to Life. The Challenges of the Contemporary Cultural Context. Proceedings of the VIIIth Assembly of the Pontifical Academy for Life (Vatican City, 25–27 February 2002).* Vatican City: Libreria Editrice Vaticana, 2003.

Seifert, Josef. *The Philosophical Diseases of Medicine and Their Cure. Philosophy and Ethics of Medicine, Vol. 1: Foundations.* Dordrecht: Springer, 2004. (Philosophy and Medicine, Volume 82)

Seyfer, Tara L.. "An Overview of Chimeras and Hybrids." *The National Catholic Bioethics Quarterly* 6.1 (2006): 37–49.

Shallo, Michael W.. *Lessons in Scholastic Philosophy.* London: Hirschfeld Brothers Limited, 1916.

Sherwin, Susan. *No Longer Patient: Feminist Ethics and Health Care.* Philadelphia, PA: Temple University Press, 1992.

Shewmon, D. Alan. "'Brainstem Death,' 'Brain Death' and Death: A Critical Re-Evaluation of the Purported Equivalence." *Issues in Law & Medicine* 14.2 (1998): 125–145.

Shewmon, D. Alan. "The Brain and Somatic Integration: Insights into the Standard Biological Rationale for Equating "Brain Death" With Death." *Journal of Medicine and Philosophy* 26.5 (2001): 457–478.

Shewmon, D. Alan. "Mental Disconnect: 'Physiological Decapitation' as a Heuristic for Understanding Brain Death." In: Sánchez Sorondo, Marcelo (Ed.). *Working Group on The Signs of Death. The Proceedings of the Working Group 11–12 September 2006.* Vatican City: Pontificia Academia Scientiarum, 2007. pp. 292–333. (Pontificiae Academiae Scientiarum Scripta Varia 110)

Shewmon, D. Alan. "Constructing the Death Elephant: A Synthetic Paradigm Shift for the Definition, Criteria, and Tests for Death." *Journal of Medicine and Philosophy* 35.3 (2010): 256–298.

Shuster, Michèle, Janet Vigna & Matthew Tontonoz. *Biology for a Changing World.* Third Edition. New York: W. H. Freeman, 2018.

Singer, Peter. *Animal Liberation: A new Ethics for our Treatment of Animals.* New York: New York Review/ Random House, 1975.

Singer, Peter. *Practical Ethics.* Third Edition. Cambridge: Cambridge University Press, 2011.

Singer, Peter & Helga Kuhse. "Ethics and the Limits of Tolerance." *The Journal of Medicine and Philosophy* 19 (1994): 129–145.

Solomon, Robert C. "Reflections on the Meaning of (Fetal) Life." In: Bondeson, Willi-am B., Tristram H. Engelhardt, Stuart F. Spicker & Daniel H. Winship (Eds.). *Abortion and the Status of the Fetus.* Dordrecht, Boston & Lancaster: Reidel, 1983. pp. 209–226. (Series "Philosophy and Medicine", Volume 13)

Spaemann, Robert. "Begotten, Not Made." *Communio: International Catholic Review* 33 (Summer 2006a): 290–297.

Spaemann, Robert. "Töten oder sterbenlassen?" *Aufklärung und Kritik* 13 (2006b): 80–88. (Sonderheft 11)

Spencer, Mark K.. "A Reexamination of the Hylomorphic Theory of Death." *The Review of Metaphysics* 63.4 (2010): 843–870.

Spike, Jeffrey. "Brain Death, Pregnancy, and Posthumous Motherhood." *Journal of Clinical Ethics* 10.1 (1999): 57–65.

Steinberg, Avraham & M. Hersch. "Decapitation of a pregnant sheep: a contribution to the brain death controversy." *Transplantation Proceedings* 27.2 (1995): 1886–1887.

Sullivan, Scott M.. "The development and nature of the ordinary/extraordinary means distinction in the Roman Catholic tradition." *Bioethics* 21.7 (2007): 386–397.

Sulmasy, Daniel P. "Dignity and Bioethics: History, Theory, and Selected Applications." In: The President's Council on Bioethics (Ed.). *Human Dignity and Bioethics. Essays Commissioned by the President's Council on Bioethics.* Washington, DC: The President's Council on Bioethics, 2008. pp. 469–501. Available at: https://bioethicsarchive.georgetown.edu/pcbe/reports/human_dignity Accessed: March 28, 2022.

Sutovsky, Steve Peter & Won-Hee Song. "Post-fertilisation sperm mitophagy: the tale of Mitochondrial Eve." *Reproduction, Fertility and Development* 30.1 (2017): 56–63.

Sykes, Nigel & Andrew Thorns. "The use of opioids and sedatives at the end of life." *The Lancet Oncology* 4.5 (2003): 312–318.

Thomson, Judith Jarvis. "A Defense of Abortion." *Philosophy and Public Affairs* 1.1 (1971): 47–66.

Tollefsen, Christopher. "The New Natural Law Theory." *Lyceum* 10.1 (2008): 1–8. Available at: http://lyceumphilosophy.com/?q=node/97 Accessed: March 28, 2022.

Tooley, Michael. "Abortion and Infanticide." *Philosophy & Public Affairs* 2.1 (1972): 37–65.

Urry, Lisa A., Michael L. Cain, Steven A. Wasserman, Peter V. Minorsky & Jane B. Reece. *Campbell Biology.* Eleventh Edition. New York: Pearson, 2017.

Utz, Arthur F. & Gerda Hieronimi. *Grundsätze der Sozialpolitik. Solidarität und Subsidiarität in der Altersversicherung*. Stuttgart: Seewald Verlag, 1969. (Veröffentlichungen des Instituts für Gesellschaftswissenschaften Walberberg, Band II)

van Dyke, Christina. "The End of (Human) Life as We Know It: Thomas Aquinas on Persons, Bodies and Death." *The Modern Schoolman* 89.3–4 (2012): 243–257.

Vanezis, Peter. "Diagnosing Death and Changes after Death." In: Vanezis, Peter & Philip Beh (Eds.) *Essential Forensic Medicine*. Hoboken, NJ: Wiley, 2019. Chapter 16.

Verberne, Eline A., Emma de Haan, J. Peter van Tintelen, Dick Lindhout & Mieke M. van Haelst. "Fetal methotrexate syndrome: A systematic review of case reports." *Reproductive Toxicology Review* 87 (2019): 125–139.

Vij, Krishan. *Textbook of Forensic Medicine and Toxicology: Principles and Practice*. Fifth Edition. London: Elsevier, 2014.

Vreeman, Rachel C. & Aaron E. Carroll. "Medical Myths." *British Medical Journal* 335 (2007): 1288–1289.

Wallace, William A.. "Aquinas's Legacy on Individuation, Cogitation, and Hominization." In: Gallagher, David M. (Ed.). Thomas Aquinas and His Legacy. Washington, D.C.: The Catholic University of America Press, 1994, pp. 173–193. Gallagher's book has been republished 2018 as volume 28 of the series "Studies in Philosophy and The History of Philosophy" edited by The Catholic University of America Press.

Walker, Adrian J.. "Altered Nuclear Transfer: A Philosophical Critique." *Communio: International Catholic Review* 31.4 (2004): 649–684.

Walzer, Michael. *Just and Unjust Wars. A Moral Argument with Historical Illustrations. Fifth Edition*. New York: Basic Books, 2015.

Warren, Mary Anne. *Moral Status: Obligations to Persons and Other Living Things*. Oxford: Clarendon Press, 1997.

White, Robert J., Maurice S. Albin & Javier Verdura. "Isolation of the Monkey Brain: In vitro Preparation and Maintenance." *Science* 141.3585 (1963): 1060–1061.

White, Robert J., Maurice S. Albin & Javier Verdura. "Preservation of Viability in the Isolated Monkey Brain Utilizing a Mechanical Extracorporeal Circulation." *Nature* 202.4937 (1964): 1082–1083.

White, Robert J., Maurice S. Albin, George E. Locke & Eugene Davidson. "Brain Transplantation: Prolonged Survival of. Brain after Carotid-Jugular Interposition." *Science* 150.3697 (1965): 779–781.

White, Robert J., Lee R. Wolin, Leo C. Massopust, Norman Taslitz, & Javier Verdura. "Cephalic exchange transplantation in the monkey." *Surgery* 70.1 (1971): 135–139.

White Robert J., Heinz Angstwurm & Ignacio Carrasco de Paula (Eds.). *Working Group on the Determination of Brain Death and its Relationship to Human Death. 10–14 December 1989.* Vatican City: Pontificia Academia Scientiarum, 1992. (The Pontifical Academy of Sciences, Scripta Varia 83) Available at: https://www.pas.va/en/publications/scripta-varia/sv83pas.html Accessed: March 28, 2022.

Wiggins, David. *Sameness and Substance.* Oxford: Oxford University Press, 1980.

Williams, Thomas. "Human Freedom and Agency." In: Davies, Brian & Eleonore Stump (Eds.). *The Oxford Handbook of Aquinas.* Oxford: Oxford University Press, 2012. pp. 199–208.

Index[1]

[1] Note: Page numbers followed by 'n' refer to notes.

C. Erk, *The Ethics of Killing*,
https://doi.org/10.1007/978-3-031-07183-6

The manufacturer's authorised representative in the EU is Springer
Nature Customer Service Centre GmbH, Europaplatz 3, 69115 Heidelberg,
Germany. If you have any concerns regarding our products, please
contact ProductSafety@springernature.com

Printed and bound by CPI Group (UK) Ltd, Croydon, CR0 4YY
30/09/2026
02225490-0012